ATMOSPHERIC PHYSICS FROM SPACELAB

ASTROPHYSICS AND SPACE SCIENCE LIBRARY

A SERIES OF BOOKS ON THE RECENT DEVELOPMENTS

OF SPACE SCIENCE AND OF GENERAL GEOPHYSICS AND ASTROPHYSICS

PUBLISHED IN CONNECTION WITH THE JOURNAL

SPACE SCIENCE REVIEWS

VOLUME 61

PROCEEDINGS

ATMOSPHERIC PHYSICS FROM SPACELAB

PROCEEDINGS OF THE 11th ESLAB SYMPOSIUM,
ORGANIZED BY THE SPACE SCIENCE DEPARTMENT
OF THE EUROPEAN SPACE AGENCY,
HELD AT FRASCATI, ITALY, 11–14 MAY 1976

Edited by

J. J. BURGER and A. PEDERSEN

ESA Space Science Department (ESLAB)

and

B. BATTRICK

ESA Scientific and Technical Publications Branch

D. REIDEL PUBLISHING COMPANY

DORDRECHT-HOLLAND / BOSTON-U.S.A.

ISBN-13:978-94-010-1530-1   e-ISBN-13:978-94-010-1528-8
DOI: 10.1007/978-94-010-1528-8

Published by D. Reidel Publishing Company,
P.O. Box 17, Dordrecht, Holland

Sold and distributed in the U.S.A., Canada and Mexico
by D. Reidel Publishing Company, Inc.
Lincoln Building, 160 Old Derby Street, Hingham,
Mass. 02043, U.S.A.

TABLE OF CONTENTS

PREFACE

The European Space Agency (ESA) is presently developing Spacelab,
a general-purpose re-usable laboratory to be flown in the cargo
bay of the Space Shuttle Orbiter being developed in parallel by
the National Aeronautics and Space Administration (NASA).
The first flight of Spacelab is scheduled for 1980.

Spacelab's ability to support large, heavy instruments, to provide high levels of electrical power, and to give scientists direct
access to their experiments in space will provide unique opportunities for advanced studies in many disciplines of space science.
One of the fields in which considerable discussion on the novel
opportunities is taking place is atmospheric physics. The possibilities of using active sounding techniques and carrying out coordinated investigations with instruments with large apertures
and long focal lengths on a global scale offer good prospects for
obtaining fundamentally new information on the stratosphere, mesosphere and lower thermosphere. It is this build-up of interest
that prompted ESA's Space Science Department (ESLAB) to devote its
eleventh annual symposium to 'Atmospheric Physics from Spacelab'.

Because the discussions in the international scientific community
on this subject are now continuing apace, speed of publication of
the symposium proceedings has been given high priority and we
have therefore opted for direct reproduction of author's type-
scripts. We hope that these proceedings will serve as a timely
review of the scientific aims and experimental techniques of
atmospheric physics in the light of the opportunities furnished
by Spacelab.

The Editors

OPENING ADDRESS

by Head of Space Science Department, ESA.

The Space Science Department of ESA (formerly ESLAB) exists in
order to provide internationally acceptable projects scientists
for ESA's scientific programme. In order to attract scientists of
sufficient calibre and to maintain their competence in a field
of fast changing experimental techniques Space Science Department
operates its own internal research programme. This is organised
around four scientific Divisions which have evolved over the last
ten years in response to changes of direction in the overall ESA
scientific programme. At present these Divisions are the Cosmic
Ray Division, the Space Plasma Physics Division, the High Energy
Astrophysics Division and the Astronomy Division. It can be seen
that a very broad range of disciplines is covered by only twenty-
three scientists.

Within the broad range covered there is, however, no atmospheric science. ESRO some years ago decided that it should not
attempt to compete in certain fields and from that time planetary
science and atmospheric science were deliberately excluded. (ESRO
did, however, fly one outstandingly successful atmospheric experiment, the S80 mass spectrometer, on its ESRO IV spacecraft.) But
when that decision was made the idea of Spacelab in Europe had not
been born. It would now seem inconceivable to exclude atmospheric
science from a space programme which may have to depend to a large
extent on Spacelab in the 1980's.

Although it is clearly not possible to switch people and
laboratory research as easily or speedily as it is to switch planning of the ESA programme, we felt that this year the annual
ESLAB Symposium should be essentially educational from the point
of view of an SSD staff member.

We thought it would be useful to listen as you debated what
science should be done in the atmosphere, how this should be done
experimentally and what use could be made of Spacelab. Judging
from the size and expertise of the audience which has assembled
it would appear that you believe a great deal can be achieved and
we look forward to your deliberations.

Our ESLAB symposia in past years have taken various forms.
Sometimes we discussed results in fields where we had internal
expertise. In this framework we particularly encouraged the intercorrelation of data from different experiments and different
spacecraft. On occasions we discussed and planned what we hoped
to achieve with future spacecraft - for example the Vienna symposium last year was based on GEOS within the International
Magnetospheric Study. On at least one earlier occasion we held
a symposium on a subject in which we had no expertise - that was
Infra Red Physics - and to-day this subject forms the core of
our Astronomy Division. I would like to think that we could go
the same way with Atmospheric Physics but scientists cannot be
converted overnight to be experts in other fields and the budge-
tary constraints these days make the employing of new staff
exceedingly difficult. It appears that we will have to steer
gradually toward an in-house expertise.

In recent years major advances have been made in understanding the relationships between the ionosphere and motions of the
neutral atmosphere. Satellite results are demonstrating clearly
the response of the neutral atmosphere to changes in geomagnetic
activity. These latter changes are in turn related to the behaviour of the sun. It must be evident then that in future none
of these areas can be treated in isolation but we must consider
a linked system of sun - interplanetary medium - magnetosphere -
atmosphere/ionosphere. Although since the beginning of the space
age a great deal of money has been invested in research in the
field of sun/earth relationships, and many correlations between
parameters such as solar wind velocity and geomagnetic activity
have been established statistically, the actual physics of many
situations has not been described.
Nowhere is this more evident than in the topical subject of sun/
weather relationships. It would be unwise for any open-minded
physicist to dismiss the considerable observational evidence that
solar activity affects the motions of the earth's atmosphere.
However, until the physical processes are uncovered it will clearly be difficult to convince meteorologists of the reality of the
link.

During this symposium we expect to talk about atmospheric
constituents and their time variations (in response to the sun
or Concorde or whatever), about the dynamics of the atmosphere,

about the experimental tools to be employed and finally how these
tools can be most advantageously deployed on Spacelab.

Concerning Spacelab itself there are still many unknowns
such as how much flights will cost the experimenter, how long
will be the lead time from experiment acceptance till the data
is obtained and how serious will be the chemical and electromagnetic contamination problems. While it is clear that even at this
late time not all the scientific community is convinced that
Spacelab was the right way to go, it is equally clear that most
scientists are gearing their thoughts and efforts toward making
the best use of that new facility when it becomes available.
It would appear at this stage that Spacelab will provide even
better opportunities for atmospheric scientists than for those
in other fields.

We hope in this symposium to hear how eagerly you anticipate
these opportunities and how you plan to grasp them. I welcome you
to the 11th ESLAB Symposium and wish you a pleasant and profitable
stay in Frascati.

D. Edgar Page

LIST OF PARTICIPANTS

ACKERMANN, M.   Institut d' Aéronomie
                                 3 Avenue Circulaire
                                 B-1180 Brussels , Belgium

AERTS, E.V.N.   idem

AMMAR-ISRAEL, A.   CNES DGP/PR
                                 129 Rue de l' Université
                                 Paris 7, France

ANDEREGG, M.   Space Science Department, ESA, ESTEC
                                 Noordwijk, The Netherlands

ANDUIZI, C.   Centro Richerche Aerospaziale
                                 Rome, Italy

BERNARD, R.   CRPE-CNET
                                 38-40 Rue du General Leclerc
                                 92134 Issy les Moulineaux, France

BANKS, P. M.   University of California San Diego
                                 Department of Applied Physics and
                                 Information Science
                                 UCSD, La Jolla
                                 California 92037, USA

BATTRICK, B.   ESA Sci. Tech. Publications Branch
                                 c/o ESTEC, Noordwijk, The Netherlands

BJÖRN, L.   Uppsala Ionospheric Observatory
                                 S-75590 Uppsala 1, Sweden

BLAMONT, J.E.   Service d' Aéronomie CNRS
                                 BP3 Verrières-le-Buisson 91370, France

BLANC, M.   CRPE-CNET
                                 38-40 Rue du General Leclerc
                                 92134 Issy les Moulineaux, France

BONETTI, A.   Istituto Fisica Generale
                               University of Florence, Italy

BOWHILL, S.A.   University of Illinois
                               Department of Electrical Engineering
                               Illinois 61801, USA

BURGER, J.J.   Space Science Department, ESA, ESTEC,
                               Noordwijk, The Netherlands

CARLI, B.   Institute of Physics
                               University of Florence
                               Via S. Bonaventura 13
                               50145 Firenze-Quaracchi, Italy

CHAPPEL, C.R.   NASA Marshall Space Flight Center
                               2803 Downing Court
                               Huntsville, Alabama 35801, USA

COLETTI, A.   Istituto di Fisica dell' Atmosfera
                               CNR, Rome, Italy

DIREITINHO TAVARES, C.   Servico Meteorologico Nacional
                               Rua Saraiva Carvalho 2
                               Lisbon, Portugal

DOMINGO, V.   Space Science Department, ESA, ESTEC,
                               Noordwijk, The Netherlands

DONAHUE, T.M.   University of Michigan
                               Atmospheric and Oceanic Science
                               Department
                               Ann Arbor, Michigan 48105, USA

DURNEY, A.   Space Science Department, ESA, ESTEC,
                               Noordwijk, The Netherlands

ECKHARDT, K.   Dornier System, Germany

FARROW, J.B.   HSD Space Division
                               Gunnels Wood Road
                               Stevenage
                               Hertfordshire, England

FEUERBACHER, B.   Space Science Department, ESA, ESTEC,
                               Noordwijk, The Netherlands

FIOCCO, G.   CNR
                               Laboratorio Plasma Spazio
                               CP 27
                               00044 Frascati, Italy

FITTON, B.   Space Science Department, ESA, ESTEC,
                              Noordwijk, The Netherlands

FLASCHE, A.   DFVLR-BPT
                              Linder Hoehe
                              5 Cologne 90, Germany

FUA, D.   CNR-LPS, Rome, Italy

GAUTIER, D.   Observatoire de Meudon
                              Groupe Plantès
                              92190 Meudon, France

GRARD, R.J.L.   Space Science Department, ESA, ESTEC,
                              Noordwijk, The Netherlands

GREGORY, J.   Middle Atmospheric Program
                              Department of Physics
                              University of Saskatchewan
                              Saskatoon, Canada

GUPTA, S.P.   Institut für Physikalische
                              Weltraumforschung
                              Heidenhofstrasse 8
                              D-78 Freiburg, Germany

HASKELL, G.   ESA HQ
                              114 Av. Charles de Gaulle
                              92522 Neuilly sur Seine, France

HARRIES, J.E.   National Physical Laboratory
                              Teddington
                              Middlesex TW11 OLW, England

HOUGHTON, J.T.   University of Oxford
                              Department of Atmospheric Physics
                              Clarendon Laboratory
                              Oxford OX1 3PU, England

HUNT, G.   Meteorological Office
                              London Road
                              Bracknell
                              Berkshire, England

ISAKSEN, I.S.A.   Institute of Geophysics
                              University of Oslo
                              Blindern
                              Oslo 3, Norway

JONES, D.   Space Science Department,ESA, ESTEC,
                                   Noordwijk, The Netherlands

KNOTT, K.   idem

KRANKOWSKY, D.   Max Planck Institut für
                                   Kernphysik
                                   PO Box 103980
                                   6900 Heidelberg, Germany

LÄMMERZAHL, P.   idem

MARSDEN, R.G.   Space Science Department, ESA, ESTEC,
                                   Noordwijk, The Netherlands

MARTELLI, G.   University of Sussex
                                   Brighton, England

MARTIN, D.H.   Queen Mary College
                                   London, England

MEGIE, G.   Service d' Aéronomie CNRS
                                   Verrières-le-Buisson 91370, France

MEREDITH, L.H.   Office of Naval Research
                                   223 Old Marylebone Road
                                   London NW1 5TH, England

MULLER, C.   Institut d' Aéronomie
                                   3 Avenue Circulaire
                                   B-1180 Brussels, Belgium

MURGATROYD, R.J.   Meteorological Office
                                   London Road
                                   Bracknell
                                   Berkshire, England

NAGY, A.   University of Michigan
                                   Atmospheric and Oceanic Science
                                   Department
                                   Space Physics Research Laboratory
                                   Ann Arbor, Michigan 48105, USA

PAGE, D. E.   Space Science Department, ESA, ESTEC,
                                   Noordwijk, The Netherlands

PARESCE, F.

University of California
Space Science Laboratory
Berkeley, California 94720, USA

PEDERSEN, A.

Space Science Department, ESA, ESTEC,
Noordwijk, The Netherlands

PERRON, C.

idem

PEYTREMANN, E.

ESA HQ
114 Av. Charles de Gaulle
92522 Neuilly sur Seine, France

RASCHKE, E.

Meteorologisches Institut
Albertus Hatnus Platz
Cologne 41, Germany

REDEMANN, E.

Meteorologisches Institut der
Universität München
Theresienstrasse 37
Munich 2, Germany

RENGER, W.

DVFLR
Institut für Physik der Atmosphäre
8031 Oberpfaffenhofen, Germany

REVAH, I.

CRPE-CNET
38-40 Rue du General Leclerc
92134 Issy les Moulineaux, France

ROTHWELL, P.

University of Southampton
Department of Physics
Southampton S09 5NH, England

RUPPERSBERG, G.

DFVLR
Institut für Physik der
Atmosphäre
8031 Oberpfaffenhofen, Germany

SANDERSON, T.

Space Science Department, ESA, ESTEC,
Noordwijk, The Netherlands

SCHANDA, E.

Institute of Applied Physics
University of Bern
Sidlerstrasse 5
CH-3012 Bern, Switzerland

SCHMIDT, F.H.   Royal Netherlands Meteorological
                                 Institute
                                 Utrechtseweg 297
                                 De Bilt, The Netherlands

SCHMIDTKE, G.   Institut für Physikalische
                                 Weltraumforschung
                                 Heidenhofstrasse 8
                                 D-78 Freiburg, Germany

SCHWILLA, H.   Dornier System, Germany

SIMON, P.C.   Institut d' Aéronomie
                                 3 Avenue Circulaire
                                 B-1180 Brussels, Belgium

SONA, A.   CISE, Milan, Italy

STEGMAN, J.   Institute of Meteorology
                                 University of Stockholm
                                 Fack S-10405, Stockholm, Sweden

DOS REIS TEIXEIRA   Servico Meteorologico Nacional
                                 Rua Saraiva Carvalho 2
                                 Lisbon, Portugal

THOMAS, L.   Science Research Counsil
                                 Appleton Laboratory
                                 Ditton Park
                                 Slough
                                 Bucks. SL3 9JX, England

TORRES PEREIRA NEVES,
L.A.   Servico Meteorologico Nacional
                                 Rua Saraiva Carvalho 2
                                 Lisbon, Portugal

VERMEER, S.   ESA Sci. Tech. Publications Branch
                                 c/o ESTEC, Noordwijk, The Netherlands

VIDAL-MADJAR, A.   LPSP-CNRS
                                 91370 Verrières-le-Buisson, France

VILLIER, G.   CNRS, France

WENZEL, K.P.   Space Science Department, ESA, ESTEC,
                                 Noordwijk, The Netherlands

SESSION 1

THE ATMOSPHERE AND IONOSPHERE BELOW 150 km

SOME OUTSTANDING PROBLEMS IN THE NEUTRAL AND IONIZED ATMOSPHERE
BETWEEN 60 AND 150 KM ALTITUDE

L. THOMAS

S.R.C., Appleton Laboratory, Ditton Park,
Slough SL3 9JX, Berks., U.K.

ABSTRACT

Processes operating in the neutral and ionized parts of the
atmosphere in the height range 60-150 km are considered with
particular attention being paid to neutral composition, thermal
balance, the production/loss and roles of excited species, and the
positive and negative ion composition. Current ideas in these
areas are described and attention is drawn to the major outstanding
problems.

1. INTRODUCTION

The characteristics of the neutral and ionized parts of the
atmosphere in the height range 60-150 km are largely determined by
the absorption of solar radiations of wavelength less than about
2000 Å. Consequently, adequate information on solar flux intensities in this spectral region is an essential requirement for
quantitative studies.

The neutral composition of the region is controlled by the
photodissociation of O_2, H_2O and NO, and by the production of NO,
chiefly by ionization processes. For the first two of these gases
the mesosphere corresponds to the lowest heights at which photodissociation is important, and the constituents resulting from
these dissociations have chemical lifetimes which are significant
but less than several hours. Consequently, major diurnal and other
temporal variations are observed in the oxygen-hydrogen constituents
in the region between about 60 and 85 km. For nitric oxide, pre-
dissociation of the δ bands makes the greatest contribution to its

J. J. Burger et al. (eds.), Atmospheric Physics from Spacelab, 3–18. All Rights Reserved.
Copyright © 1976 by D. Reidel Publishing Company, Dordrecht-Holland.

dissociation in the mesosphere but no marked day-to-night change
is expected between about 70 and 110 km. For this and other constituents having long photochemical lifetimes, transport processes
have a major influence in determining the spatial distributions.
Photodissociation and photoionization processes, and the
accompanying resonance scattering and fluorescence effects, together with chemical reactions of both neutral and ionized species,
can give rise to excited atmospheric species. It is found that
the 60-150 km height region is particularly rich in such species.
In addition to the interest in the production and loss mechanisms
of the excited atoms and molecules, increasing attention is being
paid to their potential roles in ionospheric and other atmospheric
processes [1].

The absorption of solar radiations of wavelengths less than
about 2000 Å, and the emission of infra-red radiation from ground-
state atomic oxygen and vibrationally excited CO_2, chiefly determine the temperature gradient in the lower thermosphere. In
addition to the relevance of this gradient to the downward conduction of heat, and therefore the thermal balance of the region
under consideration, its contribution to the vertical motions of
constituents arising from diffusion is also of interest.

The ionization produced by solar radiations has been the most
widely studied aspect. Rocket-borne mass-spectrometer observations
have shown the presence of a variety of positive ions [2]. At
heights below about 85 km water cluster ions, $H^+ \cdot (H_2O)_n$, predominate
under normal conditions, but at greater heights NO^+ and O_2^+ are
the major ions with layers of metal ions, such as Fe^+ and Mg^+,
also being present. These measurements in rockets, together with
laboratory measurements of chemical rate coefficients, have in-
spired considerable interest in the positive-ion chemistry. At
the lowest levels of the height range under consideration account
has to be taken of the presence of negative ions but the composition is still uncertain [2].

Although increasing attention has been paid in the past decade
to the measurement of solar fluxes in the spectral region below
2000 Å, there are still considerable uncertainties both in absolute intensities and in the degree of variability. In addition
to the difficulties arising directly out of these uncertainties,
other major problems are encountered in studies of the neutral
composition, thermal balance, excited species and ion chemistry
of the height region 60-150 km. The purpose of the present
article is to draw attention to some of these problems.

2. NEUTRAL COMPOSITION AND THERMAL BALANCE

A discrepancy of a factor of about 3 has existed in the solar fluxes between about 1400 Å and 2000 Å measured with a rocket-borne spectrograph by Detwiler et al. [3], and a rocket-borne spectrometer by Parkinson and Reeves [4]. In a report of further rocket-borne spectrometer measurements Heroux and Swirbalus [5] state that their recent results are in good agreement with those of Parkinson and Reeves. They also point out that gross differences in solar flux probably cannot be attributed to different levels of solar activity. Indeed, the degree of variability is itself a matter of some controversy since Prag and Morse [6] and Heath [7] have provided evidence of substantial changes within periods of 27 days, whereas Hinteregger [8] has presented data for a 27-day period in 1974 when the solar flux between 1555 and 1791 Å showed little change, although the 10.7 cm flux varied markedly. The absolute intensity is important in determining the photodissociation rates of O_2, H_2O and NO and, particularly, the production rate of 0 atoms above about 85 km. The latter is relevant to the energy budget of the lower thermosphere and mesosphere since the excess between the photon energy and the O_2 dissociation energy, 8×10^{-12} ergs for the ground vibrational state, is available for atmospheric heating. In addition, the $O(^1D)$ atoms produced by O_2 absorption of the Schumann-Runge continuum are quenched below 150 km and the excitation energy represents an additional source of atmospheric heating.

It has been customary in theoretical models of mesospheric and thermospheric composition to take account only of transport of constituents arising from vertical diffusion. Following Colegrove et al. [9], the velocity for constituent i has been expressed as the sum of terms representing the molecular diffusion and eddy diffusion contributions

$$W_i = - D_i \left[\frac{1}{n_i} \frac{\partial n_i}{\partial h} + \frac{1}{H_i} + \frac{1}{T} \frac{dT}{dz} \right]$$

$$- K \left[\frac{1}{n_i} \frac{\partial n_i}{\partial h} + \frac{1}{H_{ave}} + \frac{1}{T} \frac{dT}{dz} \right] \tag{1}$$

where D_i representa an average molecular diffusion coefficient for constituent i diffusing in several atmospheric gases, n_i and H_i the concentrations and scale height ($^{kT}/m_i g$) of the constituent i, T the atmospheric temperature and H_{ave} the mean atmospheric gas scale height ($kT/\overline{m}g$).

Figure 1 shows a comparison of various models of eddy diffusion coefficient with the height variation of the coefficient

for atomic oxygen diffusing in nitrogen, and the corresponding
height variations of atomic oxygen derived by George et al. [10].
This serves to show the importance of the values chosen for the
eddy diffusion coefficient. The corresponding influence of the
eddy diffusion coefficient values on the height distribution of
NO in the mesosphere has been similarly demonstrated by
Brasseur and Nicolet [11].

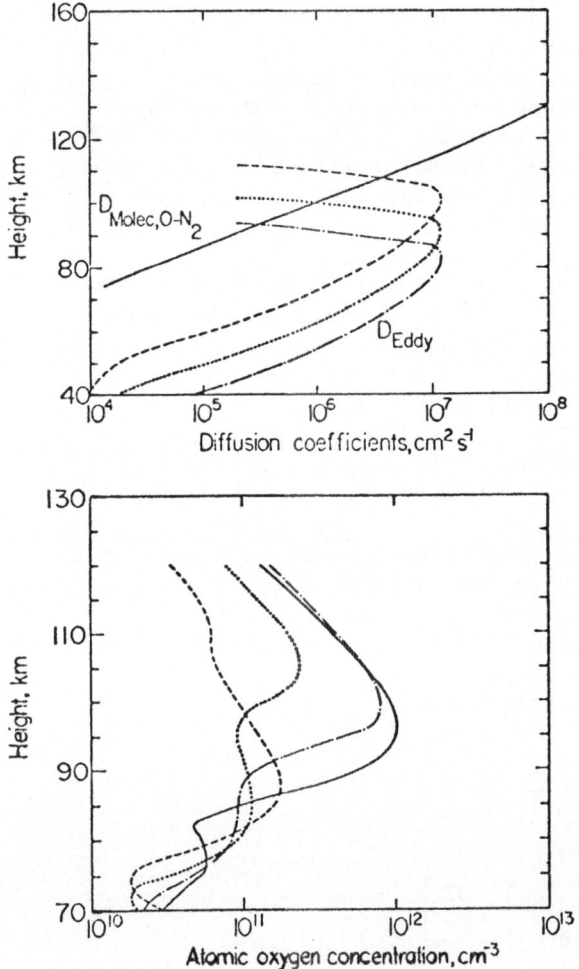

Figure 1   Height distributions of atomic oxygen concentrations
            computed for different diffusion models [10].

In their interpretation of 5577 Å, O(^{1}D-^{1}S) airglow data from
the OGO 6 satellite, Donahue and Carignan [12] have drawn attention
to the importance of the temperature gradient terms in the molecular

and eddy diffusion velocities in deriving an acceptable rate of $O(^3P)$ atom concentration for a model in which the $O(^1S)$ state is excited by the three-body reaction of oxygen atoms. These workers, like Breig [13], have demonstrated the dependence of the values assumed for K on the solar flux values adopted. Donahue and Carignan have shown that the oxygen distribution can be reconciled with a constant eddy diffusion coefficient of about 5 x $10^5 cm^2 s^{-1}$ if the temperature gradient in the thermosphere is between 10 and 20 K/km. This range was deduced using the Parkinson and Reeves [4] values of solar flux and is consistent with those found with rocket-borne Pitot tube experiments and with incoherent scatter systems, but is substantially larger than the maximum value of 10 K/km indicated in the Jacchia [14] model. The use of the Detwiler et al. [3] values of solar flux and the correspondingly higher value of eddy diffusion coefficient, 1.6 x $10^6 cm^2 s^{-1}$, yielded gradients between 30 and 50 K/km.

The question of the thermospheric temperature gradient is also relevant to the downward conduction of heat, produced by the photo-dissociation of O_2 and ionization production and loss processes, to the mesopause region where the energy is radiated by infra-red active constituents. It has been suggested by Johnson and Wilkins [15] for the Earth's atmosphere, and Shimizu [16] and McGovern and Burk [17] for planetary atmospheres, that molecular conduction is not sufficiently rapid but must be assisted by eddy conduction. For this form of conduction the heat flux F in the presence of a temperature gradient is given by

$$F = K C_p \, \rho \left(\frac{dT}{dz} + \Gamma \right) \tag{2}$$

where C_p is the specific heat at constant pressure, ρ the density and Γ the adiabatic lapse rate.

Hunten [18] has pointed out that (2) represents an upper limit to the heat flux. In addition, he has argued on the basis of Booker's analysis [19] that in order to maintain eddy mixing a power dissipation, P, per unit area is required, given by

$$P = K H \frac{g}{TR_{i_c}} \left(\frac{dT}{dz} + \Gamma \right) \tag{3}$$

where R_{i_c} represents a critical Richardson number for which turbulence can just be maintained against dissipation into heat.

Hunten has suggested that even within the uncertainties involved, it seems that P is at least equal and probably greater than F. The resolution of this question will require detailed information on solar flux inputs, improved data on the associated thermospheric temperatures and a complete treatment of the radiative processes. This will need to include the details of energy transfer from $O(^1D)$ to N_2 [20], the resonance transfer of N_2 vibrational energy to CO_2 [21], the possibility of quenching of vibrationally excited CO_2 by O atoms [22], and also the far infrared emission from $O(^3P)$ [23].

3. EXCITED SPECIES

3.1 Airglow emission and excitation mechanisms

The range of excited species that are encountered in the 60-150 km region is illustrated in the height distributions of dayglow emissions shown in Fig. 2. Except for $N(^2D)$, the production and quenching mechanisms for the corresponding excited

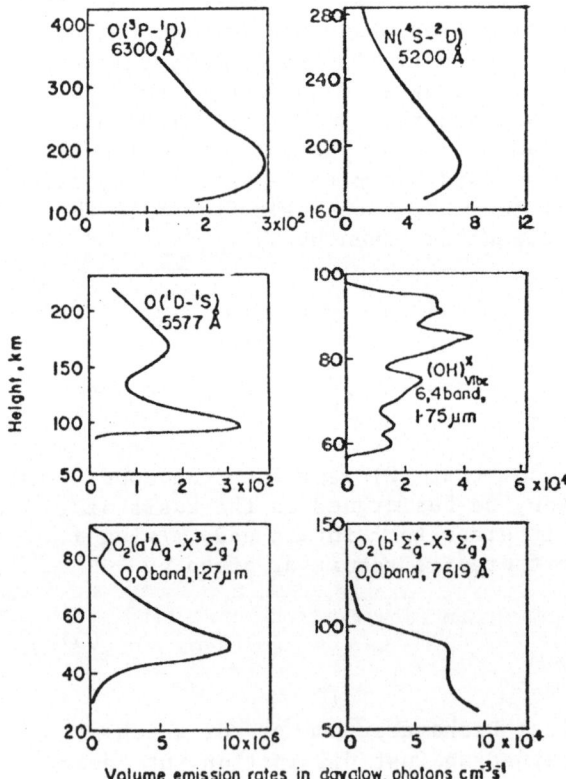

Volume emission rates in dayglow, photons $cm^{-3}s^{-1}$

Figure 2   Height variations of volume emission rates in the day-glow of 6300 Å [24], 5200 Å [25], 5577 Å [26], 1.75 μm [27], 1.27 μm [28], and 7619 Å [29].

states are reasonably well understood; the sources at nighttime are not so well established. For $N(^2D)$ the rates of these processes are still the subject of considerable interest. It is well known that the primary importance of this excited atom is that it provides the major source of NO [30, 31] in the mesosphere

$$N(^2D) + O_2 \rightarrow NO + O \qquad (4)$$

As a result it plays a principal role in both the neutral and ionized chemistry of the region under consideration, and the production and loss processes need to be firmly established. Table 1 shows the main processes identified by Strobel et al. [32], in a recent study directed towards the interpretation of $N(^2D)$ and NO data derived from Atmospheric Explorer measurements by Rusch et al. [25], and the twilight NO emission data of Feldman and Takacs [33].

Table 1 : Production and loss of $N(^2D)$ (Strobel et al. [32])

Production

$$N_2 + e \text{ (fast)} \rightarrow N + N(^2D) + e \qquad (5)$$

$$N_2 + h\nu \text{ (800-1000 Å)} \rightarrow N + N(^2D) \qquad (6)$$

$$NO^+ + e \rightarrow N(^2D) + O \qquad (7)$$

$$N_2^+ + O \rightarrow NO^+ + N(^2D) \qquad (8)$$

$$N_2^+ + e \rightarrow N + N(^2D) \qquad (9)$$

$$N^+ + O_2 \rightarrow O_2^+ + N(^2D) \qquad (10)$$

Loss

$$N(^2D) + O_2 \rightarrow NO + O \qquad (11)$$

$$N(^2D) + O \rightarrow N + O \qquad (12)$$

$$N(^2D) + NO \rightarrow N_2 + O \qquad (13)$$

$$N(^2D) + e \rightarrow N + e \qquad (14)$$

From their analysis, Strobel et al. have concluded that most of the dissociative recombinations of NO^+ ions must produce $N(^2D)$ atoms and that $N(^2D)$ is quenched by atomic oxygen with a rate coefficient of $1 \times 10^{-12} \text{cm}^3\text{s}^{-1}$, about one-half the value reported

from laboratory measurements by Davenport et al. [34]. The
results of the analysis also indicate that for heights below 150 km
the production mechanisms (5)-(8) represent the primary sources of
$N(^2D)$; only the first two loss processes are important below 150 km
and it is only below 140 km that the reaction with O_2 represents
the major loss process for the excited atoms.

3.2 Roles of excited states

It has long been realised that the internal energies of
excited species can strongly influence the rate of a chemical reaction, and even the reaction mechanism. In addition to this
chemical influence, excited species can play roles arising from
the transfer of their excitation energies to other constituents
during collisions. Finally, excitation can permit photoionization
by an extended range of wavelengths, as in the case of the $O_2(^1\Delta_g)$
molecule in the D region (Hunten and McElroy [35]). It seems
likely that excited species have a much greater influence in
atmospheric and ionospheric processes that is presently
appreciated.

Considerable progress has been made in laboratory measurements of quenching reactions for electronically excited states,
such as $O(^1D)$, $O(^1S)$, $N(^2D)$ and $O_2(^1\Delta_g)$ but the kinetics of
vibrationally and rotationally excited species is still not well
developed. In order to illustrate the importance of the state of
excitation of reactants it is useful to indicate the roles of
excited species in the areas of energy balance and chemical rate
coefficients.

(i)   Energy balance   The quenching of $O(^1D)$ atoms in the
height range under consideration is largely by collision with
molecular nitrogen, the observations of Slanger and Black [20]
demonstrating that vibrational excitation of the N_2 molecule
occurs for 35% of the collisions

$$O(^1D) + N_2 \rightarrow (N_2)^* + O(^3P) \tag{15}$$

The excited molecules cannot radiate and are not readily
quenched in collision with other nitrogen or oxygen molecules.
However, the following processes might play important roles

$$(N_2)^* \quad + O \quad \rightarrow N_2 + O + \text{translational energy} \tag{16}$$

$$(N_2)^*_{v=1} + CO_2 \rightarrow N_2 + CO_2(001) \tag{17}$$

McNeal et al. [36] have indicated a rather larger value for
the rate coefficient of reaction (16) than previously
accepted, and Breig et al. [37] have shown that this is
probably a dominant loss process for vibrationally excited

nitrogen in the lower thermosphere. Its inclusion results
in a significant reduction in the theoretical vibrational
temperature which has a bearing on the energy transfer to the
electron gas. Taylor and Bitterman [21] have shown that
owing to a natural resonance the transfer of energy into the
001 vibrational mode of CO_2 by reaction (17) is rapid, and
consequently infra-red emission of the ν_3 band of CO_2 would
then represent an energy loss.

(ii) Enhancement of rate coefficients Attention has already
been drawn to the reaction between $N(^2D)$ atoms and O_2 molecules to form NO molecules. Following a suggestion by
Norton and Barth [30] and Nicolet [31], Slanger et al. [38]
found that the rate coefficient, $1.4 \times 10^{-11} cm^3 s^{-1}$, was
several orders of magnitude larger than that for the $N(^4S)$
atoms.

Another reaction in which excitation has been found to
have a profound effect on the rate coefficient is the ion-
atom interchange reaction

$$O^+ + (N_2)* \rightarrow NO^+ + N \qquad (18)$$

Schmeltekopf et al. [39] have found that the rate coefficient
is markedly dependent on the vibrational temperature of N_2.

4. ION CHEMISTRY

Although the predominance in the normal D region of water
cluster ions, $H^+ \cdot (H_2O)_n$, below about 82 km in daytime, and 86 km
at nighttime, is well established there is still some doubt about
the order n of the ions. Fragmentation of weakly-bound ions
can occur through thermodynamic break-up at the shock layer because of increased temperatures, or through collisions resulting
from the draw-in electric fields [40]. Concerning the formation
of the cluster ions, the major problem has been to identify the
reaction scheme beginning with NO^+, the major ion produced during
photoionization. A major requirement of this scheme is that it be
rapid in view of the rapid loss of water cluster ions by dissociative recombination in the middle D region [41]. Following
suggestions by Ferguson [42] and others it is believed that the
initial production of $NO^+ \cdot H_2O$ results from clustering of NO^+ to
N_2 and CO_2 followed by reactions with H_2O

$$NO^+ \quad + N_2 + M \rightleftarrows NO^+ \cdot N_2 + M \qquad (19)$$

$$NO^+ \cdot N_2 + CO_2 \quad \rightarrow NO^+ \cdot CO_2 + N_2 \qquad (20)$$

$$NO^+ \quad + CO_2 + M \rightleftarrows NO^+ \cdot CO_2 + M \qquad (21)$$

$$NO^+ \cdot CO_2 + H_2O \quad \rightarrow NO^+ \cdot H_2O + CO_2 \qquad (22)$$

Recent measurements by Johnsen et al. [43] using a drift-tube technique have yielded values of the rate coefficient for the forward reaction of (19) and of the equilibrium constant as functions of temperature. Models incorporating these results have shown that the formation of $NO^+ \cdot H_2O$ is reasonably well understood on the basis of reactions (19)-(22), the relative importance of (19), (20) and (21) in the formation of $NO^+ \cdot CO_2$ depending on the mesospheric temperature. However, the formations of the higher hydrates up to $NO^+ \cdot (H_2O)_3$, which leads via a reaction with H_2O to $H^+ \cdot (H_2O)_3$, are rather more problematic.

The question of the nature of the ambient negative ions in the D region has not been resolved, the conflict between the two sets of mass-spectrometer observations [2] still persisting. It appears from the laboratory measurements of ion-neutral reactions by Fehsenfeld and Ferguson [44] and of positive-ion negative-ion recombination by Smith et al. [45], that the general ideas based on earlier negative-ion schemes will not need to be modified significantly to incorporate the hydration of negative ions.

It is usually assumed that the daytime E region is the best understood part of the ionosphere. The source of ionization is the photoionization by solar radiations chiefly of wavelengths 800-1027 Å and 52-300 Å [46]; perhaps the greatest uncertainty is in the variability of solar fluxes, particularly those below 300 Å. The ion chemistry is reasonably well understood on the basis of relatively few reactions

$$O_2^+ + e \rightarrow O + O \tag{23}$$

$$NO^+ + e \rightarrow N + O \tag{24}$$

$$O^+ + O_2 \rightarrow O_2^+ + O \tag{25}$$

$$O^+ + N_2 \rightarrow NO^+ + N \tag{26}$$

$$O_2^+ + NO \rightarrow NO^+ + O_2 \tag{27}$$

$$O_2^+ + N \rightarrow NO^+ + O \tag{28}$$

$$N_2^+ + O \rightarrow NO^+ + N \tag{29}$$

$$N_2^+ + O_2 \rightarrow O_2^+ + N_2 \tag{30}$$

The incorporation of relevant rate coefficient data and constituent concentrations have demonstrated the major importance of the NO and N concentrations in controlling the E-region ion composition. In this respect the ion chemistry of the E region, like the neutral chemistry, requires an improved knowledge of the chemistry of odd nitrogen which, as indicated in Section 3, implies a better understanding of the production and loss processes for N(^{2}D) atoms.

For nighttime conditions, additional information on the production of ionization by scattered Lyman-α and β and Helium II radiations is required, and an added complication arises from the need to take account of transport processes. The electron and ion concentrations at nighttime are about two orders of magnitude smaller than in daytime, and this implies that the chemical time constants are correspondingly larger. It can be shown that movements can then be important in determining the distribution of ionization. The effect of such processes has been examined by Strobel et al. [47] who took account of diffusion and diurnal winds deduced from pressure gradients given by model atmospheres. A comparison of the theoretically computed ion composition with mass-spectrometer measurements showed that for heights between 115 and 140 km the nighttime sources of ionization and chemical processes alone are insufficient to account for the observations. Furthermore, a more complex wind field than that assumed is required, probably including upward propagating tidal winds and gravity waves.

Although mass-spectrometer measurements have revealed that the presence of metal ions is a consistent feature of the ionosphere between about 82 and 120 km [2], there is little quantitative understanding of the chemical processes operating. It is generally assumed that meteor ablation is chiefly responsible for the metal atoms, M, but considerations of cross sections at high energies have suggested that the simultaneous production of metal ions, M$^+$, is unimportant. Instead, the largest proportion of these ions is believed to arise from photoionization or charge transfer processes; because of the relatively low values of ionization potentials, charge transfer can occur even with NO$^+$ ions

$$M + h\nu \rightarrow M^+ + e \tag{31}$$

$$M + O_2^+ \rightarrow M^+ + O_2 \tag{32}$$

$$M + NO^+ \rightarrow M^+ + NO \tag{33}$$

Metal oxide ions, MO$^+$ and MO$_2^+$, can also be formed from the corresponding oxide by photoionization or charge transfer. It was believed that the dissociative recombination of these oxide ions might play an important part in the chemical loss of the metal ions, since the direct loss of these by radiative recombination is so slow. However, as pointed out by Ferguson and Fehsenfeld [48], the metal oxide ions MO$^+$ are rapidly reduced by atomic oxygen

$$MO^+ + O \rightarrow M^+ + O_2 \tag{34}$$

There is still some uncertainty about the fate of the dioxide ion, MO_2^+, and particularly whether it is converted to MO^+ by a reaction analogous to [34] or lost by dissociative recombination. The significance of the loss processes for MgO_2^+ has been examined by Anderson and Barth [49].

The neutral and ion chemistries of metals are intimately related as is illustrated by the reaction scheme derived by Brown [50] shown in Fig. 3. In order to emphasise the need for

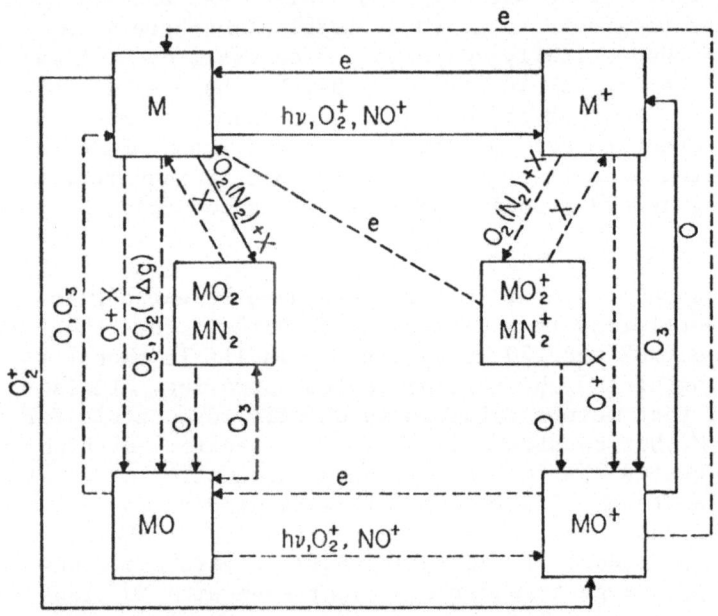

Figure 3   Schematic representation of metal atom and ion reactions [50].

laboratory measurements of reaction rate coefficients, those reactions for which no data are available in the case of Mg are shown by broken lines; a similar lack of data is found for Fe and other metal atoms. In addition to the reactions shown, Ferguson [51] has drawn attention to the possible hydration of metal ions, either directly or via the intermediate ion $M^+.CO_2$ and switching the CO_2 by H_2O.

It seems that increased information on relevant reaction rate coefficients, and minor neutral reactants, will need to be incorporated in present theoretical treatments of metal ion redistributions resulting from convergence by wind shears [52] to provide a realistic model of the narrow layers commonly observed.

CONCLUSIONS

An improved knowledge of the solar flux intensities and their variability is a common requirement for studies of the neutral and ion composition, the thermal balance and the production of excited species in the height region 60-150 km. The spectral region between about 1300 and 2000 Å is of special interest in considerations of the neutral atmosphere and the region below 300 Å deserves particular attention for E-region investigations.

The incorporation of eddy processes in the vertical transport of constituents and the possible importance of such processes in the downward conduction of heat represent major uncertainties in theoretical studies of neutral composition and thermal balance. The magnitude of the thermal gradient in the lower thermosphere is also a critical parameter for such studies.

The 60-150 km height region shows substantial concentrations of excited atmospheric species. The excitation mechanisms are reasonably well understood but the role of such species in atmospheric and ionospheric processes have attracted relatively little attention to date.

The ion composition at heights above about 85 km is well established. The major processes controlling the distributions of NO^+ and O_2^+ at such heights in daytime are reasonably well understood, although quantitative treatments require a better knowledge of neutral constituents such as NO and N. These are involved in the odd nitrogen chemistry which is being actively studied. The chemical processes controlling the distribution of metal ions at such heights has attracted little quantitative treatment, chiefly because of the difficulties encountered in laboratory measurements of the rate coefficients of relevant reactions.

At heights below 85 km the mass-spectrometer observations have probably suffered from fragmentation of weakly bound ions, and the establishment of the true ambient positive and negative-ion composition continues to be a major difficulty. The details of the formation of water cluster ions from NO^+ are still to be clarified.

ACKNOWLEDGEMENT

This review is published with the permission of the Director of the Appleton Laboratory of the Science Research Council.

REFERENCES

1.   M. N. Vlasov, J. Atmosph. Terr. Phys. 38, (1976) (In Press).
2.   R. S. Narcisi, in Physics and Chemistry of Upper Atmospherics, Ed. B. M. McCormac, Reidel, Dordrecht-Holland, 171, (1973).
3.   C. R. Detwiler, D. L. Garrett, J. D. Purcell and R. Tousey, Ann. Geophys. 17, 263, (1961).
4.   W. H. Parkinson and E. M. Reeves, Solar Phys. 10, 342, (1969).
5.   L. Heroux and R. A. Swirbalus, J. Geophys. Res. 81, 436, (1976).
6.   A. B. Prag and F. A. Morse, J. Geophys. Res. 75, 4613, (1970).
7.   D. F. Heath, J. Geophys. Res. 78, 2779, (1973).
8.   H. E. Hinteregger, J. Atmosph. Terr. Phys. 38, (1976) (In Press).
9.   F. D. Colegrove, W. B. Hanson and F. S. Johnson, J. Geophys. Res. 70, 4931, (1965).
10. J. D. George, S. P. Zimmerman and T. J. Keneshea, Space Research XII, Ed. S. A. Bowhill, L. D. Jaffe and M. J. Rycroft, Akademie-Verlag, Berlin, 695, (1972).
11. G. Brasseur and M. Nicolet, Planet. Space Sci. 21, 939, (1973).
12. T. M. Donahue and C. R. Carignan, J. Geophys. Res. 80, 4565, (1975).
13. E. L. Breig, J. Geophys. Res. 78, 5718, 1973.
14. L. G. Jacchia, Special Report 332, Smithsonian Astrophys. Observ. Cambridge, Mass. (1971).
15. F. S. Johnsen and E. M. Wilkins, J. Geophys. Res. 70, 1281, (1965).
16. M. Shimizu, in Planetary Atmospheres, Ed. C. Sagan, T. C. Owen and H. J. Smith, D. Reidel, Dordrecht-Holland, 331, (1971).
17. W. E. McGovern and S. D. Burk, J. Atmos. Sci. 29, 179, (1972).
18. D. M. Hunten, J. Geophys. Res. 79, 2533, (1974).
19. H. G. Booker, J. Geophys. Res. 61, 673, (1956).
20. T. G. Slanger and G. Black, J. Chem. Phys. 60, 468, (1974).
21. R. L. Taylor and S. A. Bitterman, Rev. Mod. Phys. 41, 26, (1969).
22. P. J. Crutzen, Q. J. Royal Met. Soc. 96, 769, (1970).

23. R. A. Craig and J. C. Gille, J. Atmos. Sci. 26, 205, (1969).
24. T. Magata, T. Tohmatsu and T. Ogawa, J. Geomagn. Geoelect. 20, 315, (1968).
25. D. W. Rusch, A. I. Stewart, P. B. Hays and J. H. Hoffman, J. Geophys. Res. 80, 2300, (1975).
26. L. Wallace and M. B. McElroy, Planet. Space Sci. 14, 677, (1966).
27. E. J. Llewellyn and W. F. J. Evans, in The Radiating Atmosphere, Ed. B. M. McCormac, D. Reidel, Dordrecht-Holland, 17, (1971).
28. W. F. J. Evans, D. M. Hunten, E. J. Llewellyn and A. Vallance Jones, J. Geophys. Res. 73, 2885, (1968).
29. L. Wallace and D. M. Hunten, J. Geophys. Res. 73, 4813, (1968).
30. R. B. Norton and C. A. Barth, J. Geophys. Res. 75, 3903, (1970).
31. M. Nicolet, Planet. Space Sci. 18, 1111, (1970).
32. D. F. Strobel, E. S. Oran and P. D. Feldman, U.S. Naval Research Laboratory Memorandum Report 3090, (1975).
33. P. D. Feldman and P. Z. Takacs, Geophys. Res. Letters, 1, 169, (1974).
34. J. E. Davenport, T. G. Slanger and G. Black, J. Geophys. Res. 81, 12, (1976).
35. D. M. Hunten and M. B. McElroy, J. Geophys. Res. 73, 2421, (1968).
36. R. J. McNeal, M. E. Whitson and G. R. Cook, Chem. Phys. Lett., 16, 507, (1972).
37. E. L. Breig, M. E. Brennan and R. J. McNeal, J. Geophys. Res., 78, 1225, (1973).
38. T. G. Slanger, B. J. Wood and G. Black, J. Geophys. Res. 76, 8430, (1971).
39. A. L. Schmeltekopf, E. E. Ferguson and F. C. Fehsenfeld, J. Chem. Phys. 48, 2966, (1968).
40. R. S. Narcisi and W. Roth, Adv. Electronics and Electron Phys., Academic Press, N. York, 79, (1970).
41. M. T. Leu, M. A. Biondi and R. Johnsen, Phys. Rev. A, 7, 292, (1973).
42. E. E. Ferguson, Rev. Geophys. Space Phys., 9, 997, (1971).
43. R. Johnsen, C. M. Huang and M. A. Biondi, J. Chem. Phys. 63, 3374, (1975).
44. F. C. Fehsenfeld and E. E. Ferguson, J. Chem. Phys. 61, 3181, (1974).
45. D. Smith, N. G. Adams and M. J. Church, Planet. Space Sci. 24, (1976), (In Press).
46. L. Heroux, M. Cohen and J. E. Higgins, J. Geophys. Rev., 79, 5237, (1974).
47. D. F. Strobel, T. R. Young, R. R. Meier, T. P. Coffey and A. W. Ali, J. Geophys. Res. 79, 3171, (1974).

48.   E. E. Ferguson and F. C. Fehsenfeld, J. Geophys. Res. 73, 6215, (1968).

49.   J. G. Anderson and C. A. Barth, J. Geophys. Res., 76, 3723,(1971).

50.   T. L. Brown, Chemical Revs., 73, 645, (1973).

51.   E. E. Ferguson, Radio Sci. 7, 397, (1972).

52.   J. D. Whitehead, J. Atmosph. Terr. Phys., 20, 49, (1971).

DISCUSSION

T.M. Donahue: The kind of temperature gradients that we infer from oxygen distributions are consistent with those obtained from incoherent scatter-observations.

HIGH-LATITUDE IONOSPHERE-ATMOSPHERE INTERACTIONS

Peter M. Banks

Department of Applied Physics and Information Science,
University of California, San Diego, USA

A review of high-latitude ionosphere-atmosphere interactions was
presented. Recent experimental data relating to ionospheric-
plasma convection were discussed in relation to the formation of
the ionospheric F-layer trough and the higher altitude plasma-
pause. With regard to thermospheric dynamics, the effects of ion
drag and Joule and particle heating were considered. Some emphasis
was given to recent measurements which provide evidence for a
horseshoe distribution of particle and Joule heating with strongest
energy input lying in the region of the polar cusp. Finally,
mention was made of the role of E-region winds in creating a high-
latitude dynamo capable of influencing ionospheric plasma
convection.

J. J. Burger et al. (eds.), Atmospheric Physics from Spacelab, 19–20. All Rights Reserved.
Copyright © 1976 by D. Reidel Publishing Company, Dordrecht-Holland.

DISCUSSION

S.A. Bowhill: What questions can a facility such as EISCAT
address in collaboration with a high-inclination Spacelab
flight?

P.M. Banks: EISCAT could provide long-term measurements of
magnetospheric activity, i.e. electric fields, auroral
precipitation, Joule heating, electric currents. Point
comparisons will be difficult and not necessarily of primary
importance. Spacelab auroral photography showing time
variations of auroral precipitation on a quasi-global scale
would be of great use in the interpretation of EISCAT results.

General comparison of point quantities should not be
completely ignored. They are simply difficult to arrange.

P. Rothwell: Can you say roughly by how much the ion temperature
near the Harang discontinuity is lower than the temperature
in the rest of the auroral oval, and also how reliable are
temperature measurements in this very disturbed region?

P.M. Banks: Ion temperatures in the vicinity of the oval are
highly variable. In general,

$$T_i = T_n + 1/3 \frac{m}{k} v_r^2$$

where m is the neutral particle mass, k is Boltzmann's
constant, v_r is the ion-neutral relative velocity and T_n is
the neutral gas temperature. Near the auroral oval, v_r is
proportional to the perpendicular electric field intensity $E_\perp$.
The Harang discontinuity is a region of electric-field
transition and $E_\perp$ is generally small compared with values seen
at earlier and later local times. Thus, T_i in the midnight
sector should be substantially lower than would be seen elsewhere in the oval. The variability of ion temperature would
be directly related to the spatial and temporal structure of
the ionospheric electric field.

SOLAR RADIATION VARIATION AND CLIMATE

V.Domingo

Space Science Department, ESA
ESTEC, Noordwijk, The Netherlands

ABSTRACT

A short review of the variability of the solar radiation
flux at different wavelengths and of the Earth's climate is
presented. Some examples of correlation between solar activity
and atmospheric parameters at different heights are shown.
The interest of studying the correlation that exists between
important atmospheric parameters (physical and chemical) with
the variations of the Sun flux at different wavelengths is
stressed.

INTRODUCTION

The purpose of this paper is to review the present knowledge
of the variability of the solar irradiance at the earth, the
variability of the terrestrial atmosphere, and the interrelation
that has been found between both.

For more extensive information on the topics treated in this
review I would recommend the reader to refer to Smith and Gottlieb
(1974) for the solar irradiance variations, Akasofu and Chapman
(1972) for solar corpuscular radiation variations and volume 16
of the GARP publications series (GARP, 1975) for climatic
variations.

The quoted observations should not be considered as an
exhaustive list, but as a representative list to the best of the
knowledge of the author. Many relevant measurements have been
made during the last few years, particularly by sun observing

J. J. Burger et al. (eds.), Atmospheric Physics from Spacelab, 21–41. All Rights Reserved.
Copyright © 1976 by D. Reidel Publishing Company, Dordrecht-Holland.

satellites and by atmospheric research satellites, but most of
the results of the analysis of those measurements are still to
appear in the literature.

SOLAR VARIABILITY

The emission of energy by the sun in very different forms
is known to suffer variations with very different time constants.
To quote two extreme cases: gravitational pulsations with a
period of 48 and 160 minutes have been observed (Weiss, 1976) and
the evolution of the sun is estimated to produce a brightening
of 1% in 5 x 10^7 years (Cameron, 1973).

In this review we are interested in the variations of the
electromagnetic radiation that constitutes the main source of
energy for the earth environment and in those of the corpuscular
radiation that reaches the earth's atmosphere after being
filtered by the magnetosphere.

Electromagnetic Radiation

Since a large fraction of the solar electromagnetic radiation
is absorbed or reflected by the atmosphere, detailed measurements
of interest for this topic are available only since the development
of space instrumentation. This fact limits the possibility of
speaking about time periodicities or time constants of more than
a few decades.

Figure 1 shows the solar flux per unit wavelength versus
wavelength. Note the spectral regions where marked variability
of the solar flux has been observed, at both ends of the
spectrum.

The variation of the solar electromagnetic radiation is
best described by reference to the individual observed phenomena.

Flares are transient phenomena with time constant of the
order of 1 to 100 minutes. Their main characteristic is to have
a very different spectrum compared to the normal solar spectrum
particularly due to the large enhancement of emission lines in
the visible UV and X-ray fluxes. In any of these spectral
regions the energy flux during the flare may be strongly enhanced
during the flare but, the integral energy flux, even in the larger
flares does not represent more than 0.1% of the solar constant.

Radio bursts are emissions of the sun in the radio wavelength
(between 10^{-4} and 600 m) with life-times of the order of 1 second
to 100 minutes. They are associated with almos any other activity

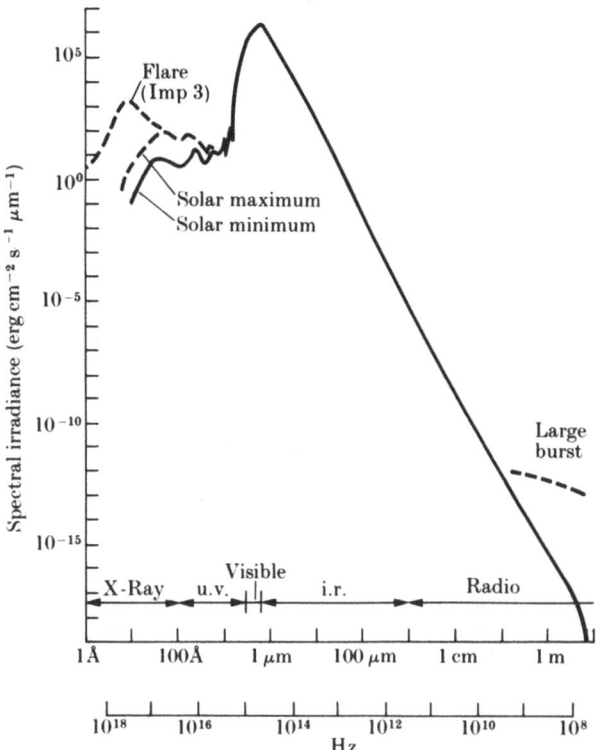

Figure 1.   The solar spectrum from 1 A to 10 m for different solar
*   conditions   (from Akasofu and Chapman, 1972, after*
*   Malitson, 1965).*

in the sun.

Faculae, plages and sunspots are transient phenomena with time
constants of the order of 1 - 100 days, that appear in active
regions of the sun. Faculae and plages, are brighter regions,
the first being only visible near the limb. Sunspots are cooler
parts of the active regions, but both, the extra brightness of
the plages and/or the reduced brighteness of the sunspots have a
relative small contribution the order of 0.1% or less to the flux
integrated over the solar disk.

Figure 2 shows the periodicity of the number of sunspots
visible on the sun of the 10.7 cm wavelength radio observation
and of the 9.6-11 Å x-ray measured by Parkinson and Pounds (1971).
The 27 - day cycle has also been observed in the far and near
ultraviolet regions by Heath (1973) on Nimbus 3 and 4. The
27-day period is caused by the solar rotation.

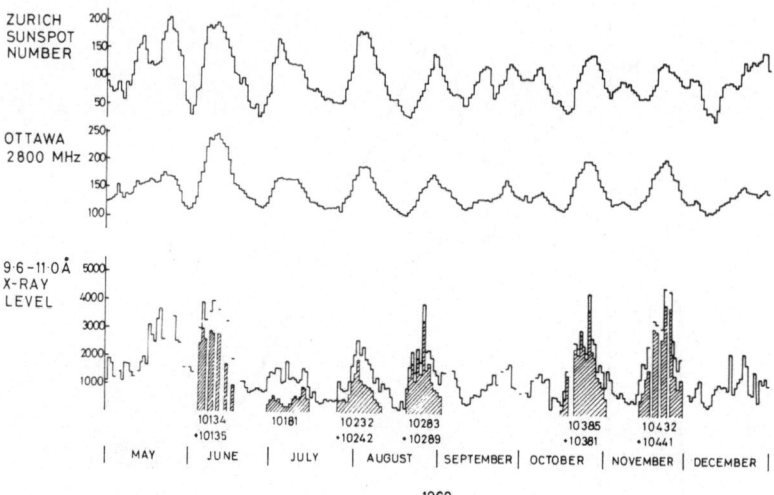

Figure 2. *Comparison of 9.6–11 A daily levels with Zürich sunspot number and Ottawa 2800 MHz, illustrating the recurrence of active longitudes (after Parkinson and Pounds, 1971).*

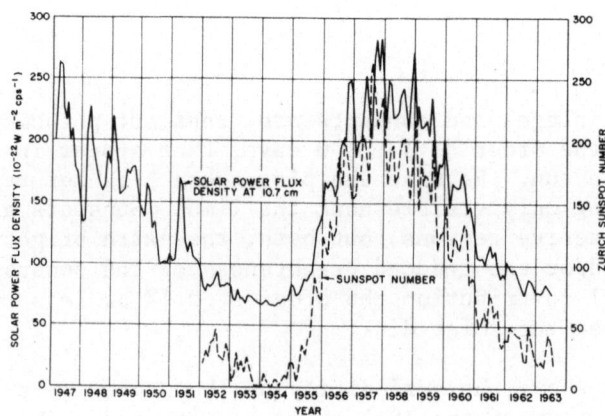

Figure 3. *The slowly varying component of solar radio emission at 10.7 cm compared with Zürich sunspot numbers (after Costelli et al., 1965).*

Table 1.   Solar Electromagnetic Radiation   Variation

Radiation	Contribution %	Variation
X-RAYS and EUV (2 - 1400 A°)	0.001	large (> 100 %)
Far UV (1400 - 2100 A°)	0.14	∿20 %
Near UV (2100 - 3300 A°)	2.44	$\lesssim$ 1 %
Visible (3300-10000 A°)	66.8	0.1
IR (10⁴ A° - 1 mm)	30.6	0.1
Long Waves (1 mm - 10 cm)	$<10^{-5}$	very large

Variations produced by: active regions, plages, sunspots, flares, bursts, storms.
27 - day period, 11 - year cycle.

The appearance and disappearance at the limb of the visible sun of the centers of activity on the surface of the sun with enhanced emission that last more than a few tens of days produce a 27 day recurrence of the enhanced radiations.

The 11 - year cycle has been established a long time ago by the observation of the number of sunspots in the photosphere of the sun   (Fig. 3).   This cycle appears as a varying number of active regions on the surface of the sun and is visible in most wavelengths.

Table 1 summarizes the main variations in a few bands mainly selected by their relationship to variability. Notice that the regions of the solar spectrum that are subject to a larger variability, i.e., x-rays, extreme ultraviolet and long waves represent a very small proportion of the total radiation flux (about 10^{-5}).

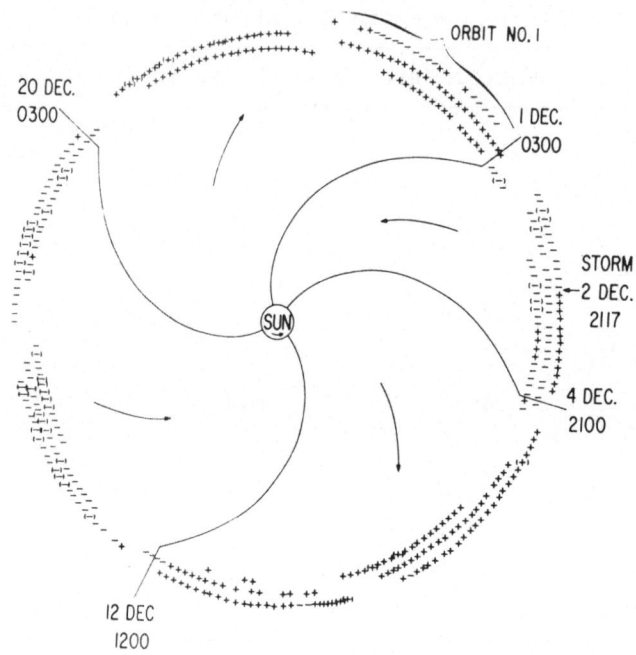

Figure 4.   *The plus signs (away from the sun) and minus signs*
            (toward the sun) at the circumference of the figure
            indicate the direction of the measured interplanetary
            magnetic field during successive 3-h intervals.   The
            inner portion of the figure is a schematic representation
            of a sector structure of the interplanetary magnetic
            field that is suggested by these observations (after
            Wilcox and Ness, 1965).

       The sun has an overall magnetic field of dipolar character.
During one 11-year sunspot cycle the north hemisphere of the
sun shows a dominant north magnetic polarity and opposite for the
southern hemisphere.   Then in the following 11 years the polari-
ties appear reversed.   Therefore the solar magnetic field
presents a 22-year cycle or recurrence, but in the emissions of
the sun, electromagnetic and corpuscular, no evidence has been
found of 22-year repetition as opposed to the 11-year cycle.

       It also has been observed that the amplitude of the 11-year
solar cycle of the sunspot number is modulated with a period of
about 80 years (Gleissberg, 1971).

Table 2. Solar Corpuscular Radiation Variation

Radiation	Variable Parameter	Variation	Phenomena
Solar wind (thermal plasma)	Density	$1-100$ cm^{-3}	Fluctuations streams
	Bulk velocity	$150-1500$ km/s	Shock waves
	Temperature	10^4-10^6 $^{\circ}$K	(27-day recurrence)
	Magnetic field	$1-100$ γ	The same and sector structure
Low energy particles	electrons($e \lesssim 1$ MeV) protons, nuclei ($e \lesssim 10$ MeV/n)	$0 \sim 10^4$ p/cm^2 sterad	Flares, streams
High energy particles	electrons($e \gtrsim 10$ MeV) protons, nuclei ($e \gtrsim 100$ MeV/n)	$0 \sim 10^4$ p/cm^2 sterad	Flares

Corpuscular Radiation

     The thermal expansion of the solar corona produces an outflow
of magnetized plasma known as the solar wind. The interaction of
the solar wind with the earth's magnetic dipole consitutes the
magnetosphere and the latter relates to the earth atmosphere through
the ionosphere and the earth magnetic field.

     Many spacecraft have measured solar wind parameters, such as
bulk velocity, density, composition, temperature, magnetic field
and the variability of these parameters.

     Most properties of the solar wind are affected by activity
in the solar corona and often suffer major disturbances that are
caused by solar storms. The passage of these disturbances past
the earth cause usually disturbances in the earth magnetosphere.

     Solar coronal disturbances are usually associated with solar
active regions that last more than one solar rotation, a 27-day
recurrence in the variation of solar wind parameters has been
observed. The magnetic field indicates the existence of dominant
structures in the coronal magnetic field for many solar rotations.
Fig. 4 shows the well k nown sector structure of the

interplanetary magnetic field.

Apart from the frequency of disturbances that follow solar storms, no 11-year solar cycle has been found in the mean values of the solar wind parameters.

The activity centres on the sun also emit energetic particles i.e. ions and electrons accelerated to suprathermal energies. Particles of relatively low energies are often released by the solar corona during periods extending over several solar rotations, thus producing the recurrent particle events with a period of 27 days.

Particles accelerated up to relativistic energies, i.e., electrons of several Mev and protons of more than 1 GeV are produced in association with large solar flares, as defined by the electromagnetic luminosity. We see that in the case of energetic particles we deal with discrete events, that - seen from the earth - have characteristic time constants that vary between a few hours and some days. Small amount of 27-day periodicity is caused by a single solar active region producing many solar flares during subsequent solar rotations.

Table 2 resumes the extreme values of the variability of the solar corpuscular radiation observed at the earth.

VARIABILITY IN THE EARTH ATMOSPHERIC PARAMETERS AND THEIR RELATION TO SOLAR VARIABILITY

Observations in the different levels of the atmosphere with precision enough to study the influence of solar variability have started only in the last few decades. For the low atmosphere, meteorological records have been kept for only a few centuries, but because of its effects on the biosphere and hidrophere, archaeological and geological records can be used to estimate certain parameters such as temperature or rainfall for much longer periods of time.

Table 3 shows a few significant examples of variations of atmospheric parameters that are or may be associated with solar irradiance variations other than the daily or annual effects.

Upper Atmosphere

It has been known for a long time that the composition and temperature of the thermosphere change drastically in association with the sunspot cycle (Fig. 5 is taken from a paper published by M. Nicolet in 1966). The excellent review by Priester et al

Table 3.   Sun/Atmosphere   Coupling

Atmospheric height(km)	Variable	Variation	Period or association
100–2000	Temperature Composition Density	700–1700 ^{o}K	11-year cycle 27 days: solar activity geomagnetic activity
60–100	Ionization	4–60 %	11 years cycle
20–100	Temperature		11 years cycle
20–40	Temperature	5–30 ^{o}C	11 years cycle
	O_3	25 %	11 years cycle
200–850 mb	Vorticity area index	10 %	solar magnetic sector boundaries
10–20	Thermopause height	5 %	11 years cycle
Ground level	(local)temperature rainfall thunderstorms		11 and 22 years
	Global climate	<0.1 ^{o}C/year	1-10^{7} years

(1967), shows the correlation that exists between the thermospheric density and the 27-day and 11-year solar cycles, and the effect that geomagnetic activity has on the density of the upper atmosphere (Fig. 6), based on results obtained by satellite drag data.

Middle Atmosphere

     included here are the stratosphere, ionosphere and lower part of the thermosphere (20 - 100 Km). This is the part of the atmosphere that is less well known from the point of view of variability and the one about wich present and near future satellite observations should produce important contributions. Schwentek (1971) reports on measurements of radio abosrption (2.61 KHz) covering almost a whole sclar cycle that show a large change in the density of the upper ionosphere and lower thermospher

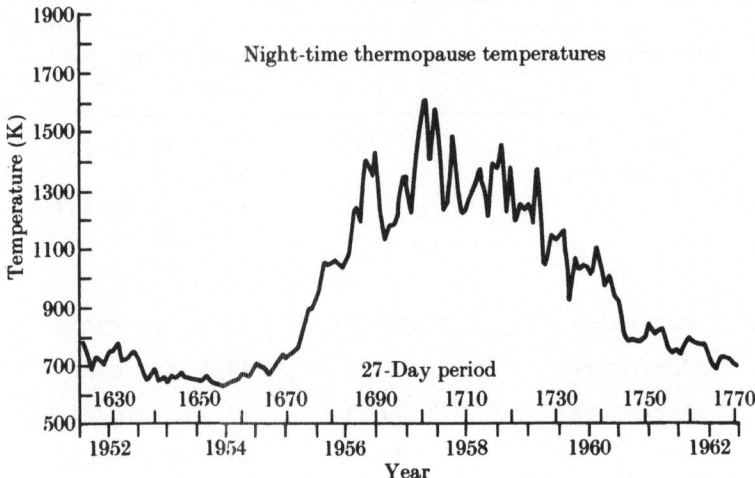

Figure 5.   *The variation of the thermopause temperature at*
night during the course of the sunspot cycle (after
Nicolet, 1964, from Akasofu and Chapman 1972).

He also reports stratosphere measurements made with radiosonde
launchings over Berlin. They show the presence of a clear
solar cycle variation of the temperature at different heights
of the stratosphere.

    The ion density in the high latitude stratosphere is known
to suffer very large variations associated with the 11-year
solar cycle (Fig. 7) caused by the intensity change of the
cosmic rays that produce the ionization (Ruderman and
Chamberlain, 1975).

    Another observed modulation is the variation of the height
of the low latitude tropopause that has been shown by Cole
(1975) to be correlated with the solar sunspot cycle. It is
interesting to notice that he finds a latitudinal effect which
is better organized when he uses the geomagnetic latitude of
the observing stations than when he uses the geographycial
latitude, thus pointing toward a magnetospheric control.

    Finally a parameter that is considered to be important
for the thermal equilibrium of the atmosphere, the O_3 density
is reported by Paetzold et al (1972) to vary between 10 and
30% at the level of its maximum density (20-30 Km) during the
sunspot cycle (Fig. 8).

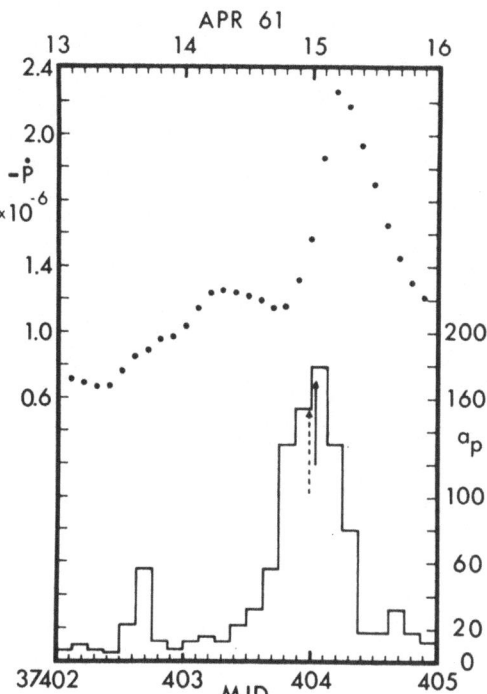

Figure 6. Increase of the atmospheric drag on the Explorer 9 satellite during a geomognetic storm on April 14-15, 1961. The rate of change of period Ṗ is plotted at 0.1 day intervals versus time and compared with the original 3-hourly a_p-indices and with the 0.2 day running mean of a_p (solid arrow) and with the 0.4 day running mean (dotted arrow), (after Priester et al, 1967).

The Lower Atmosphere

The existence of meteorological data ensures that this is the atmospheric region where most data is available by many orders of magnitude, but they are also the most complex and only a very few number of research studies show statistical evidence of solar variability influence in the atmosphere at low heights. J.W. King (1975) has put together a number of examples of meteorological parameters that show variations in the amount of rainfall and temperature that show periodicity of 11-year in some cases and of 22-years in others (Fig. 9). The lightning index obtained from the frequency of lightning strikes on the electrical power distribution system in Britain is another

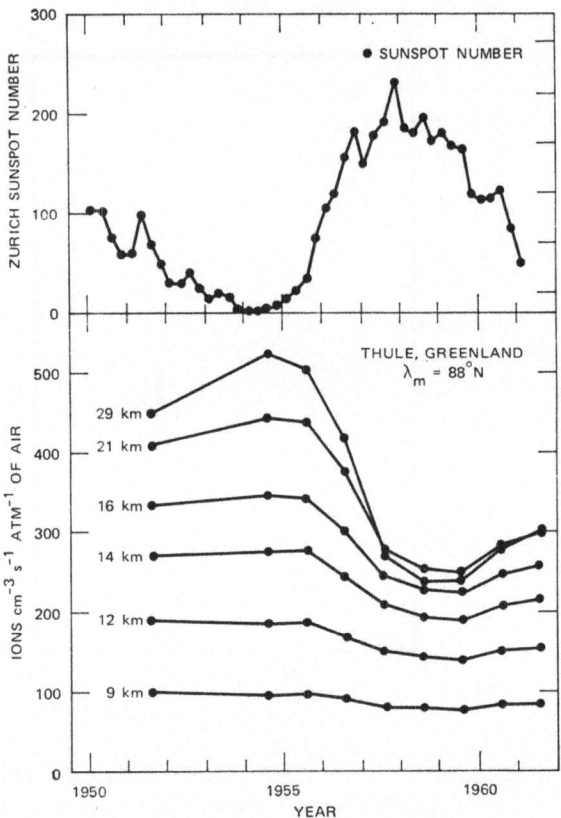

Figure 7. *Correlation of sunspot number with ionization at selected pressures over Thule, Greenland. (after Neher and Anderson, 1962, from Ruderman and Chamberlain, 1975).*

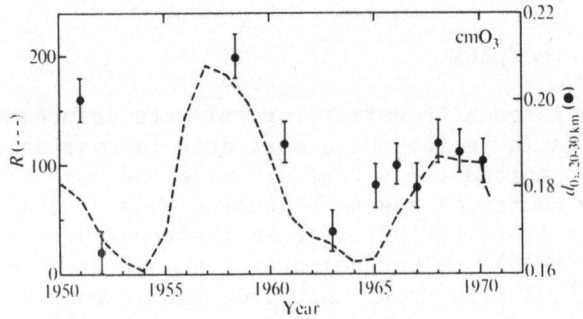

Figure 8. *Variation of the stratospheric ozone between 20 and 30 km altitude during two sunspot cycles. R, Sunspot-number (after Paetzold et al. 1972).*

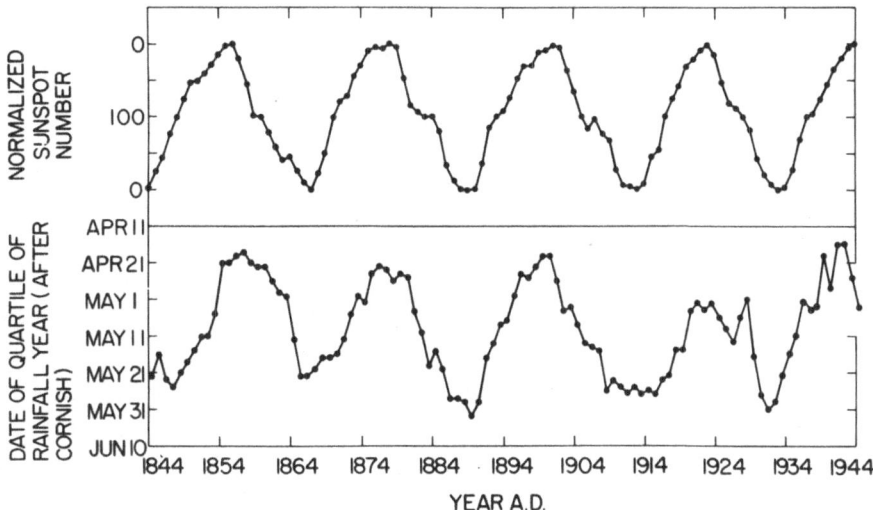

Figure 9.   *Ten-year smoothed means (lower curve) of the annual*
rainfall "quartile" (the date by which one quarter
of the annual rainfall had occured) for Adelaide,
Australia (after Cornish. The date fluctuates by
about six weeks in phase with the double sunspot
cycle plotted in the form shown in the upper curve.
(From King, 1975).

example.   The droughts every 22 years in the americhan high
plains have been reported by Roberts (1975).

Another evidence of influence of solar activity in
atmospheric conditions has been found in the correlation
between interplanetary magnetic field sector boundaries and
the atmospheric vorticity index over the northern hemisphere
(Wilcox et al. 1974, 1975). The vorticity index is found to
decrease by about 10% around the time when a sector boundary
reaches the earth (Fig.   10). In the same line of research
geomagnetic activity has been associated with troughs in
the north pacific and with the thunderstorm activity (see
review by Wilcox 1975).

So far, in the lower atmosphere, we have reffered to
periodical variations of local atmospheric parameter in the
order of years and tens of years, limitations produced by the
time of available good observations for the sensitivity involved. On the other side we have seen some correlations

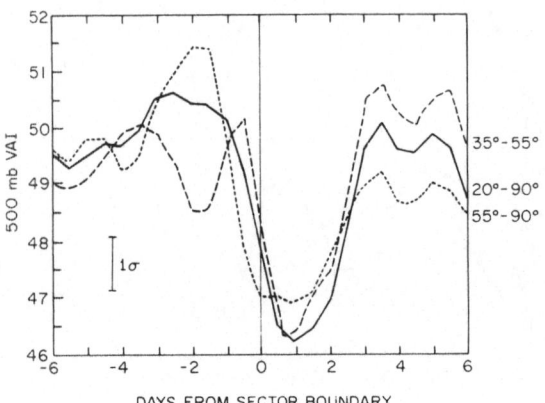

Figure 10. Superposed epoch analysis of the 500 mb vorticity area index (the area of low pressure throughs in the northern hemisphere) about times when solar magnetic sector boundaries were carried past the earth by the solar wind. The results are shown separately for the latitude zone 35⁰ N – 55⁰ N, 90⁰ N, and for the entire northern hemisphere north of 20⁰ N. (After Wilcox et al 1975).

The study of global meteorologial, archaeological and geological data gives us information on the global change of some atmospheric parameters, usually the temperature is chosen as the most representative. The observed changes at nearly all time scales between 1 and 10^7 years are very important (Fig 11).

In a review for the GARP, Flohn (1975) discussed the time scale of changes that have occured in the past, and one of the most interesting features is the existence during the past 10^5 years of climate changes that have happened with a temperature gradient of the order of 0.05°C/year for periods of some 100 years. Changes of that order would be of dramatic social consequences. In the same publication, and within the framework of present projects to study the global earth climate Oort (1975) presents results on the average temperature over the northern himisphere that show a linear variation of -0.1⁰ C/year during the five years between 1958 and 1963.

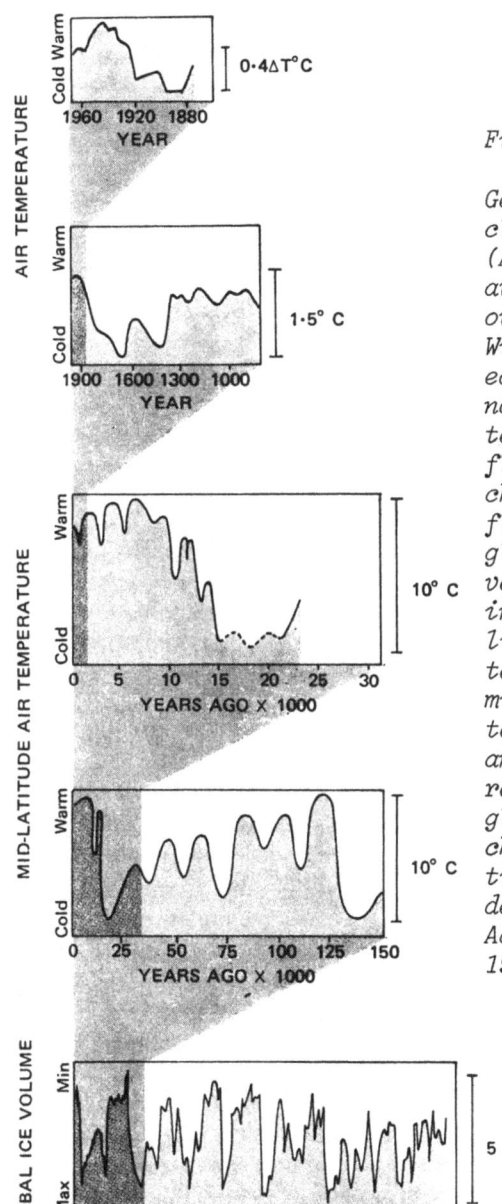

Figure 11.

*General trends in global
climate; the past million years.
(A) Changes in the five-year
average surface temperatures
over the region 0-80 N; (b)
Winter severity index for
eastern Europe; (C) Generalized
northern hemisphere air-
temperature trends, based on
fluctuations in alpine glaciers,
changes in tree-lines, marginal
fluctuations in continental
glaciers, and shifts in
vegetation patterns recorded
in pollen spectra; (d) Generalized northern hemisphere air-
temperature trends based on
mid-latitude sea-surface
temperature, pollen records,
and on worldwide sea-level
records; (E) Fluctuations in
global ice-volume recorded as
changes in isotopic composition of fosil plankton in a
deep-sea core (U.S. National
Academy of Sciences, in GARP,
1975).*

COMMENTS

The global climate of the earth can certainly be affected by
endogenous processes and amongst others, mechanisms have been
proposed that take into account changes in atmospheric dust,
caused by volcanic activity (wich would influence the reflectivity
of the atmosphere), the change in composition (this would change
the absorptivity of the atmosphere) and the change in intensity
and direction of the earth's magnetic field (which would affect
the number of charged particles entering the atmosphere).

But here we are mainly interested by the exogenous processes.
The irradiance at the exterior of the earth may change for two main
kinds of reasons. One, the mechanical properties of the sun and
the earth, seasonal and annual periodicities are of this kind, and
it has been proposed that the precession of the earth about the
sun (26.000 years) would affect the radiation effectively
received by the earth and would be one of the causes of long term
climate change in the earth (Kukla, 1975). The other kind are the
changes in the radiation flux emitted by the sun. To this category
belong the well known solar activity phenomena (plages, flares),
the 11-years, 22-years cycle, 80-years period and the apparent
disappearance of sunspots during 70 years in historical times
(Eddy,1975). In the long term, stellar theory only allows for a
change in solar flux emission of the order of 1% per 10^7 years
(Cameron 1975), but it has been proposed (McCrea, 1975) that the
Ice Ages can be explained by changes in brightness that the sun
experiment when traversing dense lanes of interstelar clouds.
in its travel around the galaxy.

We have seen that solar radiation variations produce effects
at all levels in the earth atmosphere (see Table 3). At first
glance the effects are much larger on the outer atmosphere than
at lower heights and the importance of the effects seems to
decrease with increasing depth in the atmosphere. However, one
has to bear in mind that the mass increase involved when going
down and the complexity of the dynamic phenomena in the low
atmosphere makes it much more difficult to observe small
changes superimposed on the meteorological variations.

When considering the effect of the solar radiation variations
it is important to remember the relative importance of the energy
flux carried by each of the radiations. For instance the
kinetic energy that the solar wind deposits in the earth
magnetosphere is on the average only of the order of 10^{-5} of the
electromagnetic radiation intercepted by the earth. A major
complicating factor but which may be exploited to elucidate the
physical mechanisms involved is the intercorrelation that exists
between the radiations of different wavelengts and types at its

generation at the sun. For instance in relationship to the
correlation between solar magnetic sector boundaries and atmospheric vorticity area index it is interesting to notice that
Heath et al (1975) have found that the passage of the solar
boundary by the central heliographic meridian is associated with
an increase of the far ultraviolet solar radiation (around 1200 Å).
At the other end, in the earth's atmosphere and magnetosphere,
again the internal correlations may produce misleading
associations of phenomena complicated by the time constants
that may be very different, giving rise to false time sequence
associations. As an example, the case of atmospheric effects
associated with geomagnetic activity, one can imagine physical
processes by which the external magnetosphere modified by the
solar wind, transmits changes to the ionosphere and atmosphere
by charged particles and/or magnetohydrodynamic waves, or viceversa
that a bulk movement of the atmosphere of dynamic origin, perhaps
triggered by a solar radiation change, produces a variation in
the charge or conductivity of the ionosphere and from it the
perturbation propagates to the outer magnetosphere along the
geomagnetic field lines.

I can not comment in detail the implications of the solar
variability for the physics of the atmosphere, but it is clear
that the sun earth relations will only be understood when one
will have a clear picture of the dynamics of the middle and
upper atmosphere and that the knowledge of the physics of the
atmosphere can profit very much from the study of the effects
of the solar variability at the different heights of the
atmosphere.

CONCLUDING REMARKS

The foregoing comments suggest that in order to understand
the solar-terrestrial relations and the physics of the global
atmosphere, the measurements that will be done of the physical
processes in the atmosphere, with spacelab, and in the sun with
sun observation, will or would be best complemented by a long
term study that encompasses at least the 27-day and the 11-year
time periods. The measurements that would be needed are those
that monitor:
   a) the solar irradiance constant, with a stability of at
      least 0.1 %,
   b) the variations in spectral composition with a
      precission that is superior to the variability produced
      by the 27-day period and 11-year cycle,
   c) the corpuscular radiation,
   d) atmospheric parameters at different heights, that may
      indicate the destabilization of the atmosphere due to
      a relative small triggering effect,

    e) the global atmospheric energy budget and its basic
    components.

This review has been prompted by the interest of ESA in
making a preliminary study of the feasibility of one or two
spacecraft that would simultaneously monitor the solar radiation
and earth's climate.

REFERENCES

Akasofu S. and S. Chapman, "Solar terrestrial Physics", at the
Clarendon Press, Oxford 1972.

Cameron A.G.W., Major variations in solar luminosity? Rev.
geophys. Space Phys., $\underline{11}$, 505 (1973).

Castelli J., S. Basu and J. Aarons, Solar radio emission, in
"Handbook of Geophysics and Space Environements", edited by
S.L., Valley, p. 16 - 18, McGraw - Hill Book Co. New York, 1965.

Cole H.P., An investigation of a possible relationship between
the height of the low - latitude tropopause and the sunspot
number, J. Atmos, Sci. $\underline{32}$, 998 (1975)

Eddy J.A. The Maunder Minimun: when the sun lost its spots,
EOS Trans. AGU., $\underline{56}$, 1055 (1975).

Flohn H., History and intransitivity of climate, in GARP
publication Series No. 16, page 106 (see ref. GARP, 1975).

GARP "The physical basis of climate and climate modelling",
GARP Publication series No. 16 edited by World Meteorological
Organization and International Council of Scientific Unions,
Geneva, 1975.

Gleissberg W, The possible behaviour of sunspot cycle 21,
Sol. Phys., $\underline{21}$, 240 (1971).

Heath D.F., Space observations of the variability of solar
irradiance in the near and far ultraviolet, J. Geophys. Res.
$\underline{78}$, 2779 (1973).

Heath D., J.M. Wilcox, L. Svalgaard and T.C. Duvall, Relation
of the observed for ultraviolet solar irradiance to the solar
magnetic sector structure, Solar Physics $\underline{45}$, 79 (1975).

Jacchia L.G., Static diffussion models of the upper
atmosphere with empirical temperature profiles, Smithson.
Contrib. Astrophys., 8, 215 (1965).

King J.W., Sun-wether relationships, Astronautics and
Aeronautics 3, No. 4, p. 10 (1975).

Kukla G.J., Missing link between Milankowitch and climate,
Nature 253, 600 (1975)

Malitson H.H., The solar energy spectrum, Sky Telesc., 29,
162 (1965).

McCrea W.H., Ice ages and the galaxy, Nature 255, 607 (1975).

Neher H.V., and H.R.Anderson, Cosmic rays at balloon altitudes
and the solar cycle, J. Geophys. Res., 67, 1309 (1962).

Nicolet M., The structure of the upper atmosphere,"Research in
Geophysics. Vol. 1. Sun,upper atmosphere and space", edited
by H. Odishaw, p. 243, M.I.T. Press, Cambridge, Mass.(1964)

Oort A.H., On the variability of the general circulation of the
atmosphere as deduced from aerological data, in GARP publication
series, No. 16, page 95, (see Ref. GARP 1975).

Paetzold H.K., F.Piscalar and H.Zschorner, Secular variations
of the stratospheric ozone layer over middle Europe during the
solar cycles from 1951 to 1972, Nature, 240, 106 (1972).

Parkinson J.H., and K.A. Pounds, X-ray observations of solar
active regions from OSO-5, Solar Physics 17, 146 (1971).

Priester E., M.Roemer and H.Volland, The physics behaviour of
the upper atmosphere deduced from satellite drag data, Space
Sci. Rev. 6, 707 (1967).

Roberts W.O., Relationships between solar activity and climatic
change, in "Possible relationships between solar activity and
meteorological phenomena", p. 13, NASA report SP - 366
(Washington, D.C., 1975).

Rudeman M.A., and J.W. Chamberlain, Origin of the sunspot
modulation of ozone and its implications for stratospheric
no injection, Planet. Space Sci. 23, 247 (1975).

Schwentek H., The sunspot cycle 1958/70 in ionospheric absorption
and stratospheric temperature, J. Atmosph. Terr. Phys., 33,
1839 (1971).

Smith E.V.P. and D.M. Gottlieb, Solar flux and its variations, Space Sci. Rev. 16,771 (1974).

Weiss N., Solar seismology, Nature, 259, 78 (1976).

Wilcox J.M., Solar activity and the weather, in "Possible relationships between solar activity and meteorological phenomena", p. 25, NASA report SP - 366 (Washington, D.C. 1975).

Wilcox J.M., and N. F. Ness, Quasi-static corotating structure in the interplanetary medium, J.Geophys. Res., 70, 5793 (1965).

Wilcox J.M., P. H. Scherrer, L. Svalgaard, W. O. Roberts, R.H. Olson and R.L. Jenne, Influence of solar magnetic sector structure on terrestrial atmospheric vorticity, J.Atmos. Sci., 31, 581 (1974).

Wilcox J.M., L. Svalgaard and P.H. Scherrer, On the reality of a sun weather effect, SUIPR report No. 645 (October 1975), to be publised in J. Atmos Sci.

DISCUSSION

A.C. Durney: With what accuracy is the long-term stability of the solar constant known at the present time?

V. Domingo: According to Drummond (in Advances in Geophysics, 1970), the Smithsonian Institution Astrophysical Observatory has made ground-based measurements of the solar constant over a period of 30 years which indicate a constancy of better than 1%.

A. Vidal-Madjar: A comment to point out the fact that the solar irradiance variability is not well established below 3000 Å according to the discussions that took place during the end of April '76 workshop on The Total Output of the Sun (Boulder, Colorado). As an example, following Heath and Thekaekara who presented a review of the solar irradiance from 1200 Å to 3000 Å, it seems that the variation observed over the 11 year solar cycle may be much larger than the 1% value you quoted, and may in fact be of the order of 10% to 100% between 3000 Å and 2000 Å.

V. Domingo: The 1% variation per solar rotation at 3000 Å is based on information obtained from a paper by Heath (1973).

The results that you quote are probably more recent, and I was not aware of their publication.

P.M. Banks: To what extent do variations occur in the earth-ionosphere potential during periods of solar and/or magnetospheric disturbance? Could such variations have a practical effect upon cloud nucleation or weather activity?

V. Domingo: I do not know how much investigation has been done in this respect, but it would certainly be interesting to have answers to both questions.

A.F. Nagy: Spacelab can make a contribution to solar monitoring by providing the platform for periodic <u>absolute</u> calibration of instruments that are part of a long-term monitoring programme and are carried by free-flyers.

F.H. Schmidt: I would like to remark that there is a rather well established connection between the occurrence of solar flares and subsequent blocking, i.e. "meridionalisation", of the tropospheric circulation. There also appears to be a ninety-year periodicity, connected with this link.

G. Hunt: How can you be sure that the correlations of changes in atmospheric state (such as rainfall) with solar activity, are not due to natural variations in the Earth's meteorology? We are aware, through numerical modelling, that changes in the ocean temperature are responsible for large variations in the Earth's climate, e.g. desert regions.

V. Domingo: I suppose that you refer to Dr. King's and other compilations of 11-year and 22-year periodic phenomena. Only the authors of such compilations can work out their statistical significance.

MARÉES ET ONDES PLANÉTAIRES DANS LA BASSE THERMOSPHÈRE ET LA HAUTE MÉSOSPHÈRE

R. BERNARD
CNET-CRPE ISSY LES MOULINEAUX FRANCE

RÉSUMÉ :   Les caractéristiques moyennes des marées et des ondes planétaires, et leur évolution saisonnière sont présentées, telles qu'elles sont observées par le radar météorique de Garchy et le sondeur à diffusion incohérente de St Santin. Les résultats obtenus par d'autres radars météoriques, ou d'autres sondeurs sont pris en compte pour étudier la variation en latitude des marées observées. Les principales caractéristiques des marées, diurne, semi-diurne et ter-diurnes sont étudiées en rapport avec les prévisions des théories classiques des marées. On montre que la variabilité des marées paraît liée aux variations de la circulation générale de l'atmosphère, et qu'il faut tenir compte de cette circulation pour expliquer théoriquement cette variabilité.

INTRODUCTION :

La dynamique de la haute mésosphere et de la basse thermosphère, entre 80 et 150 km peut-être observée par différentes techniques : mesures par fadings (1) ou réflexion partielle (2), mesures par laser (3), par fusées (4-5). Enfin , étude par radar météorique (6,28) ou par sondeur à diffusion incohérente.(7,20)

Nous nous limiterons ici aux résultats donnés par ces deux derniers appareils, qui apportent à la fois une mesure directe du vent et de la température (diffusion incohérente) et qui permettent une observation continue de ces paramètres. Nous utiliserons principalement les résultats du radar météorique de Garchy (France) et du sondeur à diffusion de St Santin (France) les autres sondeurs à diffusion et radars météoriques donnés en annexe servant à étendre ces mesures à d'autres latitudes ou longitudes.

J. J. Burger et al. (eds.), Atmospheric Physics from Spacelab, 43–60. All Rights Reserved.
Copyright © 1976 by D. Reidel Publishing Company, Dordrecht-Holland.

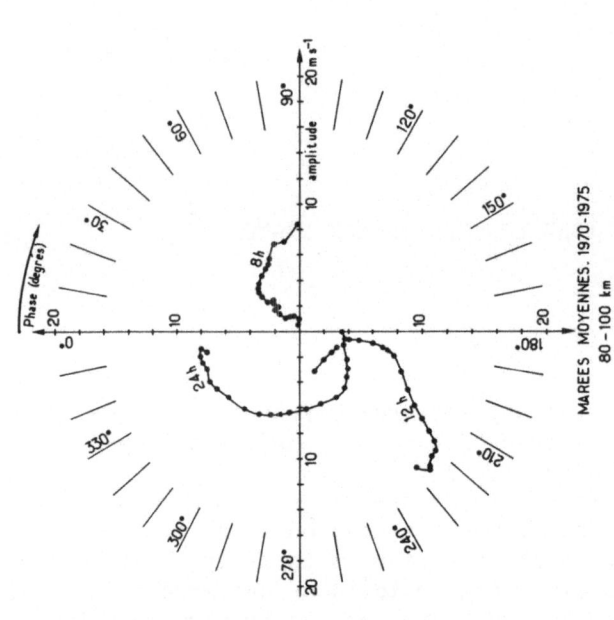

Fig 2 Amplitude et phases moyennes des marées à Garchy- l'intervalle entre deux points est 1 km. La flèche donne le sens des altitudes croissantes. La phase correspond à une variation en sin $(\omega t + \phi)$

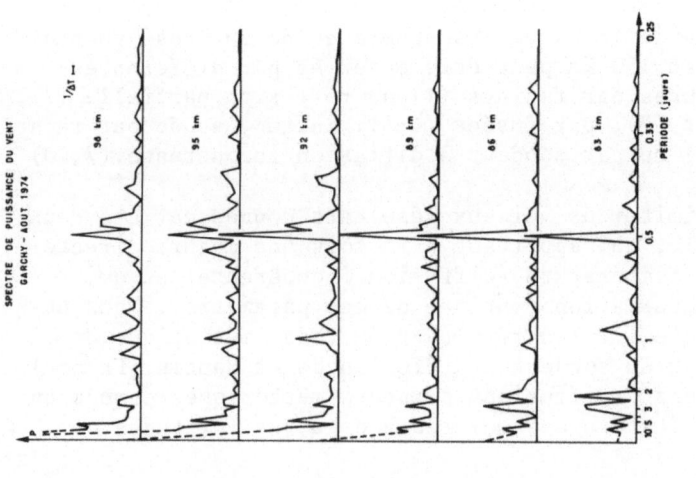

fig 1. Spectre de puissance du vent zonal -10-20 Août 19/4 Garchy France

I ANALYSE DES DONNÉES :

Une analyse harmonique peut-être appliquée aux variations complexes de la composante zonale du vent observée à Garchy (fig 1). Le spectre obtenu présente des raies caractéristiques, la plus importante étant à 12 h une autre raie est visible à 24 h ou près de 24 h, et une troisième apparait parfois à 8 h.

D'autres raies, plus variables peuvent être observées vers 2 ou 3 jours, ou à de plus grandes périodes de même qu'une large composante moyenne sur la campagne.

Nous allons étudier successivement ces différentes composantes du spectre, et leurs variations en fonction de la latitude ou de la saison.

II ÉTUDE DES MARÉES :

1°) Les composantes harmoniques de 24 h peuvent être associées aux phénomènes de marée atmosphérique maintenant bien étudiées (8). L'absorption de l'UV solaire par l'Ozone vers 50 km d'altitude peut exciter des oscillations d'ensemble de l'atmosphère, comprenant des variations du vent, de la température, de la pression et de la densité. Les conditions aux limites sur la sphère ne permettent la présence que d'une série discrète de modes, chaque mode étant caractérisé par sa structure verticale et en latitude. Certains modes peuvent se propager verticalement. La conservation de la densité volumique d'énergie de l'onde lui impose alors une croissance suivant une loi exponentielle.

$$A \simeq \exp(\alpha_0 3) \qquad \alpha_0 = 1/2 \, h_0$$

h_0 échelle de hauteur

à 90 km $\qquad T = 190° \qquad \alpha_0 = 0.09 \, km^{-1}$

On peut chercher à identifier ces modes en faisant une analyse moyenne des termes de 24 h, 12 h , 8 h, observés dans le spectre. Cette analyse a été faite à Garchy sur 5 années de mesures, représentant 330 j de mesure effective. D'autre part, les différentes campagnes ont été groupées en périodes présentant des caractéristiques analogues et assimilées aux saisons bien que ne correspondant pas aux solstices et aux équinoxes.

2°) Marée diurne
2-1 Etude moyenne.

La marée diurne moyenne à Garchy (fig 2) croit régulièrement de 3 à 7 ms^{-1} entre 80 et 90 km, puis se stabilise jusqu'à 100 km. La phase tourne régulièrement. Le vent est maximum vers l'est

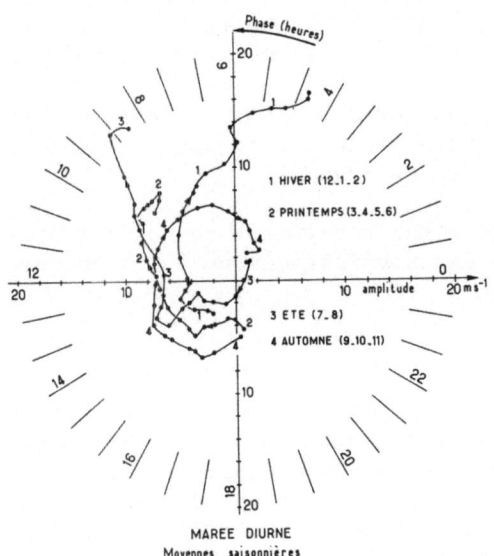

fig 3   Amplitude et phase (heure du max vers l'est) de
la marée diurne pour les 4 saisons.

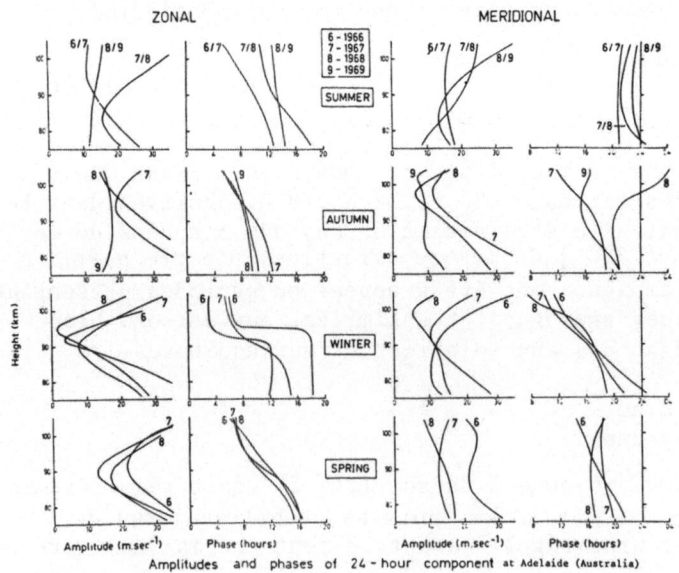

fig 4 amplitude et phase de la marée diurne à
Adélaïde (35°s) Les saisons correspondent à l'hémisphère Sud.

à 7 h à 100 km, à 18 h à 80 km: la phase se propage vers le
bas avec une longueur d'onde moyenne de 42 km.   Ceci correspond
à une marée dont l'énergie se propage vers le haut.

Le taux d'acroissement est voisin de α_0 jusqu'à 90 km,
puis tombe à α = 0,015, correspondant à un fort amortissement.

2-2   Cette structure moyenne recouvre des structures différentes suivant les saisons (fig. 3).

- En hiver et en été, la croissance reste régulière jusqu'à
100 km; il n'y a pas d'amortissement.   La longueur d'onde verticale reste voisine de 40 km.   Le maximum vers l'est apparaît en
moyenne 3 à 4 h plus tôt en hiver qu'en été.

- Le printemps montre une variation analogue, mais la croissance y est plus faible et la longueur d'onde voisine de 50 Km.

- En automne, l'amplitude reste constante entre 85 et 95 km,
puis décroit, la longueur d'onde verticale étant égale à 33 km.
Ceci correspondrait donc à une onde fortement amortie, ou a une
superposition de deux ou de plusieurs ondes.

2-3   Ces observations aux latitudes moyennes sont confirmées
par les données d'Obninsk (U.R.S.S.) et des radars météoriques
soviétiques (10-11-12).   Sans détermination d'altitude, ils observent une marée moyenne inférieure à 10 ms^{-1} avec un maximum
en août et décembre, correspondant aux résultats d'hiver et d'été
de Garchy.

A Adélaïde, 35° S les résultats sont très différents (fig. 4)
(13).   Une forte marée diurne, 20 à 30 ms^{-1}, est observée en
automne et en été, associée à une faible croissance et une grande
longueur d'onde.   La structure verticale est plus complexe au
printemps et en hiver, avec l'apparence de la superposition
d'ondes de grande et de courte longueur d'onde.   De plus la
structure est différente pour les composantes zonales et méridio-
nales.

De fortes amplitudes sont aussi observées aux basses latitudes,
à Kingston (18° N) (14), mais très variables et associées à de
courtes longueur d'onde.   Aux hautes latitudes également, la marée
diurne atteint 20 ms^{-1} à Hayes Island (85° N) (9) à College
(65° N) (15) et Molodezhaia (67° S) (10, 12).   Cependant, il faut
noter des mesures récentes faites par le C.N.E.T à Kiruna (67° N)
montrant une marée diurne faible et très variable.

2-4   En résumé, la marée diurne apparaît comme très variable
en latitude, avec une structure généralement complexe.   Il est
difficile de lui associer des modes de propagation classique,

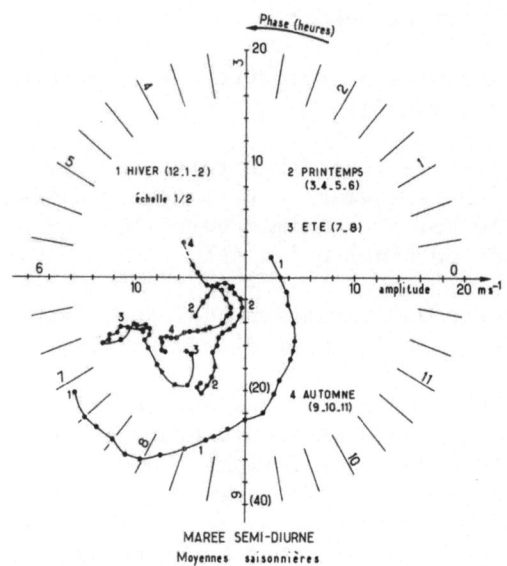

fig 5 Identique à la fig 3, pour la marée semi-diurne

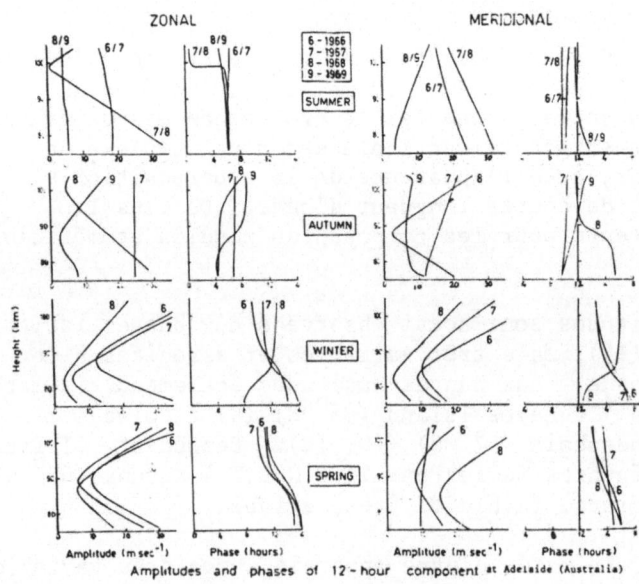

fig 6 Identique à la Fig 4 pour la marée semi-diurne

bien que la présence de fortes amplitudes à la fois aux hautes
et basses latitudes peut-être rapprochée des modes positifs et né-
gatifs définis par Lindzen (8)

3- Marée semi-diurne

3-1 Son comportement à Garchy est très différent. En moyenne
(figure 2) on observe une croissance continue jusqu'à 14 ms^{-1}
à 100 km. La phase tourne dans le même sens que la marée diurne
mais la courbe est décalée par rapport au zéro. Ceci peut se tra-
duire par la superposition d'une onde de grande longueur d'onde
verticale et d'une onde se propageant vers le bas et de longueur
d'onde verticale voisine de 40 km.

3-2 Les variations saisonnières sont très importantes (fig 5)
Une grande amplitude, jusqu'à 40 ms^{-1} à 100 km, est observée
en hiver, avec un taux de croissance $\alpha = \alpha_\circ$. La longueur d'onde
est voisine de 40 km. On peut interpréter cette oscillation comme
une seule onde progressive, dont la phase se propage vers le bas.

- En été, au contraire, l'amplitude et la phase restent
constante au-dessus de 85 km. On peut là aussi associer cette oscillation à une onde évanescente, sans propagation d'énergie.

- Le printemps et l'automne apparaissent comme des saisons in-
termédiaires, ou ces deux ondes progressive et évanescente sont
superposées. En fait, ces ondes peuvent exister, soit simultanément,
soit successivement, et la complexité de la moyenne observée pro-
vient surtout de la grande variabilité de la marée semi-diurne
durant ces saisons.

3-3 Les mesures faites aux latitudes moyennes par les sations
soviétiques, sans détermination d'altitude, ne peuvent rendre compte
de ces variations de structure. Elles observent un maximum d'amplitude en hiver et en automne et un changement de phase en
automne. Cependant, les valeurs moyennes à Obninsk correspondent
aux mesures moyennes de Garchy à 90 km (17).

La polarisation moyenne observée à Obninsk montrant un maximum vers l'est 3 h plus tôt que vers le sud, correspond bien à
la théorie des marées, qui prévoit une polarisation circulaire
Le même maximum en hiver est observé à Atlanta (35° N) (18-
19)
A Adélaide, l'amplitude atteint 3° ms^{-1} (fig6)
en hiver et au printemps. On observe également un faible accroissement et une grande longueur d'onde pour l'été de l'hémisphère
sud. Il apparait donc en été et en hiver une dissymétrie entre

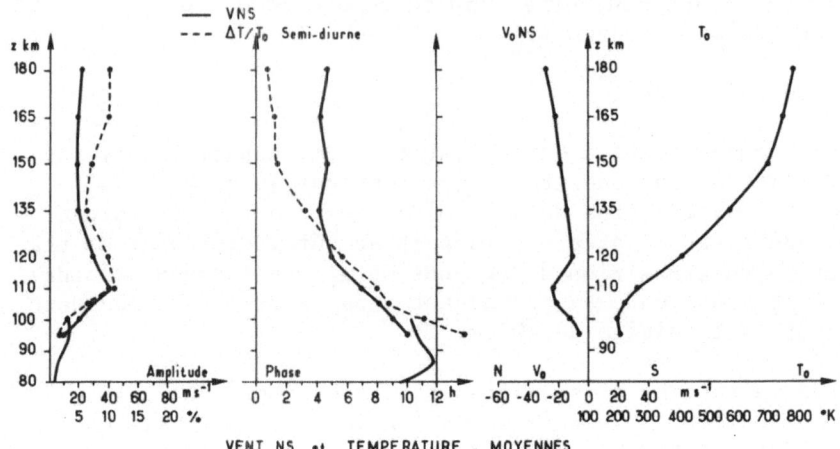

FIG 7 Composante semi-diurne du vent méridional et de la température observée à St Santin, les résultats du radar météorique de Garchy ont été portés entre 80 et 100 km pour comparaison

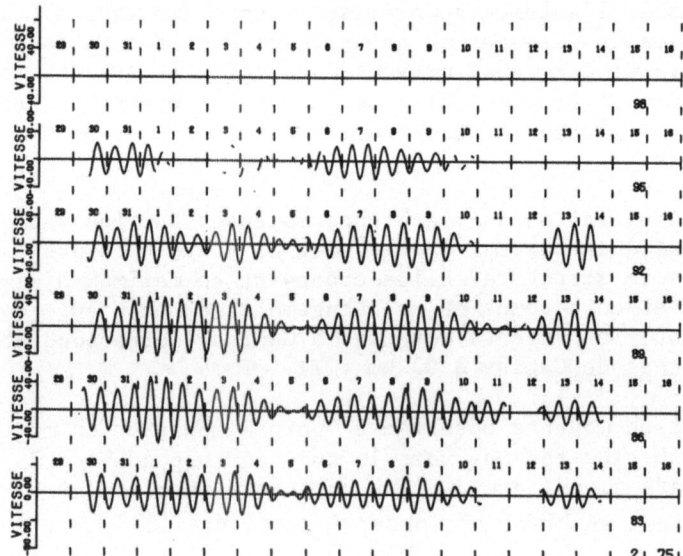

fig 8 Contribution au vent zonal des périodes comprises entre 10 et 14 h à Garchy, obtenue par filtrage numérique, à 6 altitudes.

les hémisphères nord et sud.

Les résultats aux hautes latitudes sont contradictoires, avec une forte amplitude au sud, faible au nord. La structure en latitude a été étudiée par Lysenko (9) utilisant essentiellement les données des radars soviétiques. Il trouve un maximum à 55° N. Ce maximum à 55° N correspond à des modes élevés du type S_4^2 (symétrique ou $S2^3$ (antisymétrique) et est en accord avec les courtes longueurs d'onde souvent observées.

3-4 Les modes rendent compte des observations par sondeur à diffusion au-dessus de 100 km. (fig 7) le vent et les variations de température montrent un maximum à 110 km, et une courte longueur d'onde, 50 km (T) ou 60 km (V) jusqu'à 120 km. Au-dessus, l'effet de la viscosité moléculaire et la conduction thermique expliquent la faible variation de la phase et de l'amplitude.

Les relations de polarisations au-dessous de 120 km sont en accord avec les caractéristiques des modes $S2^4$ ou $S2^5$ (7). Les résultats des autres sondeurs à diffusion, à Millstone Hill (42° N) (20) et Arecibo (21) (18° N) confirment ces résultats, mais l'absence d'accord sur la phase moyenne observée à Arecibo et à Millstone Hill ou St Santin rendent plus difficile l'interprétation en mode (22). Cette complexité est à rapprocher également de la grande variabilité jour à jour observée au-dessus de 100 km.

En raison de cette variabilité, les variations saisonnières sont peu évidentes. Cependant, on n'observe jamais la large amplitude en hiver, ni la grande longueur d'onde en été caractéristique des résultats du radar météorique. La comparaison des valeurs à 100 km obtenue par le sondeur et le radar montre cependant un accord relatif pour l'amplitude et la phase compte tenu d'un déphasage de 3 h entre les composantes EW et NS du vent, et compte tenu de la précision sur la phase moyenne à 100 h obtenu par le sondeur ($\pm$ 1 h).

3-4 Variabilité jour à jour.

En plus de la variation saisonnière de la marée semi-diurne on observe une forte variabilité à court terme, même en été ou en hiver. La fig 8 montre le résultat d'un filtrage numérique des données ne conservant que les variations dues aux mouvements de période comprise entre 10 et 14 h. On observe une forte modulation de l'oscillation semi-diurne avec une période voisine de 6 h

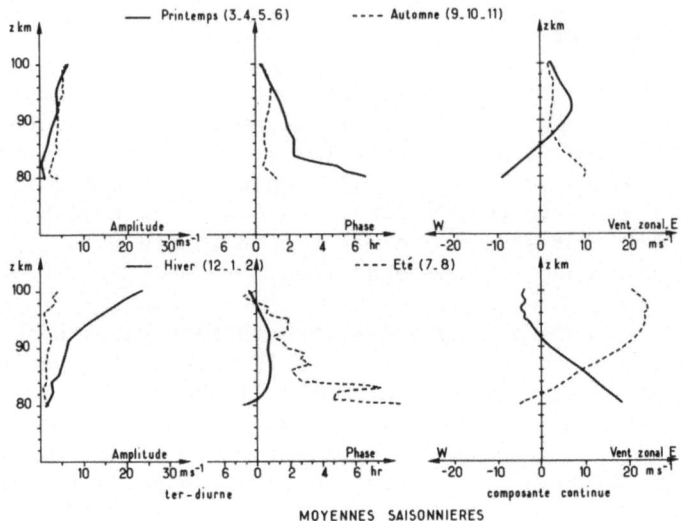

Fig 9. Moyennes saisonnières des composantes ter-diurne
et continue obtenues à Garchy.

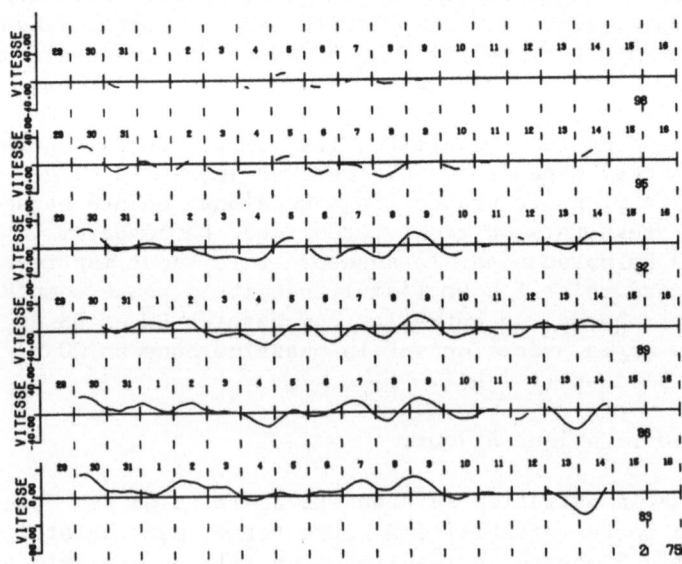

Fig 10 Identique à la fig 8 , pour les périodes supérieures à
30 h

et variable avec l'altitude. De telles variations sont fréquemment observées, et ne peuvent s'expliquer par une variabilité correspondante de l'excitation solaire.

4-1 Marée ter-diurne

L'analyse moyenne à Garchy (fig 2), montre l'existence d'une oscillation de période 8 h, faible jusque vers 95 km, puis atteignant 8 ms^{-1} à 100 km. La composante de 8 h n'est pas généralement observée et n'est pas associée à une onde de marée

Cependant, les résultats d'hiver à Garchy (fig 9) montrent que cette oscillation atteint en moyenne 18 ms^{-1} à 100 km. La croissance au-dessus de 90 km est très rapide et correspond à α = 0,13, donc à un accroissement de l'énergie de l'onde. Au-dessus de 90 km, la rotation de phase correspond à une longueur d'onde inférieure à 40 km.

Cette oscillation est totalement absente en été, et montre une amplitude faible au printemps, associée à une courte longueur d'onde, et en automne, associée à une grande longueur d'onde.

Glass et Fellous (23) suggèrent que cette oscillation semidiurne est la superposition d'une onde de marée excitée par le soleil, de grande longueur d'onde, et d'une onde de courte longueur d'onde résultant de l'interaction non linéaire entre les marées diurne et semi-diurne. Ceci peut expliquer la forte croissance au-dessus de 90 km, et la forte amplitude en hiver. Lorsque les marées diurnes et semi-diurne ont une forte amplitude.

5- En résumé, la théorie classique des marées et l'excitation solaire directe ne permet pas de rendre compte des observations, en particulier des courtes longueurs d'onde observées généralement, ni de la structure en latitude des marées diurne et semi-diurne : l'excitation solaire directe devrait générer des modes ayant un maximum à l'équateur, et de grande longueur d'onde.

D'autre part, la variabilité saisonnière n'est pas explicable par la variation de l'excitation solaire. En particulier, il faut noter que les saisons que nous avons définies dans l'étude de la circulation générale, mésosphérique ou stratosphérique.

Il apparait donc important de tenir compte, dans l'étude de l'excitation et de la propagation des marées, des couplages avec la circulation générale, ou entre les modes de marées euxmêmes. De tels couplages ont été étudiés numériquement par Lindzen et Hong (24), mais leurs résultats ne correspondent pas encore

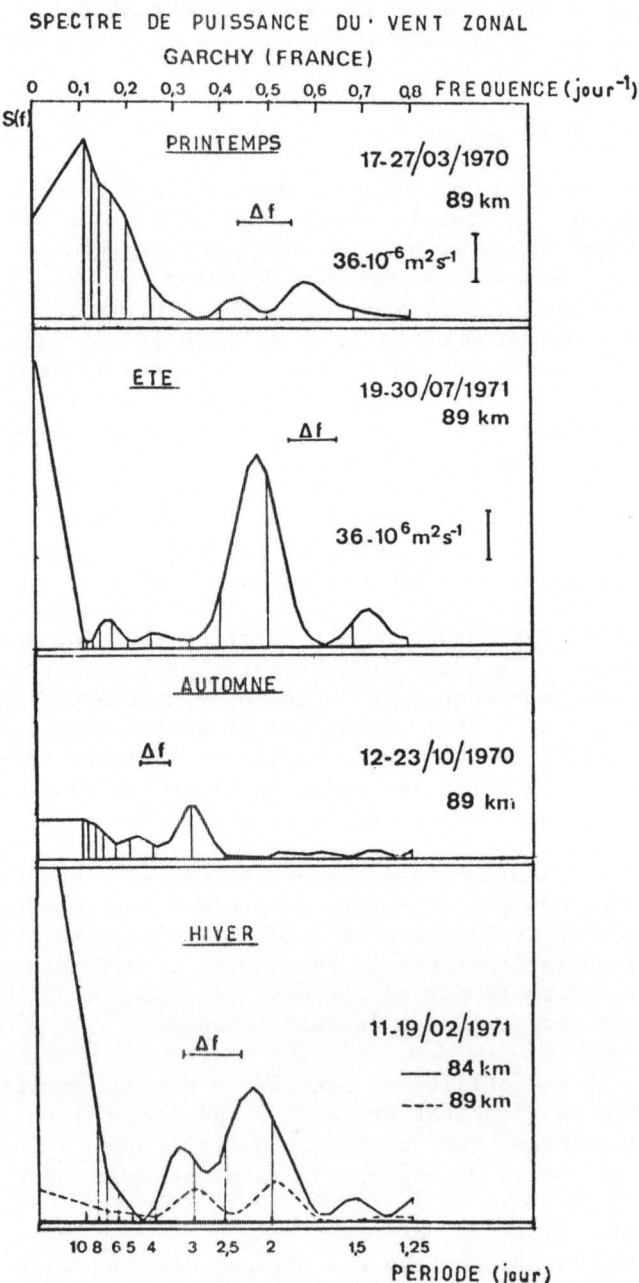

fig 11 Spectres de puissance du vent zonal pour 4
campagnes caractéristiques à Garchy

aux observations faites.

III ONDES PLANÉTAIRES ET COMPOSANTE CONTINUE

1- Un filtrage numérique éliminant les variations de période inférieure à 30 h a été appliqué aux données d'août 1974 (fig 10) On voit qu'il reste des oscillations de plus grande période, de l'ordre de deux jours, superposée à un vent moyen. La structure verticale du vent moyen zonal varie peu d'une année à l'autre et présente des variations saisonnières caractéristiques (fig 9).

- Un fort gradient, positif en été au-dessous de 90 km, négatif en hiver.

- Une circulation zonale faible au printemps et en automne qui apparaissent comme des saisons de transition entre les circulations d'été et d'hiver. Ces mêmes caractéristiques se retrouvent à Adélaide, pour les saisons de l'hémisphère sud. Le vent méridional y est en général plus faible et se caractérise par un maximum vers l'équateur en été, et vers le pole en hiver.

2- Les oscillations de période supérieure à la journée sont beaucoup plus variables que les marées. La figure II présente quelques spectres obtenus à Garchy et caractéristiques des différents régimes (25.26.27).

2.1 - Une oscillation de période voisine de 2 jours est observée systématiquement en été, moins fréquemment en hiver. Elle peut atteindre 10 à 15 ms^{-1} l'amplitude et la phase restant constante sur toute la gamme d'altitude.

Cette onde est plus rare et plus faible aux équinoxes et parait associée à la présence d'une forte circulation zonale . On a pu observer cette oscillation simultanément à Garchy et à Obninsk (26). Le décalage en phase entre ces deux stations peut correspondre à une propagation vers l'est, avec une longueur d'onde zonale supérieure à 7 000 km, les caractéristiques de cette onde sont alors à rapprocher de celles des ondes de Rossby observées dans la stratosphère.

2-2 Des ondes de période 4 à 6 jours sont également observées , surtout en hiver et au printemps. Leur variabilité en altitude est beaucoup plus grande. De telles ondes pourraient être également associées aux ondes planétaires générées dans la stratosphère, ou même la troposphère (27) bien qu'aucune corrélation n'ait été observée à Garchy entre les variations de la pression au niveau du sol et les ondes observées (26).

IV CONCLUSION

Nous avons mis en évidence les principales caractéristiques
de la circulation zonale aux altitudes météoriques. L'étude de la
variabilité saisonnière ou jour à jour des marées ou des ondes
planétaires a montré l'importance de la dynamique et de la
structure générale de l'atmosphère aux niveaux troposphé-
riques et mésosphériques pour l'explication des caractéristiques de toutes les ondes observées. Une meilleure connaissance
de cette dynamique, et des relations de couplage entre les différentes ondes observées pouvait alors permettre une meilleure
explication de ces ondes, et une analyse de la pénétration de
ces ondes dans la thermosphère, et de l'apport d'énergie qu'elles
représentent.

Principales stations de mesure

1°) Radars météoriques	Latitude	Référence	
Hayes Island (URSS)	80,5°N	55°E	9-12
Kiruna (Suède)	67,5°N	20,3°E	
College (Alaska)	65°N	148°W	15
Tomsk (URSS)	56,5°N	85°E	9
Kazan (URSS)	56°N	49°E	9.10.11.12
Obninsk "	55°N	38°E	9.10
Sheffield (Grande-Bretagne)	53°N	1°W	29
Kiev (URSS)	50°N	31°E	9.10.11
Karkhov "	49°N	36°E	9.10.11
Garchy (FRANCE)	47°N	3°E	17.23.25.26.28
Frunze (URSS)	43°N	75E	9
Dushanbee (URSS)	38,5°N	69E	9
Atlanta (USA)	34°N	84W	18.19
Kingston (Jamaique)	18°N	77W	14
Adelaide (AUSTRALIE)	35°S	139E	13
Molodezhnaya (URSS)	67°S	45E	9.12
2°) Sondeur à diffusion incoherente			
St Santin (France)	45°N		7.22
Millstone Hill (USA)	42°N		20.21.22
Arecibo (USA)	18°N		21.22

Références

(1) K. SPRENGER-R SCHMINDER   1967 JATP, 29,183-199
(2) A.H MANSON, J B GREGORY
    and D.G. STEPHENSON   1974 J. Atmos SCI 31,2207,2215
(3) KENT G.S, KEELISIDE W,
    SANDFORD MCW AND WRIGHT RWH   1972 J.ATP 34.373
(4) A WOODRUM CG JUSTUS   1968 JGR 73 467 478
(5) S.P ZIMMERMAN, NW ROSENBERG,
    AC FAIRE, D COLOMB, E.A. MURPHY, W VICKERY
    LA . TROWBRIDGE AN D. REES   1973 XVI COSPAR. KONSTANZ

(6) A. SPIZZICHINO   1972 Thermospheric Circula-
                                           tion p 117 Academic Press
                                           New-york.

(7) R. BERNARD   1974 J.A.T.P 36,1105.1120
(8) S. CHAPMAN, R.S. LINDZEN   1970 R. Reidel Publ Comp-
                                           Dordrecht

(9) LYSENKO I.A, KACHEYEV B.L, KARIMOV MK,
    NAZARENKO M.K, ORLIANSKY A.D,FIALKOYE,
    CHEBOTAREV R.P   1969 I.zv Ocean and Atm.
                                           Phys 5,9,893,902

(10) LYSENKO I.A, ORLIANSKY A.D, PORTNIAGHIN Y 1972 Phil trans
                                           Roy Soc,A 271,457,471

(11) G. M TEPTIN, V.M STROSTIN   1971 JATP,33,807,814

(12) Yu. O.U.ILJICHEV, IA LYSENKO, A.D. ORLYSANKI
     aнд Yu.I. PORTNYAGIN   1974 JATP 36,1841,1849

(13) ELFORD WG 1973 IAGA Report to com VIII

(14) A.J SCHOLEFIELD, MALLEYENE 1975 J ATP, 37,273,286

(15) J.L HOOK. 1970 Planet Sp Sci 18,1623,163

(16) S. KATO 1966 J.G.R 71,3201 ,3209

(17) M GLASS, JL FELLOUS, M MASSEBEUF,

A. SPIZZICHINO, IA LYSENKO, Yu PORTNIAGHIN

 1975 JATP,37,1077,1087

(18) R.G. ROPER 1975 IUGG Grenoble

(19) R.G. ROPER 1975 Radio Science 10,363,369

(20) SALAH J.E, JV EVANS RH WAND 1975 JATP 37,461,489

(21) J.E SALAH RH WAND, JV EVANS 1975 Rad Sci,10,347,355

(22) J.E SALAH JE,RH WAND, R BERNARD 1976 à paraitre Annales de gé

physique

(23) M. GLASS, J.L FESSOUS 1975 Space Research XV 191.19

(24 R.S LINDZEN, S HONG 1974 Journal Atm Sci 1421

(25) M MASSEBEUF , M GLASS 1973 COSPAR Konstance

(26) M MASSEBEUF 1975 JATP 37 1511-1524

(27) H.G. MULLER 1966 Plan Sci 14,1253

DISCUSSION

J.B. Gregory: In Canada, at 52°N, we see great variability in the
winter circulation at 80-100 km, including the build-up and
breakdown of the polar vortex.   We would hesitate to say that
the prevailing wind in this season is constant from year to
year.   Is this the situation in France also?

R. Bernard: Yes.   The breakdown of the polar vortex is observed
but only as a change in the gradient of the prevailing wind.
The characteristic winter structure, a negative gradient, is
always observed.   (Note that our prevailing component is
defined as an average over 10 to 15 days).

CONTRIBUTION OF INCOHERENT SCATTER RADARS TO THE STUDY OF MIDDLE AND LOW LATITUDE IONOSPHERIC ELECTRIC FIELDS

Michel Blanc and Paul Amayenc

C.R.P.E./C.N.E.T.
38-40, rue du Général Leclerc
92131 Issy-les-Moulineaux, France

1. INTRODUCTION

Electrodynamics of the earth's upper atmosphere has been one of the most important fields of interest in the development of space research. Its situation in the body of outer geophysics is quite original because of its first-order importance in coupling mechanisms, making a local understanding of electric currents and fields almost impossible: coupling between neutral atmosphere and ionospheric plasma motions by dynamo action of the neutral winds; coupling between ionospheric and magnetospheric plasma along the same magnetic field line by means of strong field-aligned conductivities, participating in particular to the equilibrium of the F layer; coupling between magnetospheric and solar-wind plasma motions by generation of the so-called convection electric fields at the magnetopause. Because of the long range of electromagnetic interactions, the global atmospheric circuit cannot be closed until all the region of space enclosed between the lower boundary of the ionosphere and the solar wind/ magnetosphere bow shock is considered.

The first quantitative and experimental description of the ionosphere electrodynamics using the dynamo theory of regular magnetic variations (Chapman and Bartels, 1940) appeared at the meeting point of geomagnetism and radioelectric studies of the ionosphere, using developments of ionized-gas physics. The only experimental element available was ground magnetic variations observed by the worldwide magnetometer network, providing

J. J. Burger et al. (eds.), Atmospheric Physics from Spacelab, 61–90. All Rights Reserved.
Copyright © 1976 by D. Reidel Publishing Company, Dordrecht-Holland.

equivalent horizontal ionospheric currents. And it is by reference to that sole basis--without an experimental knowledge of
electric fields--that the dynamo theory was improved until 1970.
Since that time, two techniques have given access to hourly-to-
diurnal variations of ionospheric electric fields. One is the
tracking of drifting whistler paths, permitting the determination
of east-west electric fields in the magnetic meridian plane
(Carpenter et al., 1972; Block and Carpenter, 1974; Carpenter
and Seely, 1975) and their mapping down to the ionosphere along
field lines. The other is the measurement of ion drift velocities
in the F region by incoherent scatter. Both are based on the assumption of frozen-in plasma motions, according to which perpendicular plasma velocities are related to electric and magnetic
fields by:

$$\vec{V} = \frac{\vec{E} \times \vec{B}}{B^2}$$

so that electric fields are linear functions of plasma motions.
The possibility of measuring north-south electric fields where
the plasma is not frozen-in, i.e. approximately below 160 km
altitude, has recently extended to the E and F_1 regions the domain over which incoherent scatter can provide electric field
data (Harper et al., 1976; Cornec, 1975). This set of data,
still very limited and needing an a priori knowledge of collision
frequencies, will not be used in this paper but seems promising
for future studies.

Electric field data, published by all incoherent scatter stations since 1970, have been analysed in mainly two different
ways. The first one is centered on a local analysis of F-layer
phenomena, for the understanding of which electric fields had
not been previously available. Such studies are concerned with
relations between electric fields, parallel plasma motions, and
vertical transport of the layer. Several mechanisms were proposed in order to explain the observed relations, as F-region
dynamo (Behnke and Hagfors, 1974) and ion drag (Taylor, 1974;
Thomas and Williams, 1975). The second one consists in the
study of large-scale features of observed fields and of their relation to large-scale generation mechanisms. Figure 1 shows
perpendicular ion drift variations on a typical quiet equinoctial
day in the F region over Saint-Santin (Blanc et al., 1976). Drift
variations can be observed with a good accuracy, permitting,

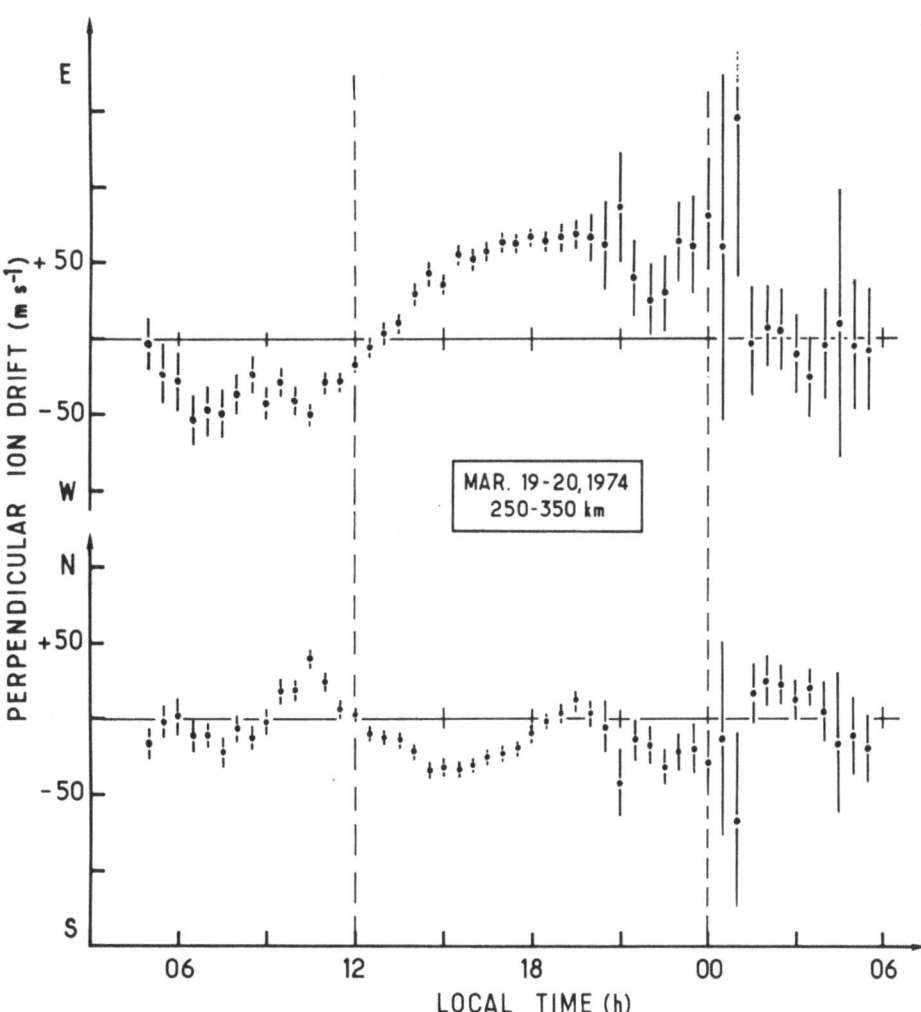

Figure 1 : West-east (top) and south-north (bottom) perpendicular F-region ion drifts for the March 19-20, 1974 incoherent scatter experiment at Saint-Santin, corresponding to a magnetically quiet day. After Blanc et al. (1975).

for instance, a harmonic analysis of the data over a 24-hour period. Together with this good local-time coverage, the locations of incoherent scatter stations from the magnetic equator to the auroral zone provide a good support, as will be appreciated later from Table I, to a study of latitude variations. Finally, as far as the longitude/local-time equivalence can be used, incoherent scatter and whistler data, taken as a whole, give access to the global electric field pattern. At middle and low latitudes, that possibility can be used to study regular daily electric field variations and to compare them to dynamo models. Section 2 reviews the present state of that work, analyses the questions it has raised in the theoretical understanding of experimental data, and proposes several directions for further improvements in both theoretical and experimental elements. In Section 3 are presented some preliminary studies of two types of variations other than diurnal which also require a global modeling: seasonal variations and some transient variations related to magnetospheric sources. Those descriptions are only at their very beginning, but should be exciting goals for future worldwide coordinated studies.

For brevity, $\underline{x}$ and $\underline{y}$ subscripts will refer respectively to south-north and west-east directions perpendicular to $\vec{B}$. According to (1), one should recall that northward (eastward) ion drift velocities correspond to westward (northward) electric fields.

2. REGULAR DAILY VARIATIONS

a. Ground magnetic variations

Diurnal variations of ionospheric currents have been studied for a long time by means of the worldwide magnetometer network, allowing for a continuous time coverage at nonpolar latitudes. At the time when no measurements of neutral winds in the thermosphere were available, the interest was first centered on the possibility of deducing neutral wind patterns from magnetic data. Later on, when experimental measurements of these winds as well as a theoretical description of their diurnal behavior by means of atmospheric tides became possible, the reciprocal method was used: assuming the neutral wind pattern including its vertical structure, one could compute the resulting horizontal currents. Figure 2 shows the main result obtained by Matsushita (1969) and Tarpley (1970). Using the (1, -1) diurnal tidal mode, they could correctly reproduce the S_q current system (shown on bottom) with their computed dynamo model (top). These authors also showed that winds associated with semidiurnal and terdiurnal modes are much less efficient in generating electric fields and nonzero integrated currents than the (1, -1) mode previously

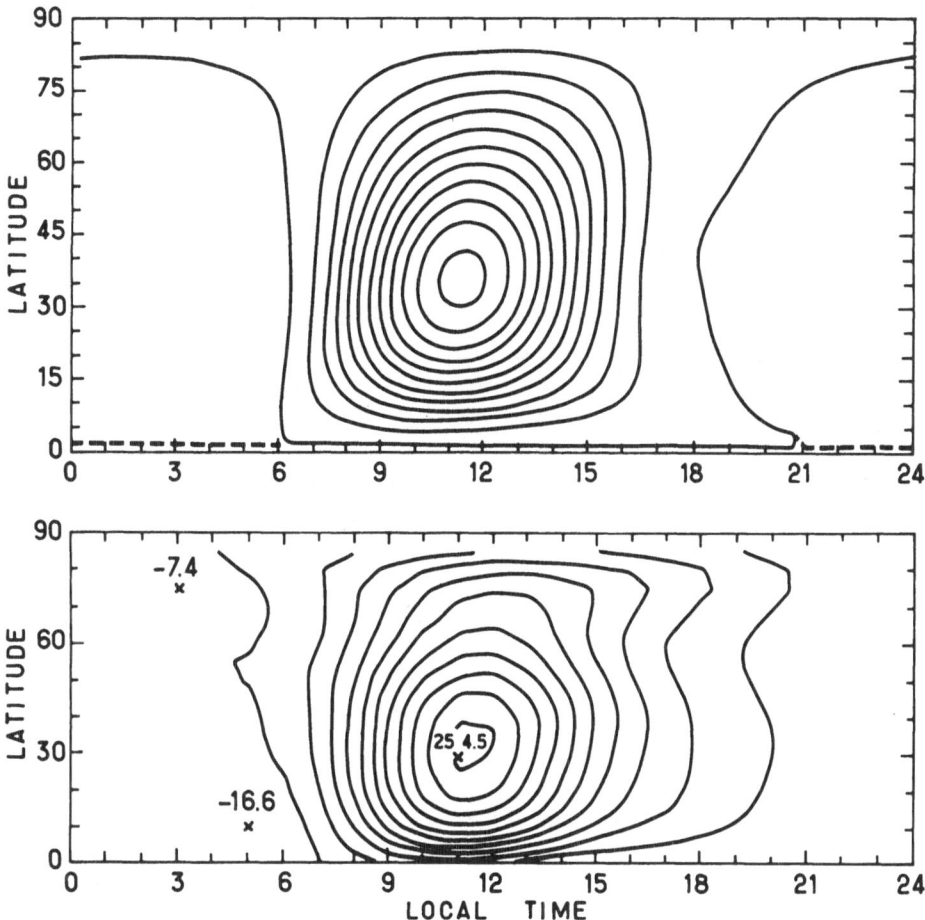

Figure 2 : Top: Current function with a 10 kA contour spacing for the (1, -1) diurnal tidal mode. Currents flow counterclockwise during daytime. Bottom: S_q current function deduced by Matsushita (1969) from IGY geomagnetic data with a 25 kA contour spacing.

mentioned, because of their strong vertical phase variation.

Therefore, by the 1970's it seemed that a satisfactory point was reached in the description of diurnal variations of electrodynamic parameters of the ionosphere. Assuming reasonable electromotive forces $\vec{V}_n \times \vec{B}$, one could explain the observed integrated horizontal current $\vec{J}$. However, the dynamo models produced values of the electric field $\vec{E}$ as well as of $\vec{J}$, but nothing comparable to the S_q current system deduced from magnetometer networks could provide an experimental reference of electric fields.

b. Incoherent scatter contribution

Since 1970 perpendicular drifts have been gathered for a large number of days of measurements by incoherent scatter observatories. Their various locations in latitude give access to the gross features of the latitudinal dependence of regular daily electric-field variations once this type of variation has been recognized for each observatory.

As an example, Figure 3 from Behnke and Harper (1973) shows a plot of $V_{\perp x}$ (positive poleward) for various winter

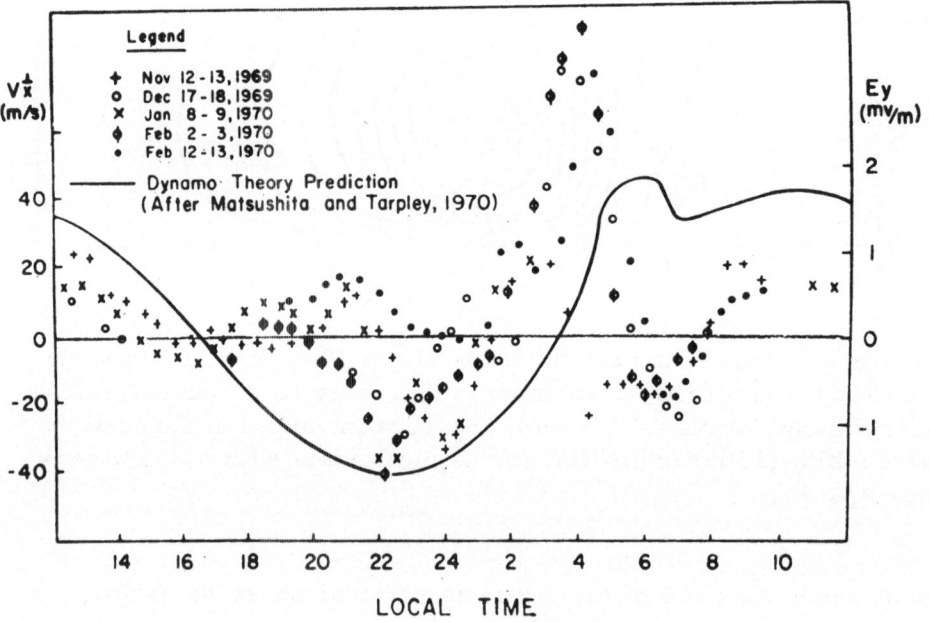

Figure 3 : South-north ion drift (positive northward) and corresponding electric field (positive eastward) as deduced from F-region incoherent scatter experiments at Arecibo, compared to the predictions of the E-region dynamo. After Behnke and Harper (1973).

days. A good reproductibility from day to day appears, except in the 20-01 L. T. Sector. In particular, a clear regular variation is observed over all daytime hours and a strong poleward drift occurs in the vicinity of 04 L. T. The presence of regular patterns has been also reported by Woodman (1970, 1972) for Jicamarca results, by Evans (1972) and Kirchhoff and Carpenter (1975) for Millstone Hill results, by Taylor (1974) for Malvern results, and by Blanc et al. (1976) for Saint-Santin, though these patterns are significantly different from one station to another.

Using presently available data for various locations (see Table I below), Richmond (1976) made a first attempt to model

TABLE I.

Station Name	Geographic Latitude	Geographic Longitude	Apex Latitude	Apex Longitude
Millstone Hill[1]	- -	- -	63°	2°
Whistlers[3]	-75.5°	-80.5°	60°	3°
Millstone Hill[2]	42.6°	-71.5°	55°	2°
Malvern	52.1°	- 2.3°	50°	77°
Saint-Santin	44.1°	2.0°	40°	78°
Arecibo	18.5°	-66.8°	31°	5°
Jicamarca	-12.0°	-76.9°	1°	- 8°

[1] North-south drifts

[2] East-west drifts

[3] Eights and siple stations (change of apex latitude sign to transfer to the northern conjugate point)

Geographic and apex coordinates of the various incoherent scatter and whistlers data sources used by Richmond (1976) to model the electrostatic potential.

the ionospheric electrostatic potential prevailing on quiet days in the plasmasphere. In order to increase the latitude coverage of his representation, he included whistlers measurements of the

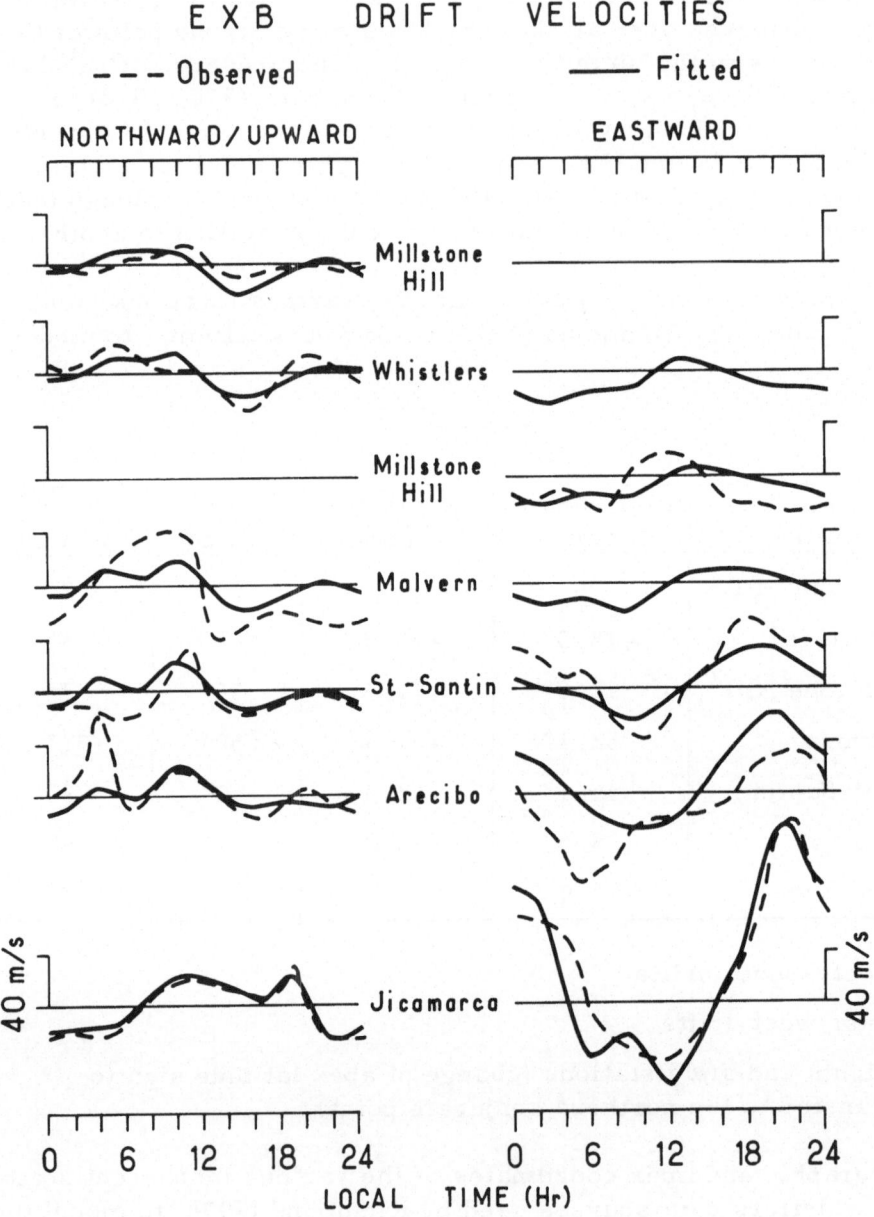

Figure 4 : Comparison between observed quiet-day F-region plasma drifts (see Table I) and drifts obtained by a least squares fit of the electrostatic potential shown in Figure 5. After Richmond (1976).

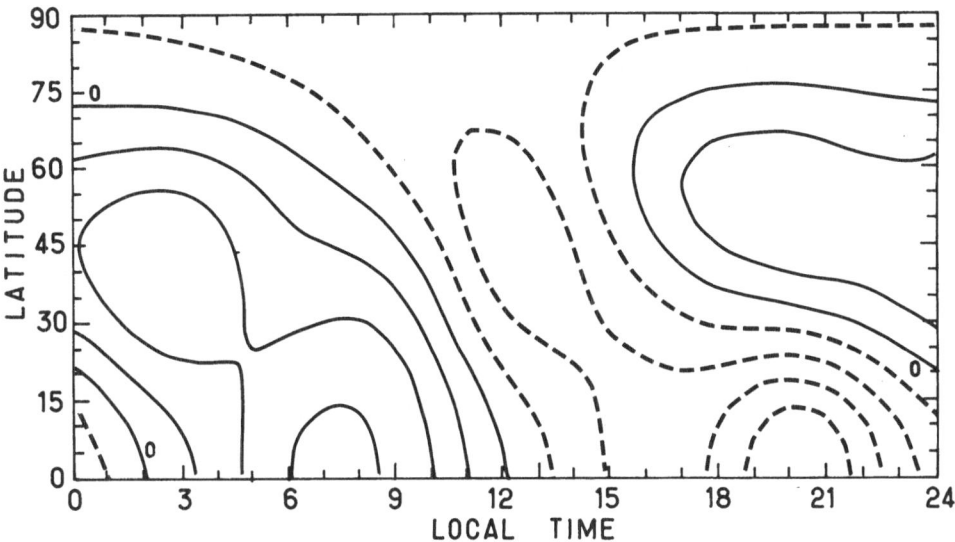

Figure 5 : Quiet-day F-region electrostatic potential pattern in apex-latitude/local-time coordinates with 1 kV contour interval. Solid (dotted) contours represent positive (negative) potentials. The maximum potential of 3.3 kV occurs at the equator around 07 L. T., while the minimum of -5.0 kV occurs at the equator around 20 L. T. After Richmond (1976).

electric field (Carpenter and Seely, 1975) at 63° apex latitude. Because of the measurement technique at Millstone Hill, east-west and north-south drifts for that station have been regarded as representative of two different latitudes. The electrostatic potential Φ , function of the apex colatitude θ (Van Zandt et al., 1972) and the longitude or local time φ , is represented as a finite series of spherical harmonic functions, symmetric about the equator, whose coefficients are adjusted in order to obtain the best fit of the regular daily variations observed in each station. The resulting fit, shown in Figure 4, is found to reproduce fairly well the gross features of all observations. The corresponding potential map, shown on Figure 5 should be very close to the potential responsible for the flow of the S_q current system, though the criteria used here for the selection of quiet days are less drastic than those applied to magnetic data for selection of S_q.

One of the most striking features of that potential map is the presence of two extrema, positive at 07 L. T., negative at 20 L. T., at each end of the equatorial electrojet. If one thinks of it as a highly conductive ribbon imbedded in regions of lower conductivities, it is not so surprising that strong polarizations appear at the points where it is abruptly connected to the night-side ionosphere. Hence, it is a satisfactory point that a global modeling based on the observations points out that feature. Conversely, one can expect that a correct modeling of the electrojet will be of first importance for any theoretical calculation attempting to satisfactorily reproduce the electrostatic potential.

As previously indicated, theoretical dynamo calculations have been found able to reproduce the S_q current system. However, no similar agreement has been found when, since 1970, measured electric fields have been available. For each station strong discrepancies appear with most models, and, when one of them appears nevertheless to reproduce the gross features of the data, it is not the same model which reproduces the data of various stations. These discrepancies can be appreciated from Figure 3, in which Behnke and Harper (1973) compared their data with the Matsushita-Tarpley (1970) model (continuous curve). In Figure 6, after Blanc et al. (1976), are plotted for $V_{\perp x}$ (Figure 6a) and $V_{\perp y}$ (Figure 6b) the comparison between four dynamo calculations (curves 1 to 4) and the quiet-day behavior at Saint-Santin (curve 5). Details of the discussion can be found in their paper. It becomes quite evident that none of the theoretical predictions can account for the observations with sufficient agreement.

Once this discrepancy between theory and experiment was clearly felt, the need for a modification of theoretical models which could benefit from recently performed neutral winds and electric fields measurements appeared as a logical consequence. Richmond et al. (1976) recently followed that direction. Using the computer program written by Tarpley (1970) with slight modifications related to numerical problems, they refined the neutral wind source by using a combination of diurnal and semidiurnal modes and by normalizing the tidal modes to incoherent scatter measurements of Amayenc (1974) and Salah et al. (1975). Their results are shown in Figure 7 and are compared with incoherent scatter and whistler electric-fields data. Coherence between theory (full curves) and experiments (dashed and dotted curves) remains approximate. The author's conclusion is that "iono-

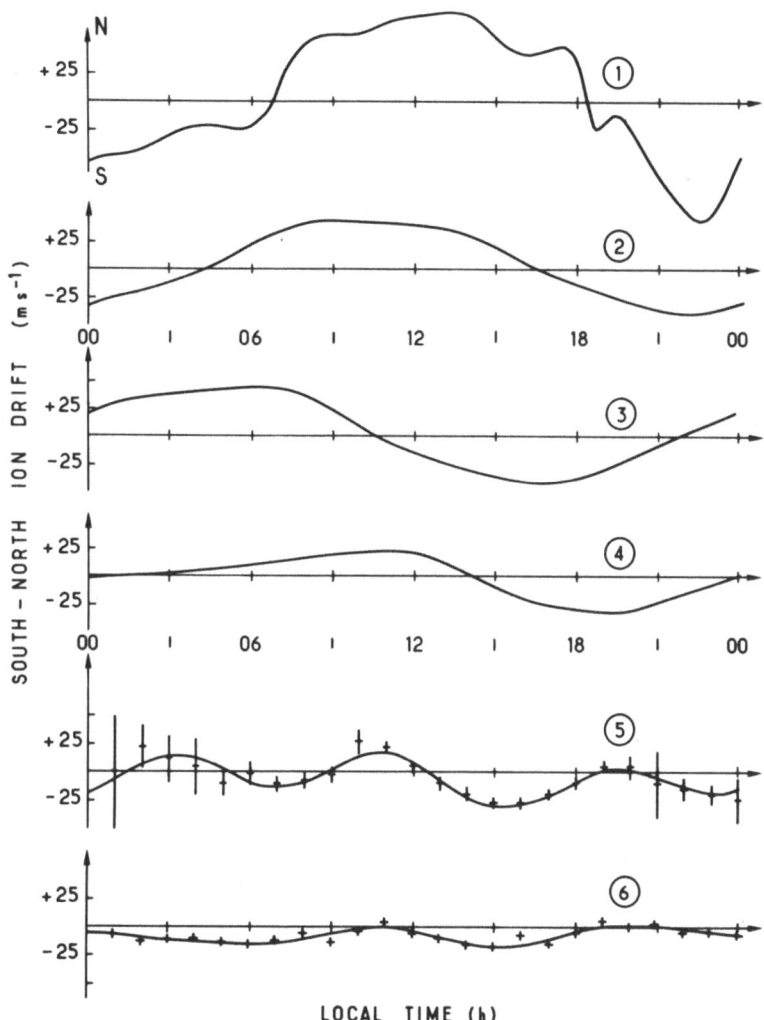

Figure 6a : Comparison between theoretical south-north ion drift from dynamo models and experimental results obtained at Saint-Santin. Theoretical curves are taken from (1) the Stening dynamo model (1974); (2) the Matsushita-Tarpley dynamo model (1970); (3) the Matsushita plasmaspheric model (1971); and (4) the Maeda semi-empirical model (1963). (5) is the data for the March 19-20, 1974 experiment reported in the 00-24 L. T. time range. (6) is a "median" model taken from equinox and winter results. After Blanc et al. (1975).

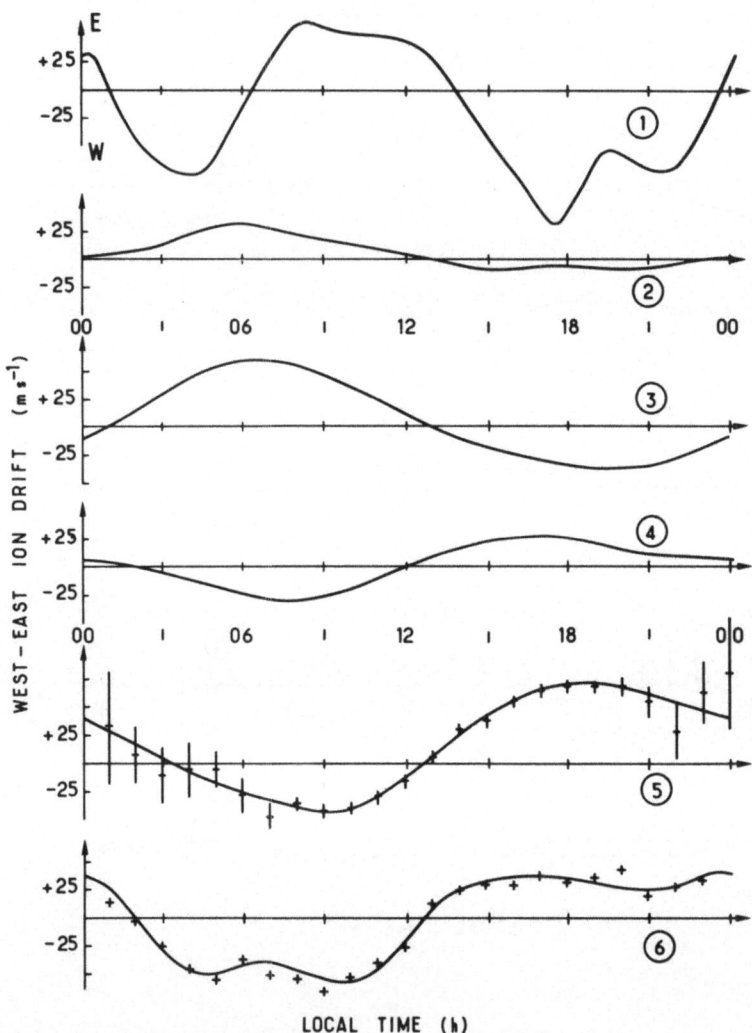

Figure 6b : Same as Figure 6a but for the east-west component of ion drift.

E X B DRIFT VELOCITIES

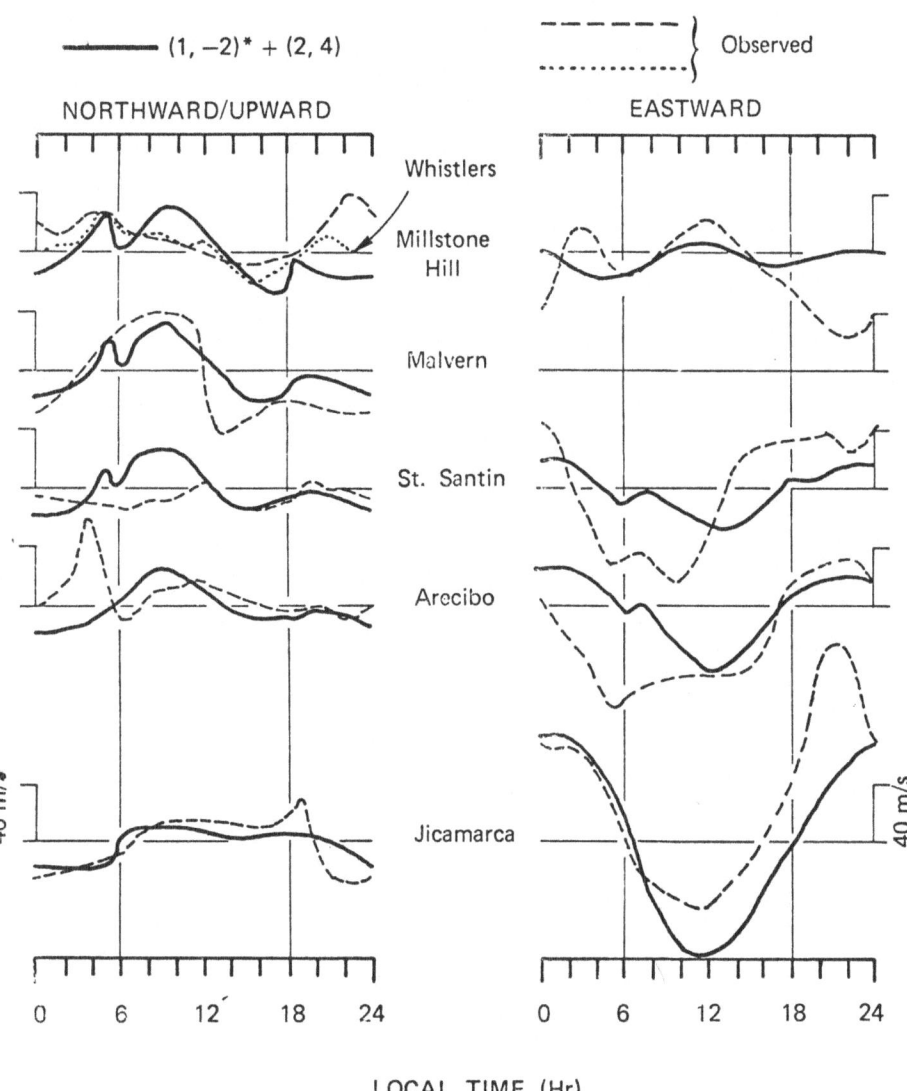

Figure 7 : F-region E x B drift velocities perpendicular to the geomagnetic field at various latitudes as a function of local time, computed from the combined diurnal and semi-diurnal tidal modes of neutral wind (solid lines) and observed at various locations (see Table I). After Richmond et al. (1976).

spheric winds can probably also explain observed drifts" (as well
as observed electric currents) and that further observations, allowing for a correct handling of the day-to-day variability of neutral winds and a more precise definition of the regular daily variations of electric fields, should help to explain the remaining
discrepancies.

c. Present state of theory/experiment confrontation

Before trying to define what could be the next steps in the design of theoretical models of the quiet electrodynamic state of
the ionosphere, one must examine both the internal structure of
theoretical models and the ways by which they are compared to
experimental data.

Figure 8 shows the elementary scheme of comparison, as it
must presently be defined to provide a significant test of the validity of a given electrodynamic model of the ionosphere. It is pre-
cisely that structure which has been used for the first time by
Richmond et al. (1976).

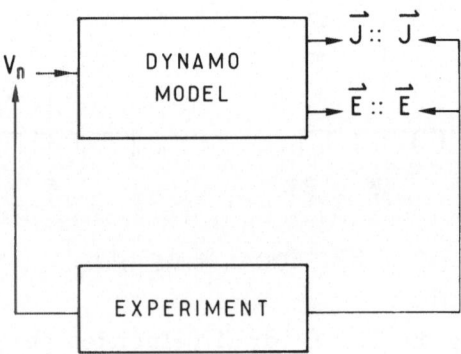

Figure 8 : Comparison scheme to be used in a significant test
of an ionospheric dynamo model.

The electrodynamic model submitted to the test is presented as a black box with one input, a global neutral wind pattern V_n and two outputs, global electric fields and current distributions. All three elements are related to the body of experimental data: i) the neutral wind model is defined so as to give a good representation of neutral winds measured at different locations; ii) thus the resulting $\vec{E}$ and $\vec{J}$ distributions are critically compared to measured field $\vec{E}$ (incoherent scatter + whistlers) and currents $\vec{J}$. Such a comparison scheme is expected to provide information on the way in which the internal structure of the black box must be modified, provided that the body of experimental data is coherently selected.

To insure the coherence of the data from various origins, two main procedures appear available:
1. to define an average for each type of data after having selected them with the same criterion. Such a procedure aiming for instance at selecting quiet magnetic periods is somewhat difficult because the number of quiet days of incoherent scatter observations of $\vec{E}$ and $\vec{V}_n$ is and will remain small compared to the number of days involved for instance in the definition of the S_q current system;
2. to collect all data $\vec{V}_n$, $\vec{E}$, and $\vec{J}$ measured on the same day and to use them in the computations. Providing such a collection of data is one of the goals of coordinated incoherent scatter observation campaigns. One such campaign, scheduled in May 1975, has been devoted to the study of quiet-day electric fields and winds so that the interpretation work could eliminate, as far as the quality of the data allows, the uncertainties due to inhomogeneous averages or day-to-day variability.

The internal structure of the black box is schematically presented in Figure 9 for the case of a modeling of nonauroral and nonpolar regions. The central element (4) is a map of integrated ionospheric conductivities defining the domain in which the two laws of electrodynamics (Ohm's law + divergence-free currents) are solved. This resolution can be performed to give currents $\vec{J}$ (6) and electric fields $\vec{E}$ (7) provided that the following elements are introduced: i) realistic conductivities Σ (2), ii) a current source (3) computed from conductivities Σ and neutral winds $\vec{V}_n$(1), and iii) realistic boundary conditions at high and low latitudes (5).

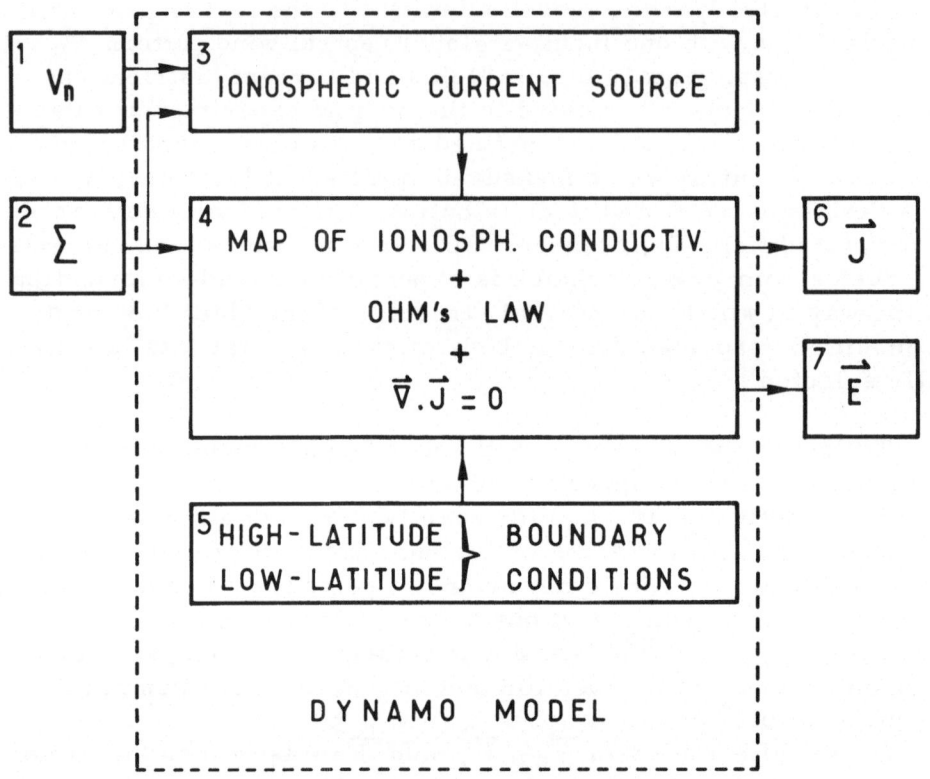

Figure 9 : Input and output parameters and internal structure
of a theoretical model describing globally the ionosphere electrodynamics at nonpolar latitudes.

Compared to the present state defined as the model of Richmond et al. (1976), several elements which could be the subject
of future improvements are now analyzed.

d.   Directions for further improvements

Ionospheric conductivities and winds should be introduced
taking account of two elements. First, as suggested by Rishbeth
(1971) and numerically tested by Heelis et al. (1973) and Matuura
(1974), F-region conductivities and winds may play a significant

role especially in the vicinity of the magnetic equator. A striking support to this idea has been provided by Heelis et al. (1973). Figure 10 shows their results comparing F-region vertical ion drifts, i. e. eastward electric fields measured at Jicamarca (dotted curve after Woodman, 1970, 1972) with corresponding theoretical results without (full curve) and with (dashed curve) inclusion of F-region neutral-wind effects. The improvement

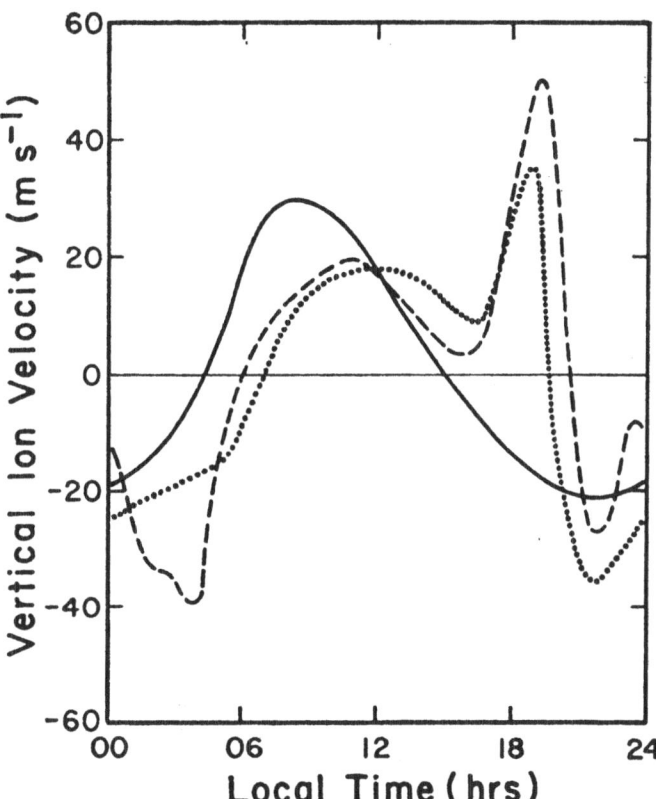

Figure 10 : Local-time variation of vertical ion velocity at 300 km above the equator computed theoretically from tidal E-region electric field only (——), tidal E-region and F-region polarization fields (---), and compared with vertical drift velocity average from 1968-1969 Jimarca observations of Woodman (1970) around equinox ($\cdots$). After Heelis et al. (1974).

in comparison in the latter case is very important including in
particular the strong 19 L. T. and 21 L. T. positive and negative
peaks which are characteristic and permanent features of Jimarca data.

Second, the same particular magnetic-field geometry which
accounts for the influence of the so-called F-layer dynamo influence at the equator is responsible for the equatorial electrojet. This narrow strip of very strong conductivity has not yet
been self-consistently introduced in a global dynamo model to
fit magnetic data though Schieldje et al. (1973) introduced it without taking account of its current interchanges with other latitudes. This kind of work is therefore yet to be done either
through direct inclusion of the electrojet in the conductivities
map or through an adequate modification of the low latitude boundary condition.

The last improvement possibilities and probably the most
difficult concern high latitude boundary. Until now, all dynamo
models simply ignored the complex problem of coupling with the
outer magnetosphere and closed their calculations at the poles
where a unique value of the potential was assumed to be obtained.
However, this procedure is far from actual physical representation. First, a magnetospheric source imposed by the solar-wind
magnetosphere coupling (Axford and Hines, 1961; Dungey, 1961)
is projected onto polar regions. It is this same source which
drives on quiet days the S_q^p current system evidenced by Nagata
and Kokubun (1962) from high-latitude magnetograms. Therefore, quiet-day electric fields are a superimposition of at least
two fields, one due to magnetospheric convection, the other due
to atmospheric dynamo. But the effect of ring current particles
is expected to confine the first type of field at high latitudes.
This so-called "shielding effect" (Vasilyunas, 1970, 1972; Wolf,
1970; Jaggi and Wolf, 1973) must also interact with dynamo fields
leading to an important weakening of the coupling between the dynamo currents prevailing poleward and equatorward of the equatorial boundary of the auroral zone (considered roughly as the
ionospheric projection of the ring current inner boundary). This
effect is worth taking into account for instance by considering
that boundary as the poleward boundary of the dynamo model and
by choosing there a correct boundary condition. However, it is
not sure that shielding mechanisms are correctly described by
present theories because it is extremely difficult to take into
account nonadiabatic processes in the time and space variation

of energetic particle distributions in the magnetosphere. Checking the effect of various boundary conditions on electric field and current global distributions might be an indirect way for improving the description of those complex mechanisms.

On the experimental side, the problem lies in the definition of the "regular daily variations." Progress should be achieved mainly in two directions. First, the daily variations are certainly season-dependent in a highly complex way due particularly to the distinction between the rotation and the magnetic axes of the earth and to the complexity of the earth's magnetic field morphology: resulting annual variation must be experimentally investigated. Second, many random-like features of the data, mostly corresponding to short-term variations and which strongly contribute to the day-to-day variability, must be physically understood. A few contributions to that direction of progress are presented in the next section.

3. VARIATIONS OTHER THAN DIURNAL

a. Seasonal variations

It is at Jimarca, where the line-of-sight velocity in the F-region was directly interpretable as an electrodynamic drift, that the first information on seasonal variations was obtained by Woodman (1970). Three of the five seasonal periods he considered are shown in Figure 11: winter solstice from June 1 to August 8 (top), spring equinox (middle), and summer solstice (bottom). The general behaviors appear very similar for all seasons: nearly constant upward drifts during daytime, followed by nearly constant downward drifts at night. Day-to-day variations for each season are of the same order of magnitude as seasonal changes. Among those changes, Woodman emphasized the quasi-disappearance at winter solstice of the strong post-sunset upward drift, which is a very striking feature of the data for all other seasons. That study did not discriminate between quiet and disturbed days and, moreover, Woodman mentions that there is no clear general relation between disturbed days and days of large observed drifts.

Using Saint-Santin data obtained from October 1973 to July 1975, we have performed seasonal averages over quiet days selected from the K_p index, the results of which are presented in

RONALD F. WOODMAN

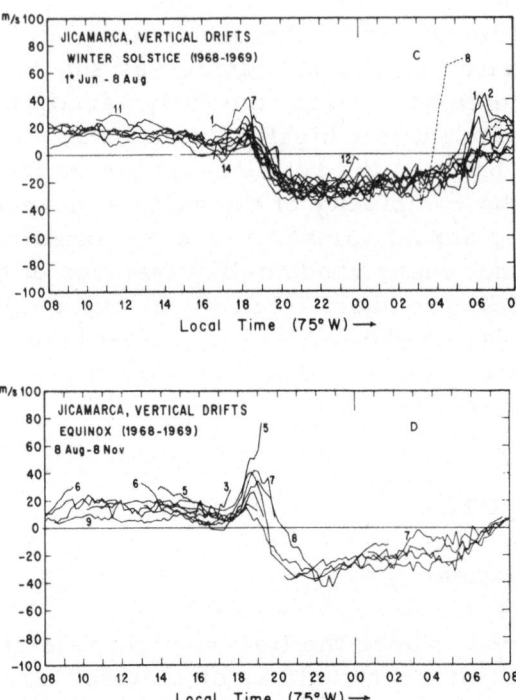

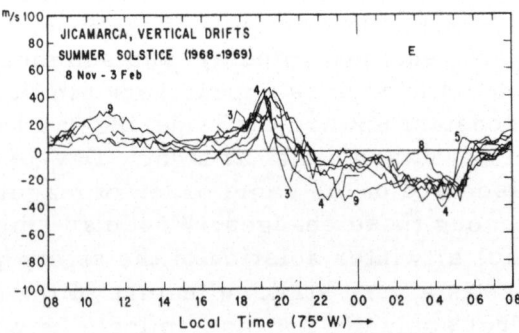

Figure 11 : Daily variations of vertical F-region ion drifts
observed by incoherent scatter at Jicamarca for different
seasons. After Woodman (1970).

Figure 12 for northward (Figure 12a) and eastward (Figure 12b)
perpendicular drifts. Error bars correspond to the standard
deviation of the indicated average, calculated from the experimental errors on individual points of measurements. Each
seasonal average includes typically five periods of 24-hour measurements. If one excepts the morning sector in summer, common features appear: diurnal oscillation of west-east drifts,
reversing to eastward between 12 and 13 L. T. and to westward
in the early morning hours around 03 or 04 L. T., with amplitudes in the range of 40 to 60 m/s; weaker values of north-south
drifts, displaying no comparable diurnal oscillation, but tending
during daytime hours to northward in the morning, and then to
southward in the afternoon, with the same reversal times as for
east-west velocities. Among the seasonal changes, most striking
are the summer behavior between 06 and 10 L. T., departing
eastward and southward from drifts observed in other seasons,
and the apparent intensification of the east-west diurnal variations during winter up to 60 or -60 m/s.

b. Relations between short-term fluctuations and magnetospheric
   sources

Determining the contribution of magnetospheric sources to
the generation of electric fields and currents at middle and low
latitudes is of first importance for our purpose: as seen in Section 2d, the question of high-latitude boundaries of the dynamo
model implies a correct understanding of the shape of the driving
field of magnetospheric convection as well as of mechanisms
governing the dynamics of energetic particle belts in the outer
magnetosphere. One of the best ways of providing information
on those mechanisms is to study the transient response of the
ionosphere-magnetosphere system to various magnetospheric
events. Obviously, the shape of the transient response itself,
i. e. its latitude/local-time dependence, gives access to some
characteristics of that system which are not accessible by the
sole study of its steady states. Moreover, provided that correlations with variations of high-latitude and magnetospheric parameters are clearly identified, the experimental study of transient
features can help to separate more easily than in the steady state
magnetospheric and dynamo fields contributions.

Contributions of incoherent scatter to that topic up to 1974
have been reviewed by Testud et al. (1975). Here we restrict

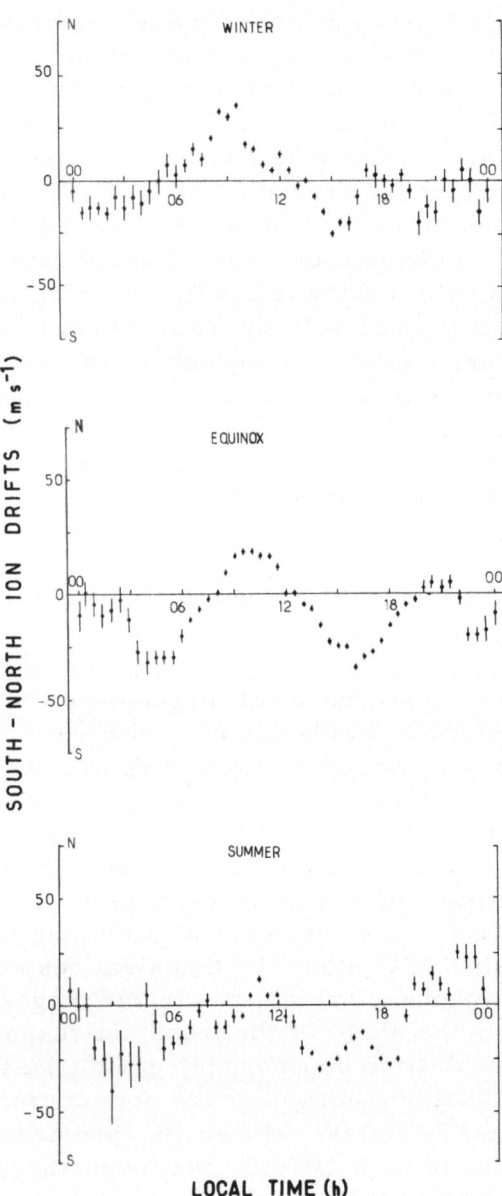

Figure 12a : Seasonal average of daily variation of south-
north perpendicular ion drifts as deduced from Saint-Santin
incoherent scatter observations for quiet magnetic conditions.

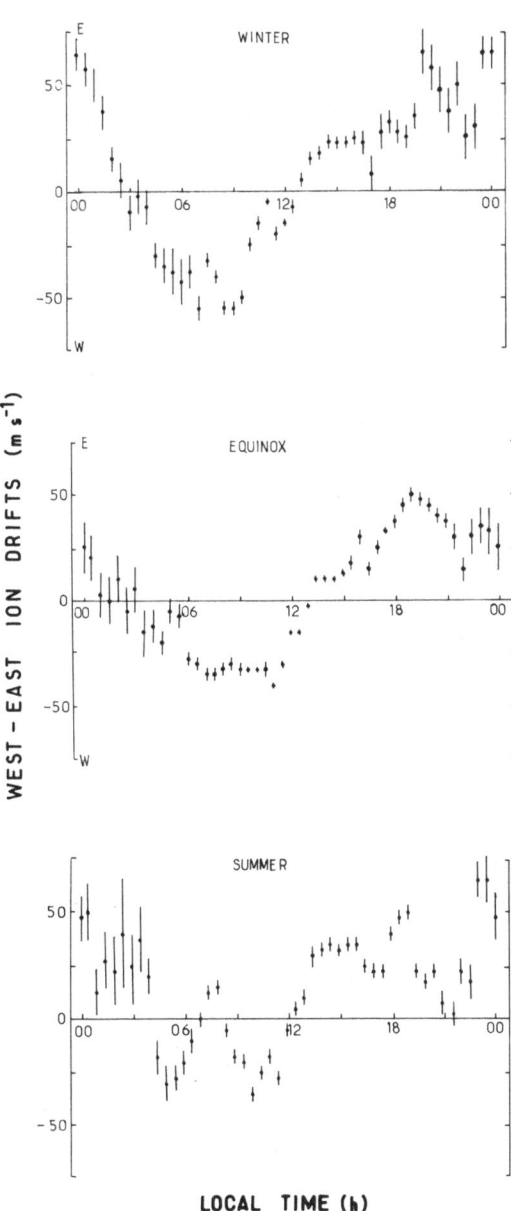

Figure 12b : Same as Figure 12a but for the west-east component of perpendicular ion drifts.

ourselves to two striking examples of observations, showing two
different types of analyses and leading to very different interpre-
tations.

Figure 13, arranged from Evans (1972), shows vector diagrams of E-region (top curves) and F-region (bottom curves)
ion drift variations during daytime. Left-hand-side diagrams
are averages of the drifts representative of magnetically quiet
conditions. Vector velocities rotate 360° clockwise in 12 hours
with amplitudes close to 25 m/s. Right-hand-side diagrams
show the drifts observed on May 14, 1969. After 14.00 EST,
F-region drifts display a very strong westward shift, corresponding to a 7 mV/m northward electric field. That strong perturbation appeared in simultaneity with the sudden commence-
ment at 14.30 EST of a strong magnetic event, which can be

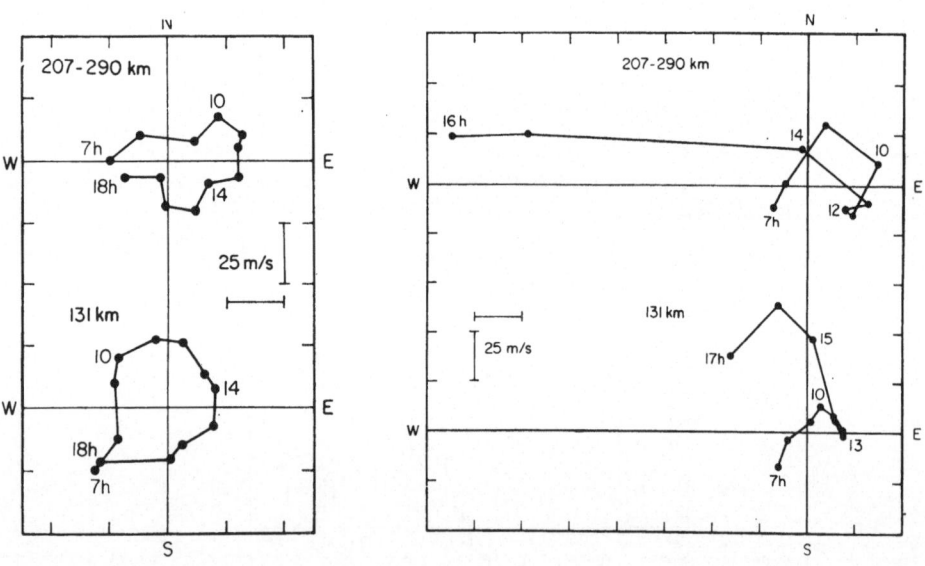

Figure 13 : Vector diagrams of hourly (EST) horizontal ion
drifts observed by incoherent scatter at Millstone Hill in
E and F regions. Left: average for quiet magnetic conditions. Right: May 14, 1969 results where a strong sudden
magnetic activity starts at 14.30 EST. Arranged from Evans
(1972).

recognized for instance by inspection of the auroral electrojet
indices (Testud et al., 1975). Such an association with onsets
of magnetic events has also been observed on some occasions
at Saint-Santin (Blanc et al., 1976). A systematic study of these
types of U. T. coincidences between electric field variations at
midlatitudes and magnetic events, particularly magnetic substorms, should help to evaluate the efficiency of the penetration
of electric fields of magnetospheric origin within the plasmasphere, and therefore of the shielding effect due to ring current
particles.

Carpenter and Kirchhoff (1975) presented a different approach.
They simultaneously observed ionospheric electric fields at Chatanika (high-latitude station) and Millstone Hill and superposed
measured fields in constant local time, corresponding to a five-
hour U. T. difference. On the left-hand side of Figure 14, their
results are shown for a first period in which they observed northward electric fields, whereas on the right-hand side are eastward
fields measured during a different period. The scales used differ
by a factor of 10, and with that factor observed fields have the
same apparent amplitude. The authors emphasized that there

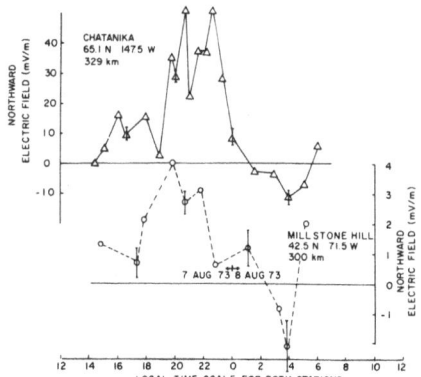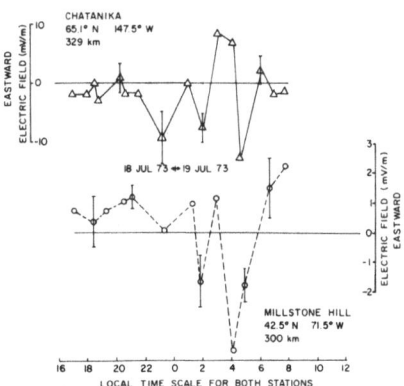

Figure 14 : August 7-8, 1973 northward (left) and July 18-19,
1973 eastward (right) electric field deduced from Chatanika
and Millstone Hill incoherent scatter experiment in each local
time. Arranged from Carpenter and Kirchhoff (1975).

appears to be a good correlation, very often peak-to-peak, between these fields. Consideration of the time scale of correlated variations, as short as 1 or 2 hours, leads us to a surprising
conclusion: short-scale features of the magnetospheric electric
field would be nearly invariant over at least 5 hours in universal
time. Such an idea is not consistent with present descriptions
of the magnetospheric electric fields, and it would be very important to test it through a larger number of observations.

Midlatitude observations of electric fields of magnetospheric
origin appear, both a priori and through first results, to be a
very promising topic. All studies until now have been made separately for each station and are in a preliminary state. For those
studies, as well as for the description of quiet-time fields, coordinated observations in all stations, followed by global comparisons, should be very fruitful.

4. CONCLUSION

Improving the coherence between electric current patterns
deduced from regular daily variations of the magnetic field at
the ground at nonpolar latitudes and dynamo theory predictions
has been one of the major goals of studies of the ionospheric
electrodynamics. By 1970, it seemed that this kind of study
was reaching a final point.

Quasi-direct measurements of ionospheric electric fields by
incoherent scatter, broadening the set of data whose coherence
with dynamo theories had to be tested, gave a new impetus to
this subject. Presently, it appears possible to model on a global
scale regular daily variations of observed electric fields at middle and low latitudes (Richmond, 1976) but a really satisfactory
agreement of them with an up-to-date dynamo model (Richmond
et al., 1976) does not seem to be observed yet. In the search
for future improvements of this comparison, both experimental
and theoretical aspects involved in its present state have been
critically analyzed.

On the experimental side, simultaneous global measurements
of all electrodynamic parameters, i.e. neutral winds and electric
currents and fields, are undoubtedly the best basis for significant
tests of theoretical models. On the theoretical side, several improvements of these models have been proposed by reference to

various specific studies:

1. At low latitudes the equatorial electrojet, as well as the F-region winds and conductivities, should be self-consistently introduced in the global circuit in a way which takes account of the exact magnetic field geometry.

2. At high latitudes, the boundary between convection-dominated fields and atmospheric-dynamo-dominated fields must be described in a physically realistic way; for instance, adequate boundary conditions on the electric potential or current function should be imposed in the vicinity of the equatorward edge of the auroral zone, considered for simplicity as the ionospheric projection of the inner edge of the ring current.

Apart from regular daily variations, seasonal variations and short-term fluctuations of electric fields deduced from incoherent scatter are few and recent, and even though the main work remains to be done, they appear very promising. In particular the study of the transient response of the global atmospheric circuit to magnetospheric events is one means to study the efficiency of the shielding effect due to ring current particles, and the penetration of the plasmasphere by electric fields of magnetospheric origin.

REFERENCES

Amayenc, P., Tidal oscillations of the meridional neutral wind at midlatitude. Radio Sci. 9, 281 (1974).

Axford, W. I. and Hines, C. O. , A unifying theory of high latitude geophysical phenomena and geomagnetic storms. Can. J. Phys. 39, 1433 (1961).

Behnke, R. A. and Hagfors,T.,Evidence for the existence of night-time F-region polarization fields at Arecibo. Radio Sci. 9, 211 (1974).

Behnke, R. A. and Harper, R. M. , Vector measurements of F-region ion transport at Arecibo. J. Geophys. Res. 78, 8222 (1973).

Blanc, M. , Amayenc, P. , Bauer, P. , Taieb, C. , Electric field induced drifts from the French incoherent scatter facility. To appear in J. Geophys. Res. (1976).

Block, L. P. and Carpenter, D. L. , Deviation of magnetospheric electric fields deduced from drifting whistler paths. J. Geophys. Res. 79, 2783 (1974).

Carpenter, D. L. and Seely, N. T. , Cross l. plasma drifts in
the outer plasmasphere; quiet-time patterns and some
substorm effects.　To appear in J. Geophys. Res. (1975).

Carpenter, D. L. , Stone, K. , Siren, J. C. and Crystal, T. L. ,
Magnetospheric electric fields deduced from drifting whistler paths.　J. Geophys. Res. 77, 2819 (1972).

Carpenter, L. A. and Kirchhoff, V. W. J. H. , Comparison of
high-latitude and mid-latitude ionospheric electric fields.
J. Geophys. Res. 80, 1810 (1975).

Chapman, S. and Bartels, J. , Geomagnetism.　Clarendon Press,
Oxford (1940).

Cornec, J. P. , Winds and electric fields in the upper E-layer
over Malvern.　Paper presented at XVI IUGG General
Assembly, Grenoble (1975).

Dungey, J. W. , Interplanetary magnetic field and the auroral
zones.　Phys. Rev. Lett. 6, 47 (1961).

Evans, J. V. , Measurements of horizontal drifts in the E and
F regions at Millstone Hill.　J. Geophys. Res. 77, 2341
(1972).

Harper, R. M. , Wand, R. H. , Zamlutti, J. and Farley, D. T. ,
E-region ion drifts and winds from incoherent scatter
measurements at Arecibo.　J. Geophys. Res. 81, 25
(1976).

Heelis, R. A. , Kendall, P. C. , Moffett, R. J. and Rishbeth, H. ,
Electrical coupling of the E and F regions and its effect on
F-region drifts and winds.　Planet. Space Sci. 22, 743
(1973).

Jaggi, R. K. and Wolf, R. A. , Self-consistent calculation of the
motion of a sheet of ions in the magnetosphere.　J. Geophys.
Res. 78, 2852 (1973).

Kirchhoff, V. W. J. H. and Carpenter, L. A. , Dominance of the
diurnal mode of horizontal drift velocities at F-regions
heights.　J. Atmos. Terr. Phys. 37, 419 (1975).

Maeda, H. , Worldwide pattern of ionization drifts in the ionospheric F region as deduced from geomagnetic variations.
Proc. Conf. on the Ionosphere, London, July 1963 (1963).

Matsushita, S. , Dynamo currents, winds and electric fields.
Radio Sci. 4, 771 (1969).

Matsushita, S. , Interactions between the ionosphere and the
magnetosphere for S_q and L variations.　Radio Sci. 6,
279 (1971).

Matsushita, S. and Tarpley, J. D. , Effects of dynamo region
electric fields on the magnetosphere.　J. Geophys. Res.
75, 5433 (1970).

Matuura, N., Electric fields deduced from the thermospheric model. J. Geophys. Res. 79, 4679 (1974).

Nagata, T. and Kokubun, S., An additional geomagnetic daily variation (S_q^p field) in the polar region on geomagnetically quiet days. Rept. Ionosphere Space Res. Japan 16, 256 (1962).

Richmond, A. D., Electric field in the ionosphere and plasmasphere on quiet days. To appear in J. Geophys. Res. (1976).

Richmond, A. D., Matsushita, S. and Tarpley, J. D., On the production mechanism of electric currents and fields in the ionosphere. J. Geophys. Res. 81, 547 (1976).

Rishbeth, H., The F-layer dynamo. Planet. Space Sci. 19, 263 (1971).

Salah, J. E., Wand, R. H. and Evans, J. V., Tidal effects in the F region from incoherent scatter radar observations. Radio Sci. 10, 347 (1975).

Schieldge, J. P., Venkateswaran, J. V. and Richmond, A. D., The ionospheric dynamo and equatorial magnetic variations. J. Atmos. Terr. Phys. 35, 1045 (1973).

Stening, R. J. Private communication (1974).

Tarpley, J. D., The ionospheric wind dynamo II solar tides. Planet. Space Sci. 18, 1091 (1970).

Taylor, G. N., Meridional F2-region plasma drifts at Malvern. J. Atmos. Terr. Phys. 36, 267 (1974).

Testud, J., Amayenc, P. and Blanc, M., Middle and low latitude effects of auroral disturbances from incoherent scatter. J. Atmos. Terr. Phys. 37, 989 (1975).

Thomas, D. P. and Williams, P. J. S., Measurements of ion drag induced by plasma velocity in the F region. J. Atmos. Terr. Phys. 37, 1271 (1975).

Van Zandt, T. E., Clark, W. L. and Warnock, J. M., Magnetic apex coordinates: a magnetic coordinate system for the ionospheric F2 layer. J. Geophys. Res. 77, 2406 (1972).

Vasilyunas, V. M., Mathematical models of magnetospheric convection and its coupling to the ionosphere. In Particles and Fields in the Magnetosphere, edited by McCormac, p. 60 (1970).

Vasilyunas, V. M., The interrelationship of magnetospheric processes. In Earth's Magnetospheric Processes, edited by McCormac, p. 29 (1972).

Wolf, R. A., Effects of ionospheric conductivity on convective flow of plasma in the magnetosphere. J. Geophys. Res. 75, 4677 (1970).

Woodman, R. F., Vertical drift velocities and east-west electric
    fields at the magnetic equator. J. Geophys. Res. 75,
    6239 (1970).
Woodman, R. F., East-west ionospheric drifts at the magnetic
    equator. Space Res. 12, 969 (1972).

DISCUSSION

S.A. Bowhill: In computing electric fields, one should include
    field-aligned currents and the asymmetrical shape of the real
    magnetic equator. Calculations by my colleagues have shown
    large effects from these considerations.

M. Blanc: To make good use of such a sophisticated model, it is
    certainly important to be free from the problem of day-to-
    day variability. Since 1975, simultaneous measurements of
    electric fields have been scheduled for the whole set of
    incoherent scatter stations, and should provide five days or
    more of simultaneous data every year. Trying to reproduce
    theoretically electric field data of individual days would be
    very exciting, and your information tells me that such work,
    in which experimenters and theoreticians could fruitfully co-
    operate, can be conducted in the near future.

SESSION 2

THE OZONOSPHERE

BASIC PROCESSES IN THE STRATOSPHERE AND THE MESOSPHERE

T.M. Donahue

Atmospheric & Oceanic Science Department,
University of Michigan, Ann Arbor, USA

The dominant transport and chemical processes controlling the distribution of minor species - particularly ozone - in the stratosphere and mesosphere were discussed. Emphasis was placed on identifying the regions in which particular reaction cycles involving OX, ClX, HOX and NOX are dominant and feedback effects that may occur when sources of these species are altered. In particular, the implications of recent measurements of height profiles for O, OH, N_2O and HCl were discussed, as well as such interactions as ClO with NO and ClO with N_2O to form $ClON_2O$. The question of the terrestrial sources and sinks of N_2O and CH_4 was addressed. Finally, the factors controlling the distribution of hydrogen species in the stratosphere and the mesosphere were related to the escape of hydrogen.

J. J. Burger et al. (eds.), Atmospheric Physics from Spacelab, 93–94. All Rights Reserved.
Copyright © 1976 by D. Reidel Publishing Company, Dordrecht-Holland.

DISCUSSION

S.A. Bowhill: Is it true to say that the introduction of chlorine
nitrate, or any other reactions, into the simple chlorine
scheme will have <u>less</u> effect on the ozone than the simple
chlorine scheme alone?

T.M. Donahue: There are no reactions that convert Cl, ClO, NO or
NO_2, OH, H, HO_2 into other chemical species I can think of
that would result in a more rapid destruction of ozone than
would otherwise occur.

R.J. Murgatroyd: (a) What effect has the incorporation of the
recent findings regarding the importance of reactions
involving chlorine nitrate had on model calculations of the
likely decreases of stratospheric ozone due to chloro-
fluoromethane emissions from the surface?

(b) In view of these developments, what are your views on
the main aspects on which research on this problem should
now be concentrated?

T.M. Donahue: (a) Preliminary results with a one-dimensional
model (at the University of Michigan) suggest that the
previous estimate of about 13% decrease if present rates of
release continue (10.5 ppb ClX) will be reduced to about 4%
in the new calculations. But these estimates are very
questionable. Present models contain about 35 ppb NO_x at
high altitude because of large $O('D) + N_2O$ rates and too
little O_3 at high altitude.

(b) It is necessary to obtain urgently better measurements
- simultaneous if possible - of the most important constituents in the stratosphere affecting this problem, e.g.
chlorine, chlorine oxides (ClO), chlorine nitrate, the
hydroxyl radical, NO_2O, etc., as the model studies clearly
require these for validation.

CHEMICAL MODELS OF THE NEUTRAL ATMOSPHERE

Ivar S.A. Isaksen
Department of geophysics, University of Oslo
Norway

1. INTRODUCTION

The main purpose of chemical models of the stratosphere
has been to describe the chemical processes which control the distribution of ozone. It is now generally
accepted that catalytic reactions with nitrogen oxides
(NO, NO_2) as originally proposed by Crutzen (1970) are
the dominant sinks of stratospheric ozone, and that we
are able to increase the amounts of nitrogen oxides in
the stratosphere through the action of high flying aircraft, to such an extent that the protecting ozone
shield may be reduced. It is further recognized that an
analogue effect on the ozone layer, from active chloride compounds (Cl, ClO) may be the result of the release of stable fluorochlorocarbons in the troposphere
(Molina and Rowland, 1974).

In order to evaluate the effect of catalytically
active nitrogen and chlorine compounds on the stratospheric ozone layer, models which in a realistic way
describe the chemical and diffusive properties of the
atmosphere have to be used. In particular, if man's
present and future effects on the ozone layer are to
be estimated, processes have to be included in the
model, with realistic values for exchange between the
troposphere and the stratosphere, the source strength
of the gases at the ground, and loss processes due to
precipitation and gas phase reactions.

In this paper a chemical-diffusive model of the
troposphere and stratosphere is discussed. Height

J. J. Burger et al. (eds.), Atmospheric Physics from Spacelab, 95–106. All Rights Reserved.
Copyright © 1976 by D. Reidel Publishing Company, Dordrecht-Holland.

profiles of gases of interest for stratospheric ozone
will be given, and the results compared with a large
number of observed profiles. This gives a measure of
how well our model describes processes in the atmosphere.

2. THE MODEL

A model where vertical eddy diffusion simulates
the transport processes of the atmosphere has been used.
It is the same as previously described by Crutzen and
Isaksen (1976). The equation expressing the time variation of a specie x takes the form:

$$\frac{d}{dt}[x] = -\frac{d}{dz}(F_x) + P_x - L_x \qquad (1)$$

F_x is the vertical flux, and P_x and L_x are the chemical
production and loss terms respectively of the specie x.
The flux is given by the expression:

$$F_x = -K_z[M]\frac{d}{dz}(\delta_x) \qquad (2)$$

where K_z is the vertical eddy diffusion coefficient,
[M] the density of air, and δ_x the mixing ratio of the
specie x. A vertical eddy diffusion coefficient which
is deduced experimentally by Ehhalt (1974) to explain
the height distribution of methane, has been applied
in the model.

For gas which is affected by heterogeneous removal
in the troposphere, a constant removal rate corresponding to a lifetime of 10 days has been included in the
loss term L_x in the lower troposphere (z<8 km). Simi-
larely, when ground level production rates, the natural or the anthropogenic, can be estimated, this source
has been included in the production term, and ground
level densities are calculated from eq.(1). When the
ground level source strength is difficult to estimate,
we simply have measured densities as lower boundary
values.

In order to illustrate how nitrogen and chlorine
compounds are distributed in the stratosphere and troposphere, a schematic presentation of sources, sinks, and
direction of transport for nitrogen and chlorine compounds are given i figure 1 and figure 2, respectively.
The main source of these compounds is at ground level.
Heterogeneous removal of nitric acid and hydrochloric
acid, as indicated by the arrows, effectively prevents
active nitrogen (NO, NO_2) or chlorine (Cl, ClO) from

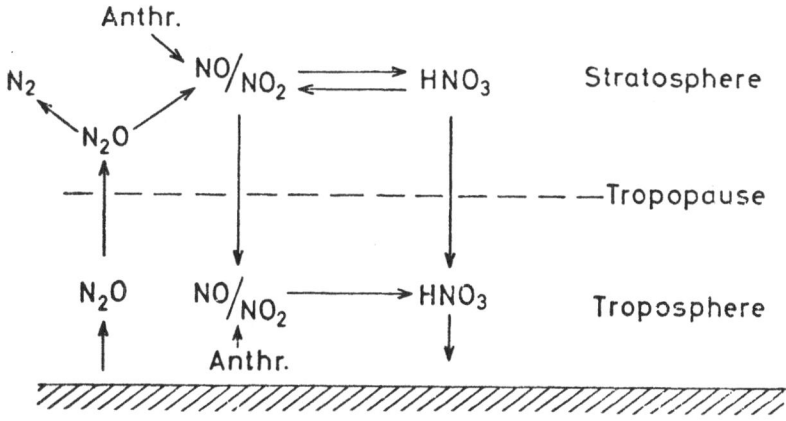

Fig.1. Schematic diagram of the nitrogen cycle in the
troposphere and stratosphere. Anthropogenic production
of NO_x,($NO+NO_2$), is included in the figure.

being transported to the stratosphere. The importance
of such processes depend of course on our choice of
removal rates. However, our findings are substantiated
by more detailed considerations of the removal processes in a model where both vertical and meridional
transport is considered (Crutzen, 1975). It is therefore obvious that ground level production of nitrogen
or chlorine compounds, has to be in the form of chemically stable compounds (N_2O, FC-11, FC-12) which are
able to penetrate into the stratosphere without being
removed in the troposphere by chemical reactions. In
the stratosphere the release of active nitrogen or
chlorine from such compounds will take place through
the action of the sun's short-wave radiation ($\lambda < 300$ nm).
Apart from precipitation processes, reactions with the
hydroxyl radical may for certain chlorine compounds
act as an effective loss process. This is the case for
chlorocarbons containing hydrogen (CH_3Cl, CH_2Cl_2, $CHCl_3$,
C_2H_3Cl, C_2HCl_3, $C_2H_3Cl_3$). The removal rates with respect to the reaction with hydroxyl range from a few
days (C_2HCl_3, C_2H_3Cl) to several months ($C_2H_3Cl_3$). It
is therefore reasonable to assume that the most long-
lived of the compounds will penetrate into the stratosphere, and act as a source for active stratospheric
chlorine (Crutzen and Isaksen, 1976). This underlines
the importance of an accurate determination of the
hydroxyl densities in the troposphere. Hydroxyl has an
equally important function in the stratosphere where

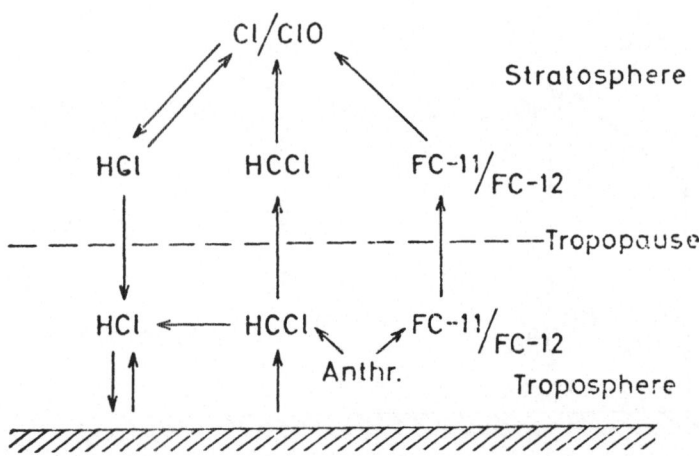

Fig.2. Schematic diagram of the chlorine cycle in the troposphere and stratosphere. Anthropogenic production (Anthr.) is included in the figure. HCCl denotes hydrochlorocarbons of the methane, ethane, and ethylene type, FC-11 denotes $CFCl_3$, and FC-12 denotes CF_2Cl_2.

it is involved in the conversion of active nitrogen and chlorine compounds to the chemically inactive compounds nitric acid and hydrochloric acid. It is interesting to notice that hydroxyl plays another role in the nitrogen than in the chlorine chemistry. High OH densities favors the conversion of NO_2 to HNO_3, and thereby reduces the effect of nitrogen oxides on the ozone layer, while it breaks up HCl to increase Cl and ClO densities which results in a more effective loss of ozone. Therefore it is necessary to have accurate estimates of the densities of hydroxyl in the stratosphere if we want realistic estimates of the effect of nitrogen and chlorine compounds.

In a recent article Anderson (1976) gives measured profiles of OH densities down to 30 km obtained at a solar zenith angle of 80°, and at 32° latitude N. Compared with our calculated profiles, his values are approximately a factor two higher. However, both measurements and calculations are subject to uncertainties, especially the calculated profiles due to poorly known rates of some of the key reactions.

3. THE DISTRIBUTION OF NITROGEN COMPOUNDS

It was shown in figure 1 that the main production
of nitrogen oxides in the stratosphere is provided by
the nitrous oxide molecules which are produced at the
Earth's surface. Since nitrous oxide chemically is
very stable in the troposphere, it has a fairly constant mixing ratio throughout the troposphere. Its
height distribution is therefore most accurately determined by adopting a fixed lower boundary value obtained from measurements.

Height profile for N_2O mixing ratios is given in
figure 3. There is generally a good agreement between
the calculated and the measured profiles. Both measure-

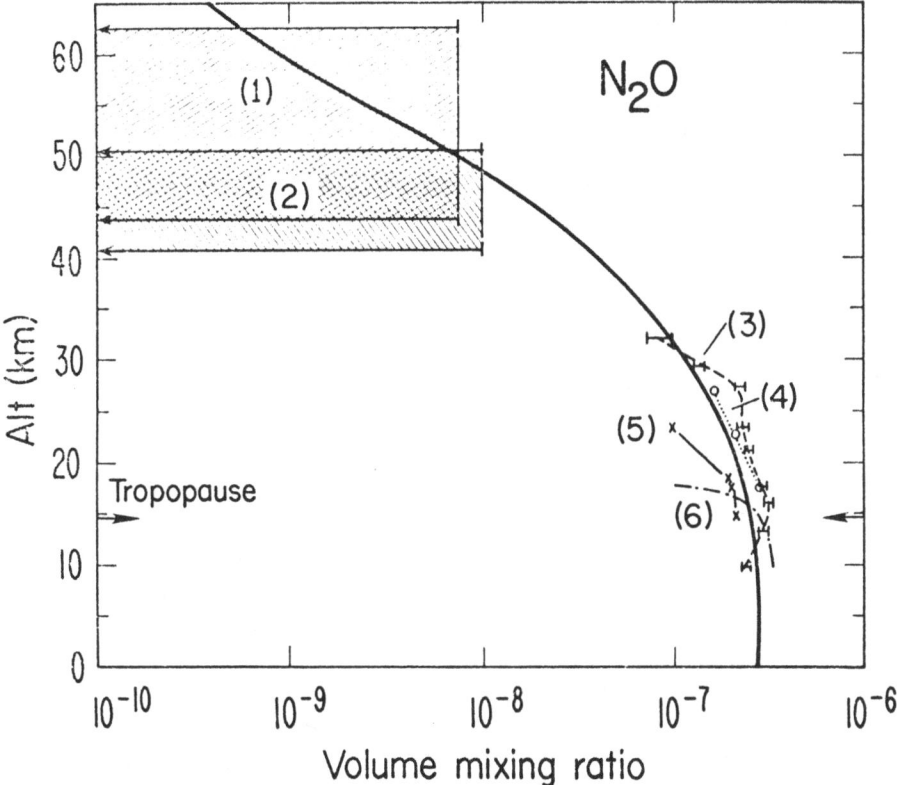

Fig.3. Calculated and measured volume mixing ratios
for nitrous oxide. The calculated profile is given by
the solid line. A constant value of 2.5×10^{-7} is used
as lower boundary conditions. For references to the
measured profiles, see Crutzen and Isaksen (1976).

ments and calculations show a slow decrease with height
in the lower stratosphere, and the upper limits given
for measurements in the upper stratosphere are in
agreement with the calculated profile.

Large attention is presently paid to the possible
future effect from the use of fertilizers on the N_2O
distribution. It has been argued that already around
year 2000 marked reductions of the ozone layer will
occur as a result of increase in the atmospheric in-
ventory of N_2O due to the use of fertilizers. It has,
however, been pointed out (Liu et al., 1976, Crutzen,
1976) that there probably is a considerable time lag,
possibly as much as several hundred years, from the
use of fertilizers to the actual release of N_2O into
the atmosphere. This will strongly reduce the threat
on the ozone layer from fertilizers in the coming cen-
tenary.

Height profile for $NO_x(NO+NO_2)$, is given in figure 4. The measurements show marked variations which
are difficult to explain from the calculations. It
should, however, be noticed that most of the measurements are either of NO or NO_2, and in order to be converted to NO_x the densities of ozone has to be known.
Variations in ozone may therefore explain some of the
discrepencies. It is further obvious that biological
or anthropogenic production at ground level will have
a negligible effect on the stratospheric profile, due
to effective removal by precipitation.

The only possible way that release of nitrogen
oxides may effect stratospheric densities of NO_x is by
the direct release in the stratosphere from high flying
aircraft (SST's, see also figure 1). The small number
of planes (<50) which will be in operation in the near
future, will have a negligible effect on the ozone
layer. If, however, the originally projected fleet of
1000 planes is in operation, the reduction in the ozone
layer is estimated to approximately 10%.

The calculated nitric acid profile shown in figure
5 is in good agreement with the observed stratospheric
profiles. The marked drop in measured mixing ratios towards the tropopause for both NO_x and HNO_3 indicate
that heterogeneous removal is effective in the troposphere, in agreement with our calculations. The mixing
ratios of HNO_3 are approximately equal to the NO_x mixing ratios below 30 km, but smaller above 30 km. This
means that most of the nitrogen oxides which are produced in the stratosphere, remain in the form of che-

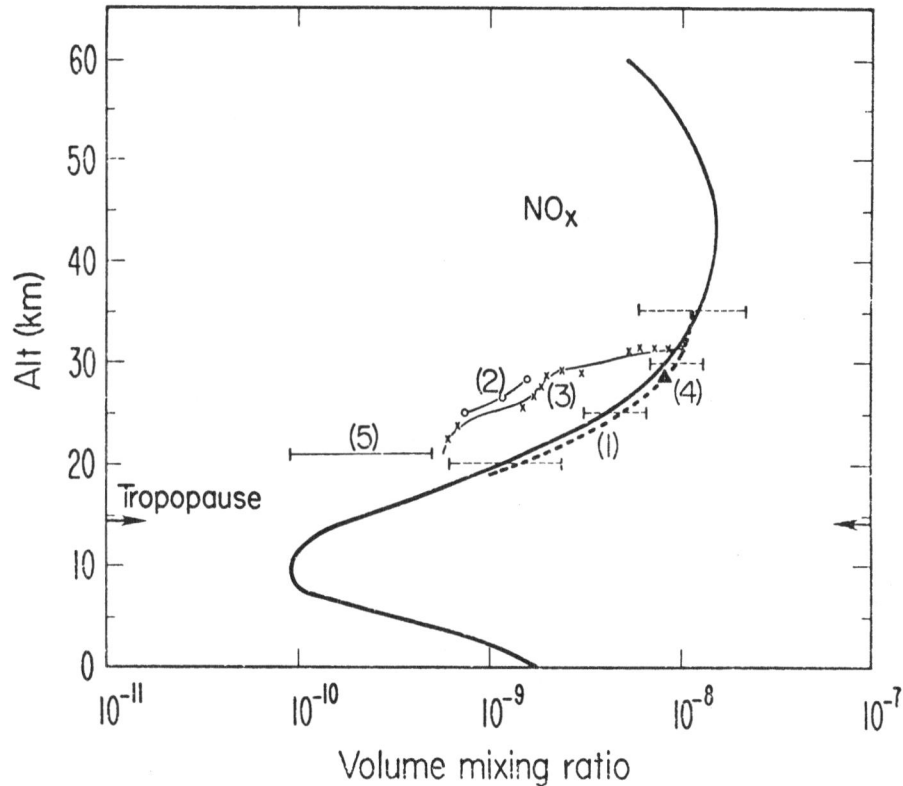

Fig.4. Calculated and measured volume mixing ratios for nitrogen oxides ($NO+NO_2$). The calculated profile is given by the solid line. Lower boundary conditions are determined from a constant flux of $2 \times 10^{+10}$ molecules cm^{-2}.

mically active nitrogen (NO and NO_2), thereby acting as catalysts for ozone. All chemical reactions of importance for the nitrogen chemistry are given by Crutzen and Isaksen (1976).

4. THE DISTRIBUTION OF CHLORINE COMPOUNDS

FC-11 ($CFCl_3$) and FC-12 (CF_2Cl_2) are probably the main sources of free chlorine in the stratosphere. Similar to nitrous oxides they are thought to be chemically stable in the troposphere, and should therefore show small variations in the troposphere. However, they are released only by anthropogenic processes, and since the release is predominantly in the Northern Hemisphere, and has increased exponentially up to now,

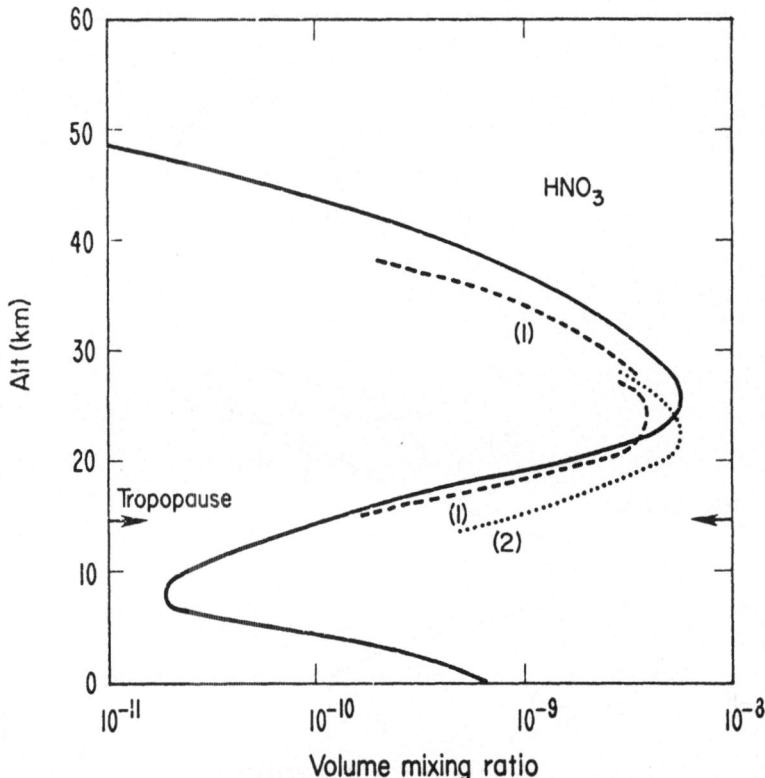

Fig.5. Calculated and measured volume mixing ratios for nitric acid. The calculated profile is given by the solid line. Lower boundary conditions are determined from a constant heterogeneous removal rate in the lowest 8 km of the troposphere, corresponding to a lifetime of 10 days.

variations between the two hemispheres and also in time have been observed.

In order to obtain the time evolvement of the distribution of FC-11 and FC-12 in the atmosphere, the estimated world release rates are used to calculate the distribution after 1950. The release rates are averaged over the globe. Mixing ratios of $CFCl_3$ and CF_2Cl_2 for 1975 are given in figure 6. The calculated mixing ratios are in remarkable good agreement with observations. In the troposphere they are well within the observed range, and in the lower stratosphere the drop in mixing ratios above approximately 15 km compares quite well the measurements. This good agreement is an in-

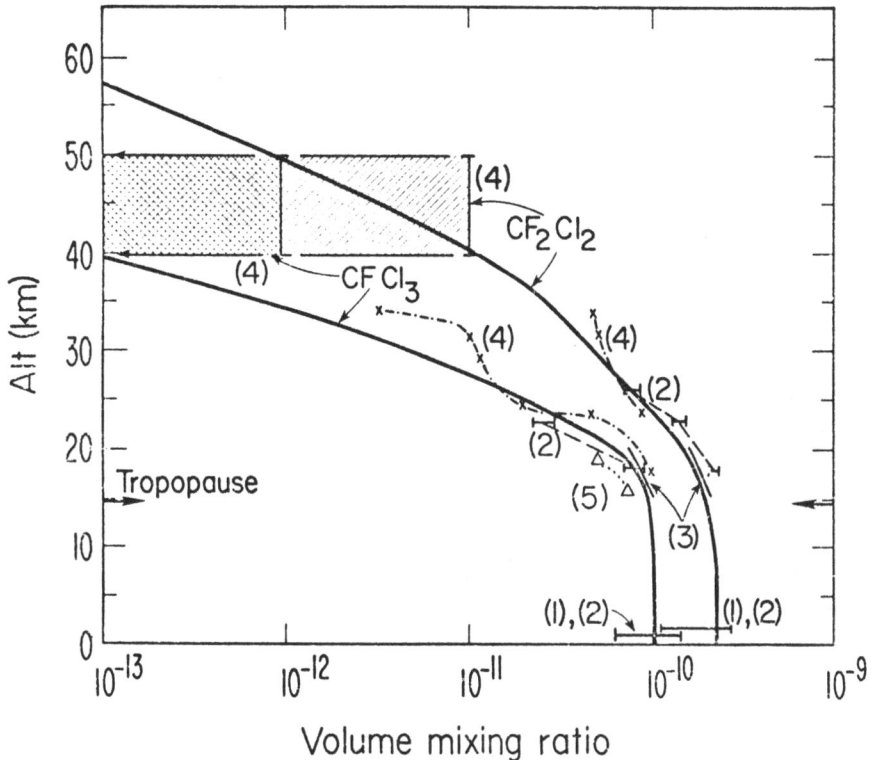

Fig.6. Calculated and measured volume mixing ratios
for CFCl₃ and CF₂Cl₂. The solid lines give the calculated profile. Lower boundary values are obtained
from global release rates, assuming a yearly increase
in release rates of 13% up to 1972, and 10% after 1972.

dication that it is not likely that any major sinks
have been left out of the calculations. The height
distribution in the stratosphere points to solar dissociation in the stratosphere as proposed by Molina
and Rowland (1974), as the main sinks.

In addition to FC-11 and FC-12 the height distribution of a large number of other chlorine containing
organic compounds, have been estimated. Those assumed
to be of anthropogenic origin are estimated in a similar way as FC-11 and FC-12, while those assumed to be
produced by nature are estimated using observed values
at the lower boundary. Of these compounds only carbon
tetrachloride (CCl₄), methyl chloride (CH₃Cl) and me-
thylchloroform (C₂H₃Cl₃) are found to be of importance

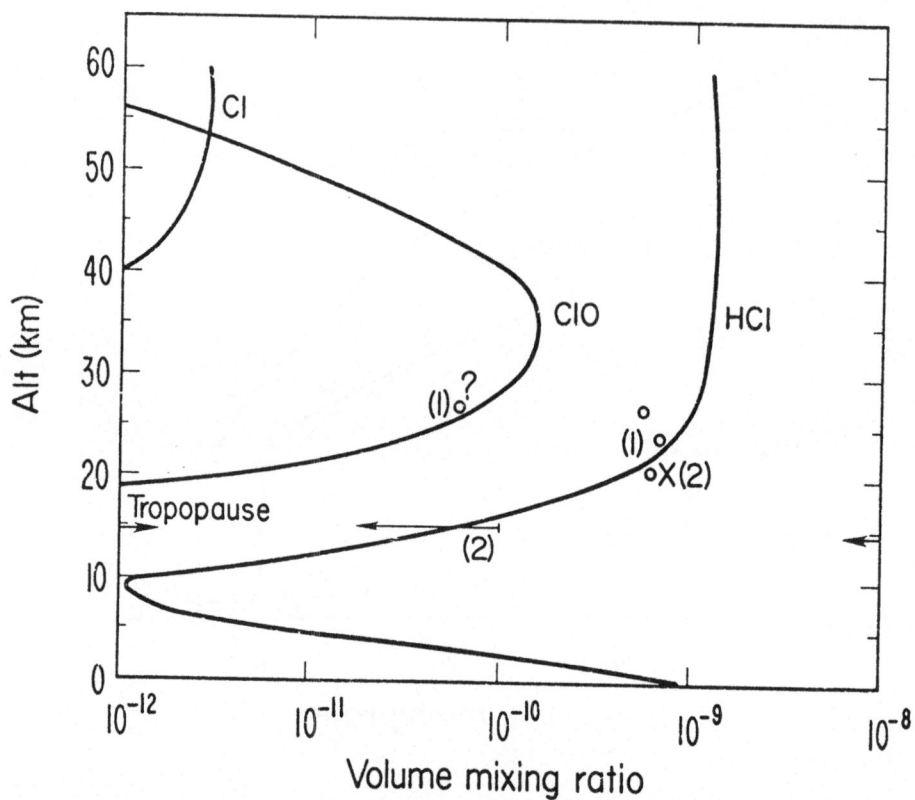

Fig.7. Calculated and measured volume mixing ratios of
HCl, ClO, and Cl assuming that the compounds are in
photochemical equilibrium. The calculated profiles are
given by solid lines. A constant mixing ratio of 10^{-9}
is used as lower boundary condition for HCl.

for chlorine in the stratosphere.

　　Based on the distribution of all the above chlorine containing species, the height profiles of hydrochloric acid, atomic chlorine and chlorine oxide are
calculated (Figure 7). Ground level production of hydrochloric acid will have a negligible effect on free
chlorine in the stratosphere, due to heterogeneous removal in the troposphere. It is important to notice
that most of the chlorine is in the form of HCl (>90%),
this strongly reduces the effect of Cl and ClO on
stratospheric ozone. The few measurements which are obtained in the stratosphere agree well with our calcu-

lated profiles.

5. REDUCTIONS IN TOTAL DENSITIES DUE TO THE RE-LEASE OF CHLORINE COMPOUNDS

The estimated reductions in total ozone due to the re-lease of chlorine compounds are given in figure 8. In order to estimate the future effect on ozone from the use of FC-ll and FC-12, two alternatives are considered. In the first case the release of FC-ll and FC-12 are stopped at the end of 1975, in the second case the release is continued three more years at the 1975 rate, and then stopped at the end of 1978. In both cases maximum reduction of ozone appears approximately ten years after the release has stopped. The maximum dif-ference between the two curves of 0.7% gives the ad-ditional effect of postponing the stop in FC-ll and

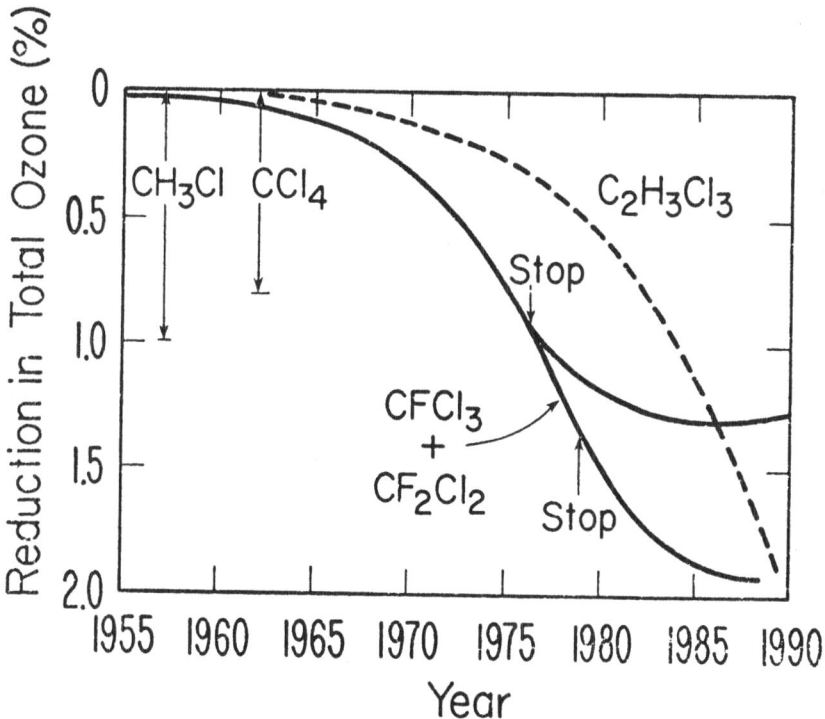

Fig.8. Reduction in total ozone due to the release of chlorine compounds (see text).

FC-12 releases with three years. A continued release on the 1975 level is estimated to result in an ozone reduction of approximately 10% late in the next century.

The reduction of ozone due to the release of methyl cloroform is based on a steady increase of 10% in the release rate. It is seen that also this compound will contribute significantly to the reduction of ozone in the future if the release rate continues to increase.

Both methyl chloride and carbon tetrachloride are in this model assumed to be of natural origin. The reduction of ozone is estimated to 1% from methylchloride, and to 0.8% from carbon tetrachloride. It should be noticed that measurements of carbon tetrachloride seem to indicate a noticable increase in the last few years. It is therefore possible that it is affected by man's activity. In that case it may have a future effect on the ozone that is more pronounced than estimated in this model.

6. CONCLUSIONS

The generally good agreement between calculated and measured profiles indicate that our model can be used to obtain realistic estimates of the future effects on the ozone layer from continued use of chlorine compounds. This is of particular interest since large natural variations in total ozone densities make it difficult to measure small, but significant variations (<5%) due to anthropogenic release of chlorocarbons.

REFERENCES

1.   P.J. Crutzen, Q.J.R. Meteorol. Soc. 96, 320 (1970).
2.   P.J. Crutzen, Private communications (1975).
3.   P.J. Crutzen, Geophys. Res. Lett. 3, 173 (1976).
4.   P.J. Crutzen and I.S.A. Isaksen, submitted to
           J. Geophys. Res. (1976).
5.   D.H. Ehhalt, Can. J. Chem. 52, 1510 (1974).
6.   S.C. Liu, R.C. Cicerone, T.M. Donahue, and W.L.
           Chameides, Geophys. Res. Lett. 3, 157 (1976).
7.   M.J. Molina, and F.S. Rowland, Nature 249, 810
           (1974).

MEASUREMENTS OF MINOR CONSTITUENTS IN THE STRATOSPHERE

M. Ackerman

Institut d'Aéronomie Spatiale
B - 1180 - Bruxelles

INTRODUCTION

The capability of spacelab to carry payloads of weight and
size an order of magnitude larger than was possible with other
vehicles will give a new dimension to investigations of the
earth homosphere from space. The use of more sophisticated
instruments will be possible as well as their association
allowing comparison between various types of measurements that
will lead to a better accuracy of the data gathered on a global
basis.

Particularly in the stratosphere, the geographic and vertical distribution of minor constituents is a key factor of the
understanding of the chemical and dynamical processes taking
place in that atmospheric region where the ozone abundance is
of crucial importance for life on earth and is at the center of
all physical and chemical stratospheric processes as discussed
by other speakers'at this symposium.

The analysis of the experimental work already performed
in this field with presently available vehicles and methods is
usefull as a guidance for the definition of spacelab borne
payload parts dedicated to this type of observations. Such an
analysis will be summarized here. Our present knowledge of the
minor stratospheric constituents originates from the alternative
interactions between model calculations and observations. These
have been obtained from various platforms: satellites, rockets,
balloons, aircrafts and the earth surface. The methods can be

J. J. Burger et al. (eds.), Atmospheric Physics from Spacelab, 107–116. All Rights Reserved.
Copyright © 1976 by D. Reidel Publishing Company, Dordrecht-Holland.

put in two different categories : those based on in situ
measurements on one hand and those based on remote sensing on
the other hand. The latter ones only can be applied from spacelab
while the other will possibly contribute indirectly in a broad
framework by providing verifications of remote measurements
gathered from space.

THE IN-SITU MEASUREMENT METHODS

These relate to local sampling or action on stratospheric
air followed by immediate or delayed application of analytical
techniques. In such cases the use of an adequate vehicle
implies balloon, aircraft and rocket flights. Samplings from
balloon have already been conducted at the end of the nine-
teenth century (Pfotzer, 1972).

Meteorological balloons are now routinely carrying
chemical ozone sondes pumping the ambient air for immediate
electrochemical analysis (Brewer and Milford, 1960). Chemi-
luminescent reactions of ozone with Rhodamine B, and ethylene
have also been applied to ozone measurements. Ridley et al
(1974) and Lowenstein et al (1975) have determined the
abundance of stratospheric nitric oxide through its luminescent
reaction with ozone respectively observed in sampling experiments on balloon gondolas and aircrafts. Even more complex
chemical procedures have been applied to the in situ detection
of carbon monoxide from aircraft (Seiler and Junge, 1969) by
observing optically mercury atoms release by reduction of
mercury oxide.

Physical processes have been applied in situ to the
determination of various minor constituents. To study the
conditions of contrail formation behind high flying aircrafts,
Dobson et al. (1945) have obtained the first evidence for the
stratospheric dryness. Methane (Bainbridge and Heidt, 1966),
freons and nitrous oxide (Schmeltekopt et al, 1975) have been
determined by means of laboratory gas chromatography on air
samples collected by balloons in pre-evacuated containers.
Cryopumping on board of balloon and rocket payloads followed
by laboratory analysis of in situ trapped samples has been
applied to the determination of several trace species. The
only information on the abundance of stratospheric H_2 is due to
this method (Ehhalt et al, 1975). Collection of trace elements
on paper filters has yielded data on nitric acid and on
halogens (Lazrus et al, 1975). Air samples have been analysed
inside an absorption cell on board of a ballon gondola by
means of a spin flip Raman laser source to detect nitric oxide

and water vapor (Patel et al., 1974). The most recently developped method is based on atomic resonance fluorescence in conjunction with a parachute borne flowthrough module (Anderson, 1975). Results have been so obtained on O and OH. The in situ measurement methods are summarized in Table I.

THE REMOTE SENSING METHODS

These are up to now essentially based on the observation of luminous phenomena in the wavelength range from the ultra-violet to the far infrared. The observation of stratospheric phenomena related to traces species is historically very old since the presence of volamic dust and of nacreous clouds is known since a long time. Ground based observations of optically active substances formed in the stratosphere and absent from the troposphere is possible at wavelengths where the lowest atmospheric layers do not inferfere. The best example appears to be ozone currently monitored by application of its ultra-violet absorption from 290 to 340 nm (Götz et al. 1934). It is however more generally suitable to observe the stratosphere from high altitude by means of instruments carried by balloons rockets or satellites. Since the molecular properties lead to a caracteristic spectrum of narrow lines for each species, the use of high resolution spectroscopic instruments ensures the specifity and the sensitivity of the method. The tangential observation of the atmosphere ensures long pathes and vertical resolution and can be made in absorption or in emission. In the first case, an extra-atmospheric light source must be used such as the sun or stars. The light intensity, I, received at the observation point equals

$$I = I_o \, e^{-\tau}$$

where I_o is the light intensity in absence of absorption and τ is the optical thickness, the product of the absorption cross section by the number of molecules on the path. In the second case the discrete emission of the atmospheric gas is observed which is related to the amount of absorber in a similar but more complex fashion. The measured intensity

$$I_E = I_o(T) \, (1 - e^{-\tau}) \, .$$

is related to the black body function $I_o(T)$ which depends on the temperature of the medium. The stratospheric emission peaks at about 10 μm.

TABLE I.   STRATOSPHERIC MEASUREMENTS [1]

CONSTITUENTS[2]	CO_2	H_2O	O_3	CH_4	N_2O	CO	H_2	(O)	(NO)	NO_2	HNO_3	HX[3]	$C_n X_y$[3]	(OH)
Range of number densities [log $n(cm^{-3})$]	13-16	11-14	10-13	9-13	8-12	9-12	9-12	5-10	8-9	7-9	7-10	7-9	6-9	6-8
METHODS OF MEASUREMENT														
REMOTE SENSING														
UV : ground based			x							x				
satellite			x											
IR : from aircraft,balloon or rocket														
emission sub-millim.waves		x	x						x	x	x			
emission middle IR		x	x	x	x	x			x	x	x	x		
absorption "														
-from ground (em.abs.)	x	x	x							x	x	x	x	
-from satellite (em.)	x		x								x?			
IN SITU														
Frost point hygrometer		x												
Chemiluminescence			x											
IR LASER on sample in situ		x												
Resonant scattering						x		x	x					x
General analytical methods applied in the labo. on in situ trapped samples	x			x	x	x	x				x	x	x	

[1] Based on actually published performance.

[2] Constituents in parenthesis present in sunlit stratosphere only.

[3] X stands for halogen.

Both emission and absorption methods require the knowledge
of molecular parameters such as energy levels, line intensities,
line widths to perform the inversion of integrated optical
depths in order to determine vertical distributions of constituents. While the absorption method is almost independent of the
temperature, the availability of the external light source limits
its use in time. On the other hand the emission method can
operate at any time but relies on the absolute determination of
intensity which depends on temperature. In addition, when the
atmospheric total density becomes to low, at high altitude $I_o(T)$
does not follow the Boltzman formula and the source function
must be determined by absorption measurements for instance.

The possibility of detecting minor stratospheric constituents by limb observations (Ackerman, 1963) began only recently
with the availability of sophisticated high altitude balloon
borne equipment dedicated to stratospheric observations. The
first measurements were on nitric acid (Murcray et al., 1968),
methane and nitrogen dioxide (Ackerman and Frimout, 1969) and
water vapor (Murcray et al., 1969). All of these were in absorption using the sun as a source as well as the more recent on
nitric oxide (Ackerman et al., 1975), chlorofluorocarbons
(Williams et al., 1975) and hydrochronic acid (Farmer et al.,
1976 and Ackerman et al., 1976).

The first stratospheric limb observations in emission
have yielded information in the submillimeter-wave range
(Harries, 1973) and in the middle infrared range (Murcray
et al., 1973 and Chaloner et al., 1975) mainly on the odd
nitrogen species NO, NO_2 and HNO_3 and on H_2O. Satellite
observation of the H_2O limb emission is now underway for the
low stratosphere from Nimbus 6 and Nimbus G will perform
measurements of the emission of several minor species by means
of several instruments.

The instruments used are of two types : the spectrometers-
interferometers and the correlation radiometers. All have to be
of large sensitivity and of high resolution in order to achieve
measurements of low concentrations as shown in figure 1, as
well as to separate the spectral features due to various gases.
Figure 2 shows as an example HCl absorption lines in the 3 μm
methane band. In order to perform occultation absorption
measurements from an orbiting vehicle the spectral scanning time
must be short which excludes the use of large path difference
interferometers. It should eventually be mentionned that ozone
is now routinely measured from the observation of ultraviolet
backscattered solar radiation.

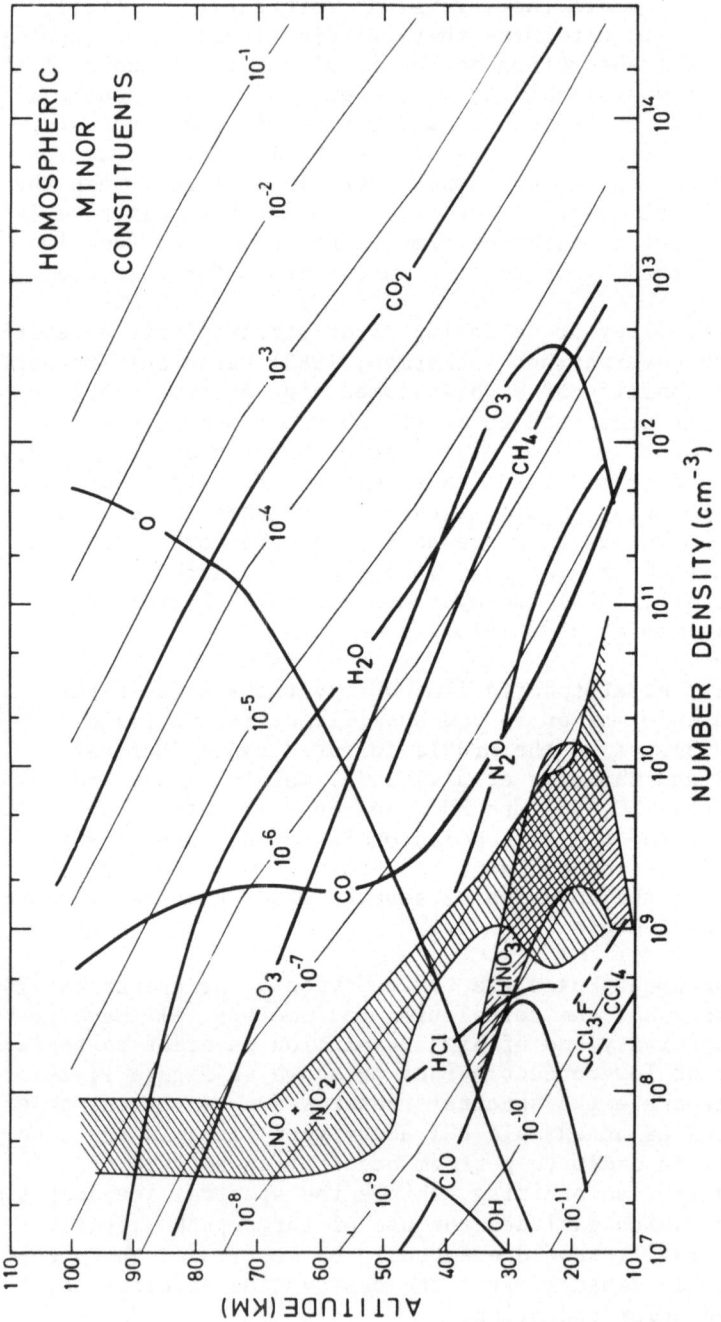

Fig. 1.- General view of the abundance of minor constituents in the stratosphere, mesosphere and low thermosphere.

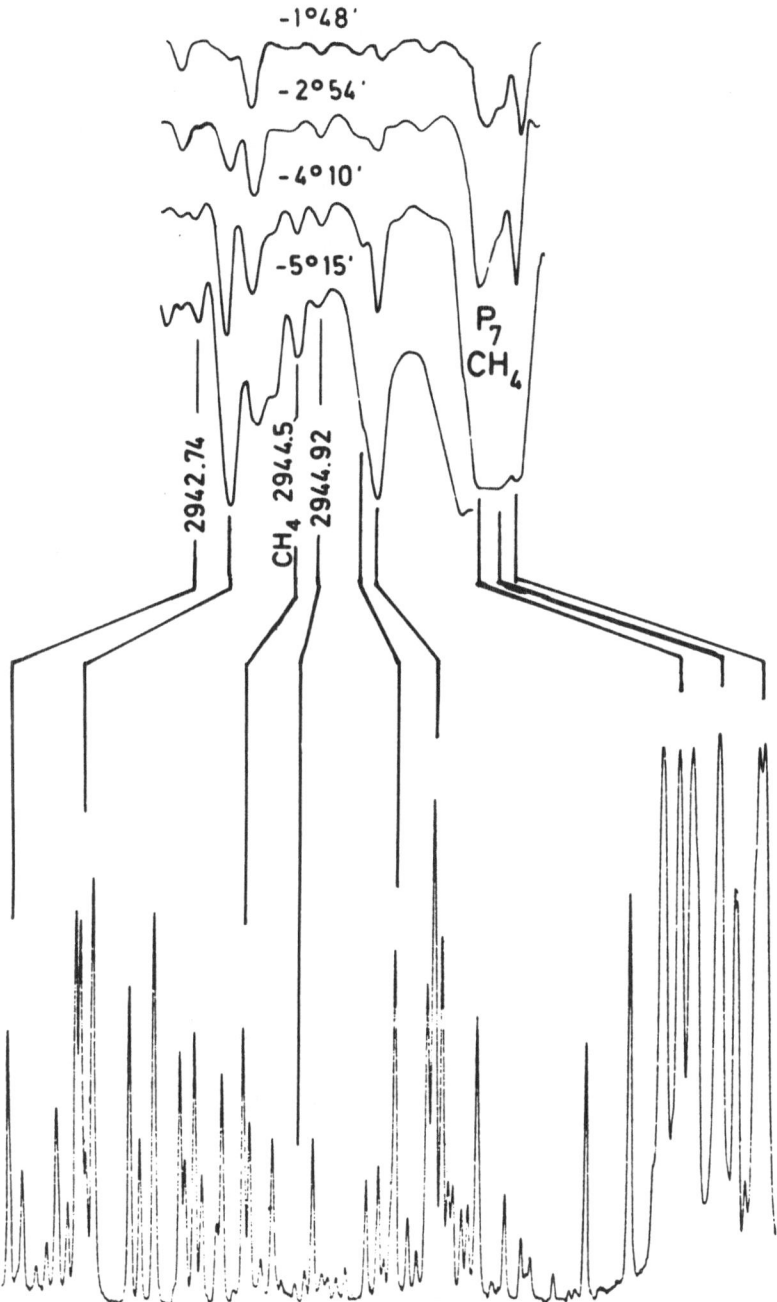

Fig. 2.- Stratospheric spectra (upper part) taken at various zenith angles in degree by means of a grille spectrometer and showing HCl absorption lines (2944.92, 2942.74 cm^{-1}) in the CH$_4$ 3μm band and laboratory spectra of CH$_4$ (R. Toth, private communication) showing the need of high resolution.

The results obtained up to now table I, by all means are
still preliminary, many discrepancies exist between the results
obtained by various methods and very few information exist on
geographic, seasonal and long term trends in the concentrations
of minor species. Of the great wealth of the electromagnetic
spectrum from the ultraviolet to submillimeter waves very little
has been inspected. The difficulties of the actual experiments
limit their number as well as their geographic and time
distribution. Such limitations will be overcome by the use of
jointly instrumented payloads on orbit allowing inter comparisons
of observations performed at any location and at any time.

REFERENCES

ACKERMAN, M. (1963), Possibilité de détection de constituants
    atmosphériques mineurs par absorption infrarouge entre
    35 et 40 km d'altitude, Aeronomica Acta B n° 1.
ACKERMAN, M. and D. FRIMOUT (1969), Mesure de l'absorption
    stratosphérique du rayonnement solaire de 3,05 à 3,7 microns
    Bull. Acad. Roy. Belgique, Cl. Sc., 55, 948.
ACKERMAN, M., J.C. FONTANELLA, D. FRIMOUT, A. GIRARD,
    N. LOUISNARD and C. MULLER (1975), Simultaneous measur-
    ements of NO and NO_2 in the stratosphere, Planet. Space Sci.
    23, 651.
ACKERMAN, M., D. FRIMOUT, A. GIRARD, M. GOTTIGNIES and C. MULLER
    (1976), Stratospheric HCl from infrared spectra, G.R.L.,
    3, 81.
ANDERSON, J.G. (1975), The absolute concentration of $O(^3P)$ in
    the earth's stratosphere, G.R.L., 2, 231.
BAINBRIDGE, H.E. and L.E. HEIDT (1966), Measurements of methane
    in the troposphere and lower stratosphere, Tellus, 18, 221.
BREWER, A.W. and J.R. MILFORD (1960), The Oxford Kew Ozone-
    sonde, Proc. Roy. Soc., A 256, 470.
CHALONER, C.P., J.R. DRUMMOND, J.T. HOUGHTON, R.F. JARNOT and
    H.K. ROSCOE (1975), Stratospheric measurements of H_2O and
    the diurnal change of NO and NO_2, Nature, 258, 696.
DOBSON, G.M.B., A.W. BREWER and B.M. CWILONG (1945), Meteo-
    rology of the lower stratosphere, Proc. Roy. Soc., 185A,
    144.
EHHALT, D.H., L.E. HEIDT, R.L. LUEB and E.A. MARTELL (1975),
    Concentrations of CH_4, CO, CO_2, H_2, H_2O and N_2O in the
    upper stratosphere, J. Atmos. Sciences, 32, 163.
FARMER, C.B., O.F. RAPER and R.H. NORTON (1976), Spectroscopic
    detection and vertical distribution of HCl in the tropo-
    sphere and stratosphere, G.R.L., 3, 13.
GOTZ, F.W.B., A.R. MEETHAM and G.M.B. DOBSON (1934), The
    vertical distribution of ozone in the atmosphere, Proc.
    Roy. Soc. London, A145, 416.
HARRIES, J.E. (1973), Measurements of some hydrogen-oxygen-
    nitrogen compounds in the stratosphere from Concorde 002,
    Nature, 241, 515.
LAZRUS, A.L., B.W. GANDRUD, R.N. WOODARD and W.A. SADBACEK
    (1975), Stratospheric halogen measurements, G.R.L.,
    2, 439.
LOWENSTEIN, M. and H. SAVAGE (1975), Latitudinal measurements
    of NO and O_3 in the lower stratosphere from 5.5° to 82°
    North, G.R.L., 2, 448.
MURCRAY, D.G., T.G. KYLE, F.H. MURCRAY and W.J. WILLIAMS (1968),
    Nitric acid and nitric oxide in the lower stratosphere,
    Nature, 218, 78.

MURCRAY, D.G., T.G. KYLE and W.J. WILLIAMS (1969), Distribution
    of water vapor in the stratosphere as derived from setting
    sun absorption data, J. Geophys. Res., 74, 5369.
MURCRAY, D.G., A. GOLDMAN, A. CSOEKE-POECHK, F.H. MURCRAY,
    W.J. WILLIAMS and R.N. STOCKER (1973), Nitric acid
    distribution in the stratosphere, J. Geophys. Res., 78,
    7033.
PATEL, C.K.N., E.G. BURKHARDT, C.A. LAMBERT (1974), Spectro-
    scopic measurements of stratospheric nitric oxide and
    water vapor, Science, 184, 1173.
PFOTZER, G. (1972), History of the use of balloons in scientific
    experiments, Space Sci. Rev., 13, 199.
RIDLEY, B.A., H.I. SCHIFF, A.W. SHAW, L.R. MEGILL, L. BATES,
    C. HOWLETT, H. LEVAUX and T.E. ASHENFELTER (1974),
    Measurement of nitric oxide in the stratosphere between
    17.4 and 22.9 km, Planet. Space Sci., 22, 19.
SCHMELTEKOPF, A.L., P.D. GOLDAN, W.R. HENDERSON, W.J. HARROP,
    T.L. THOMPSON, F.C. FEHSENFELD, H.I. SCHIFF, P.J. CRUTZEN,
    I.S.A. ISAKSEN and E.E. FERGUSON (1975), Measurements of
    stratospheric $CFCl_3$, CF_2Cl_2 and N_2O, G.R.L., 2, 393.
SEILER, W. and C. JUNGE (1969), Decrease of carbon monoxide
    mixing ratio above the polar tropopause, Tellus, 21, 447.
WILLIAMS, W.J., J.J. KOSTERS, A. GOLDMAN and D.G. MURCRAY
    (1975), Simultaneous stratospheric measurements of fluoro-
    carbons and odd nitrogen compounds, private communication
    to be published.

HIGH RESOLUTION ATMOSPHERIC EXTINCTION MEASUREMENTS FROM
THE FRENCH EXPERIMENT ON BOARD THE NASA SPACECRAFT OSO-8

A. Vidal-Madjar[o], R.G. Roble[+], W.G. Mankin[+], G. Artzner[o],
R.M. Bonnet[o], P. Lemaire[o], J.C. Vial[o]
[o] LPSP Verrières-le-Buisson - 91 - FRANCE
[+] NCAR Boulder - Colorado - 80302 - USA

ABSTRACT

The French instrument on board OSO-8 is a multichannel,
high resolution (0.02 Å) UV spectrometer, observing very small solar
areas (from 1" x 1" of arc to 6" x 2' of arc) simultaneously
in the Ca II, Mg II, Ly α, Ly β lines.

Using a classical inversion technique as described by Roble
and Hays (1972) it has been possible to invert some extinction
profiles.

The quality of the data appears to give a 500 m vertical
resolution both for O_3 and O_2. O_3 is essentially observed from
55 to 75 km and O_2 from 85 to 200 km. Due to the velocity of the
spectral scan mechanism (able to scan Ly α line in 2.56 s) it was
possible to observe the differential extinction of the Lyman
alpha line when absorbed by O_2. Furthermore data on the atomic
hydrogen absorption give a new method of evaluating the exospheric
temperature and atomic hydrogen density at each point of the
exobase.

J. J. Burger et al. (eds.), Atmospheric Physics from Spacelab, 117–128. All Rights Reserved.
Copyright © 1976 by D. Reidel Publishing Company, Dordrecht-Holland.

1. INTRODUCTION

The French instrument on board OSO 8 spacecraft was launched the 22nd of June 1975 and was placed in a circular orbit at 550 km of altitude with an inclination of 33° to the equatorial plane.

During each orbit (15 times per day) the spacecraft, pointed to an area of the solar disc, observes a sunset and a sunrise at the earth limb. This means that every day of the life of our instrument we will gather 30 extinction curves for each observation channel.

The LPSP instrument (see figure 1) is a six channel high resolution spectrometer (0.02 Å) mounted behind a high resolution Cassegrain telescope (1 x 1 second of arc).

Two channels observing the Ca H and Ca K lines at 3900 Å give us a pointing stability reference during the sunrises and sunsets since radiation at these wavelengths is not absorbed by the earth's atmosphere.

Two other channels observing the Mg H and Mg K lines around 2800 Å, give during sunrises and sunsets essentially the extinction curve corresponding to the ozone in the earth's atmosphere since in this wavelength range O_3 is by far the main absorber. A lyman alpha channel gives us the O_2 altitude distribution, but also, due to the 0.02 Å resolution of the spectrometer, the atomic hydrogen is very clearly visible on Lyman alpha profiles, giving then along the whole orbit (day side) at each point of the exobase, the exospheric temperature and the atomic hydrogen density.

Finally, the last channel centered on Lyman beta line gives us the O_2 distribution at higher altitudes, above the Lyman alpha results. Here again the hydrogen geocoronal absorption is visible at the center of the line.

These results are the main one that will be obtained in each different channel ; however because each channel covers a spectral range of the order of 20 Å, other solar lines extinctions may be observed and give other potentials of the instrument. In particular the Lyman beta channel observes the sun in several 20 Å bands due to the overlaping orders of the grating. In particular this places in our field of observation the O I lines at 1304 Å and the N I lines at 1200 Å.

In this paper we will describe the main objectives that could be reached in the different channels and give some preliminary results.

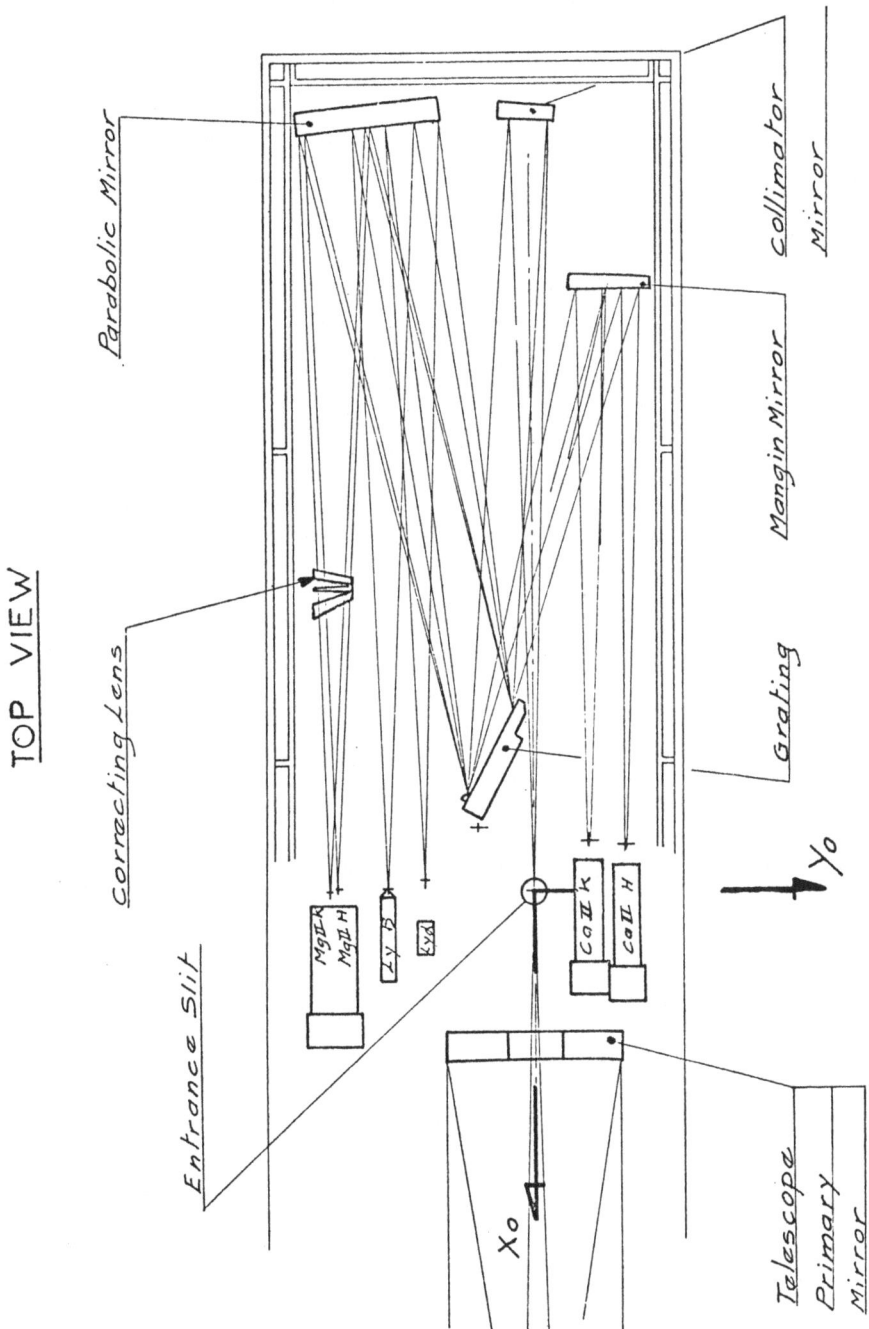

Figure 1   Optical schematic of the LPSP instrument on board the OSO 8 spacecraft.

2. THE MAGNESIUM CHANNELS (2800 Å)

At any point of the spectral range covered by these two
channels the O_3 extinction is observed at each sunrise and sunset.

An example of such an extinction curve is presented in
figure 2 showing the high quality of the data. Also having a
measurement point every 0.16 s this gives us a 500 m resolution
in altitude.

As one may see the inverted profile obtained independently
from the two magnesium channels fits almost perfectly, giving,
through that redundance, a very great confidence in the evaluated
O_3 number density. These two almost identical curves show also
the high precision obtained in the O_3 number density which is
better than 10 %. This fact will help us to observe very small
variations of the ozone concentration over the globe (in 24 hours)
and compare our map with 30 points to the ozone maps distribution
made at lower altitude (up to 55 km where our measurements start)
made with a completely independent technique (I.R. radiometry)
from Nimbus 6 spacecraft launched also in June 1975 (Gille et al,
1975). Such a comparison seems to be very promissing and will
give great confidence in both observational techniques.

It is hard to discuss very much detail from one observed O_3
distribution since the importance of new observations is to provide three dimensional informations. In effect our result (see
figure 2) agrees with other observed profiles and most of the one
dimensional theoretical models. The interesting thing will then
be, in the future, to fit our high quality three dimensional
results to three dimensional theoretical models on which they
will put in consequence strong constraints.

3. THE LYMAN ALPHA CHANNEL

As it can be seen very clearly on any Lyman alpha line
profile (see figure 3) a narrow absorption feature appears at
the center. This feature due to atomic hydrogen spread high
above the altitude of the spacecraft is visible during the whole
dayside part of the orbit. It changes only in width and depth
giving a new and promissing observational technique to measure
the exospheric temperature at each point of the orbit (a spectral
scan can be completed in 20 sec) values that could be directly
compared to model predictions or other observations. Also the
atomic hydrogen density at the exobase will be simultaneously
measured, producing a very rich set of data for understanding in
a better way that controls the hydrogen distribution at the exobase :
        - the thermal distribution alone

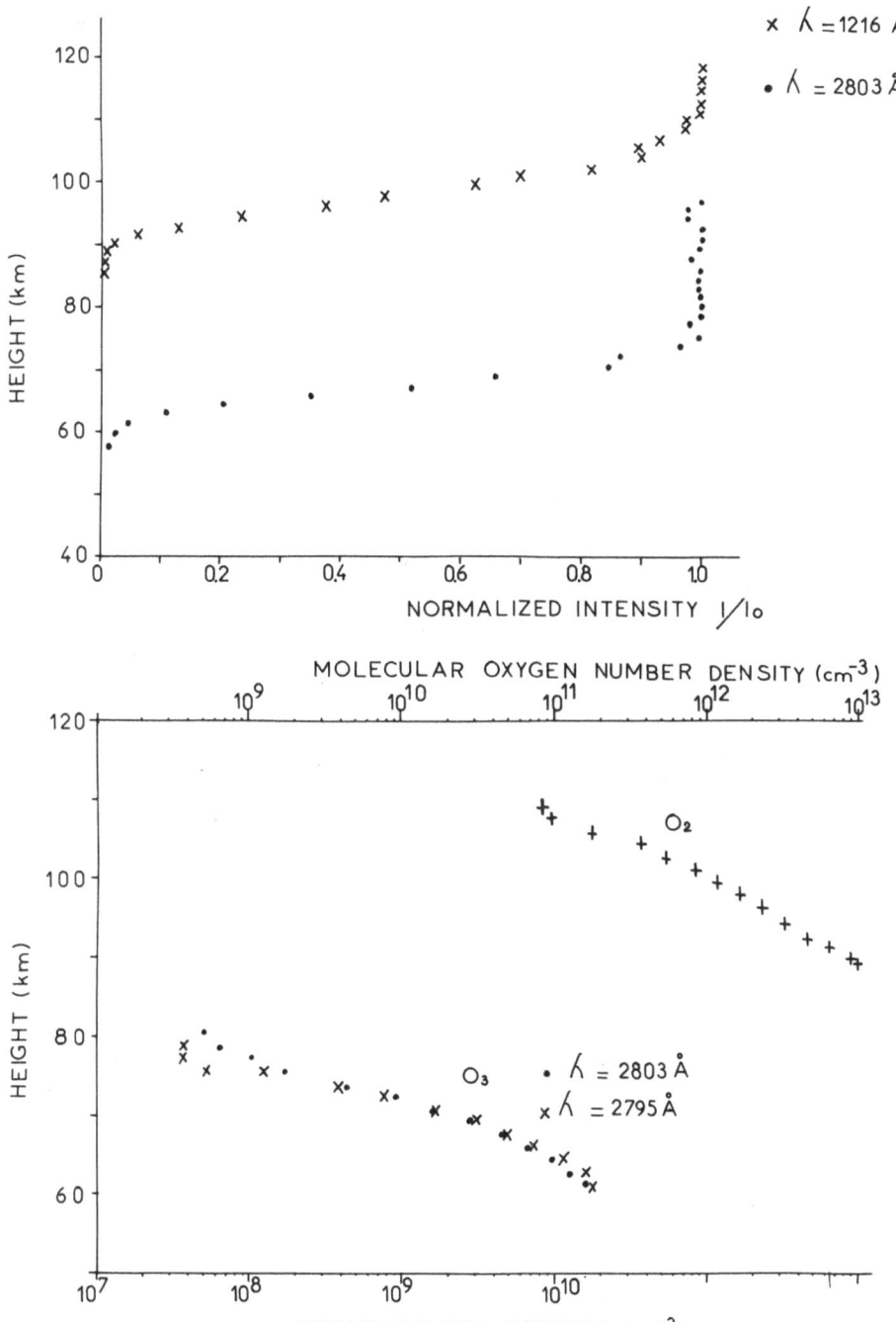

Figure 2   Extinction curves for 1216 Å and 2800 Å as observed
            from OSO 8 and the related altitude profiles of
            O_2 and O_3.

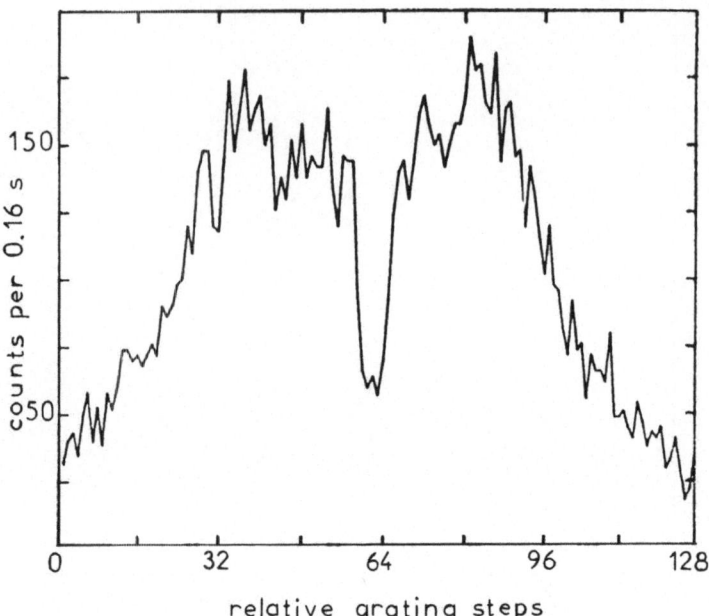

relative grating steps

Figure 3   A typical Lyman alpha profile obtained in 20 sec
             with the LPSP instrument on board OSO 8, showing
             very clearly the hydrogen geocoronal absorption
             core.

TIMES SEPARATED BY 1.28 SEC

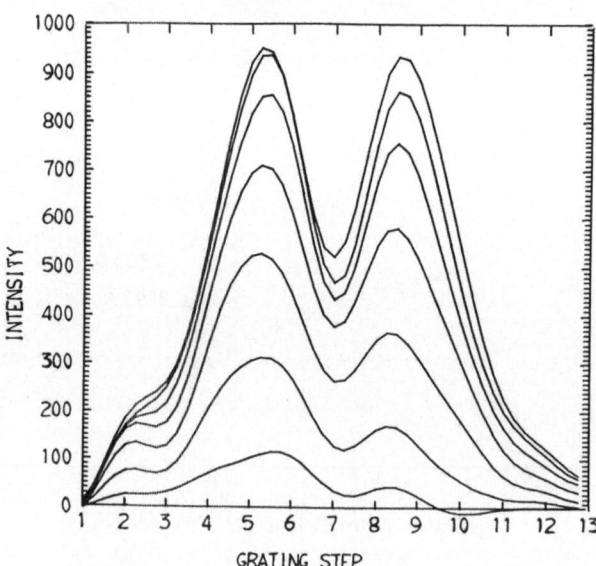

Figure 4   Extinction of the Lyman alpha profile showing a
             stronger absorption on the blue peak than on the
             red one (wavelength varies in the opposite direction
             of the grating steps).

- or some other mechanisms like perturbations due to winds or charged exchanged escape mechanism.

When the instrument is set on a given point of the Lyman alpha profile, then an extinction curve due to O_2 absorption is measured (figure 2). From the inversion of that curve the O_2 number density can be deduced between 90 and 110 km. Again one curve in agreement with many measurements and models does not provide too much new information. The interesting thing here as for O_3 will be to build up three dimensional maps of O_2, extended much higher in altitude, by using simultaneous measurements in the Lyman beta channel (see next paragraph).

But in the case of the Lyman alpha channel, it was also possible to scan the whole line (in 2.56 s) during the sunrises or sunsets showing (figure 4) directly the differential extinction over the line, since the O_2 cross section varies over the line (Ogawa, 1968). It is very clearly seen that the blue peak is absorbed faster than the red one (the grating steps increasing with decreasing wavelength) showing directly that the O_2 cross section is larger on the blue side of the line than on the red one. But from this set of measurements one may evaluate the relative variation of the cross section over the line, evaluation presented on figure 5 along with Ogawa laboratory measurements. It is clear that the relative variation is not the same and even that a bump seems to appear at 1216.1 Å in the observed cross section variation. Is this difference due to the different conditions existing in the upper atmosphere and in the laboratory ? This is possible, but since after inverting the extinction profiles for each different wavelength, assuming the same cross sections for all, we found that the O_2 number density distributions were not compatible one with the other by more than a factor of two (figure 6), it seems more likely that the difference is due to other absorbers present in the upper atmosphere. Following Weeks (1975) the most probable candidate for the overall change of the relative cross section shape is NO which is ionized by the Lyman alpha flux producing the well known D region. N_2 may be responsible for the bump around 1216.1 Å since an N_2 absorption band starts at approximately that wavelength (O_3 may also be present at high altitudes).

This result seems very exciting since it may lead to an evaluation of NO and/or N_2 and/or O_3 number densities in this altitude range.

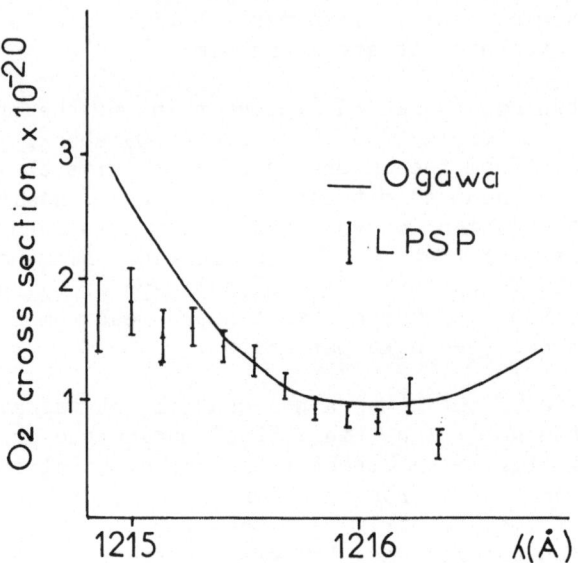

Figure 5 O_2 absorption cross sections deduced from figure 4
                extinction curves.

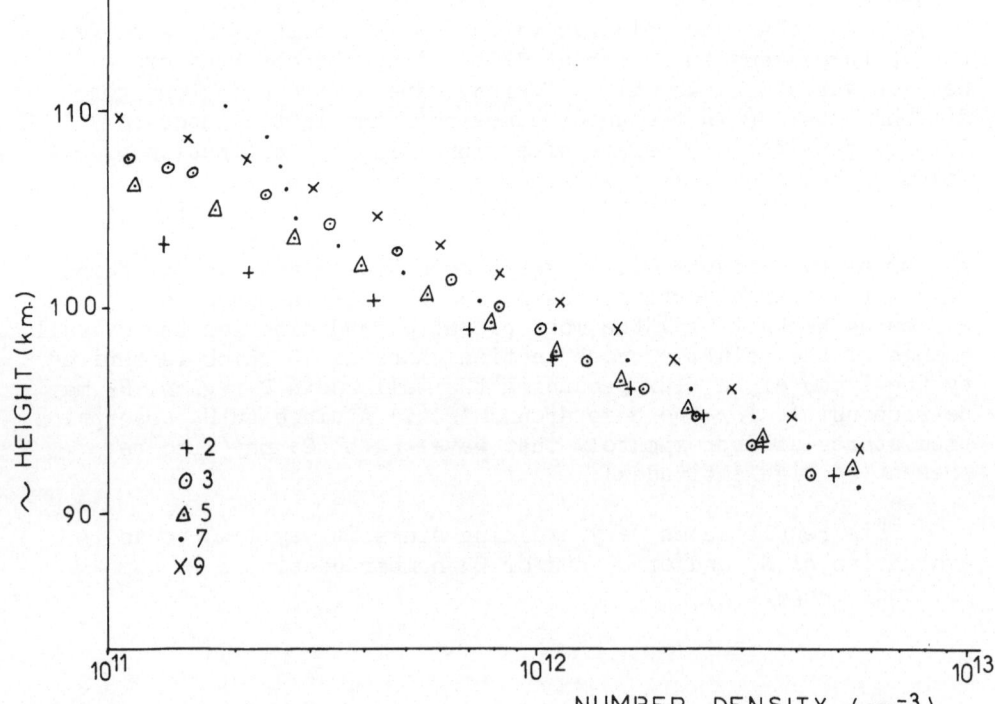

Figure 6 O_2 altitude profiles deduced from figure 4
                extinction profiles for different grating positions.

4. THE LYMAN BETA CHANNEL

From this channel like from the Lyman alpha one the results obtained will give essentially the O_2 number density but at much higher altitudes, from 150 to 200 km. The O_2 number density profile obtained simultaneously with the one from the Lyman alpha channel will give the whole thermospheric distribution profile of O_2.

This result in our set of data is very interesting since it will be the link between all our atmospheric measurements going from O_3 around 65 km up to the exospheric atomic hydrogen density and temperature.

5. SOME POSSIBLE OTHER CONSTITUENTS OBSERVATIONS WITH THE OSO 8 INSTRUMENT

In addition to the possible detection of NO and N_2 we made, we observed a very weak absorption feature in the magnesium channels around 2816 Å. If confirmed this result may be due to OH absorption above 75 km and will give us observational data of great value since this element is directly linked through photochemical equilibrium to other observed ones like O_3, O_2, H.

A simultaneous observation of such a set of constituents is certainly the best way to observe and understand the complicated photochemical equilibrium present in that part of the atmosphere.

6. CONCLUSION

The French instrument on board OSO 8 demonstrates that a multichannel spectrometer observing extinction through the earth's atmosphere can be a very powerful tool to study the three dimensional structure of the atmosphere. Essentially this is due to the simultaneous observations made in all channels giving simultaneous observations of different constituents over different altitude ranges in the atmosphere. Also the high quality, precision and rapidity of the grating mechanism give us the possibility of observing even evolutions of line profiles during sunrises and sunsets.

Such an instrument, adapted to a specific atmospheric mission, using either high resolution channels for line profiles studies or broad band detector for continuum absorption observations should be able to give a complete survey of the earth's atmosphere from 20 to 500 km with a whole set of constituents like :

O_3, O_2, O, OH, NO, H, N_2, N

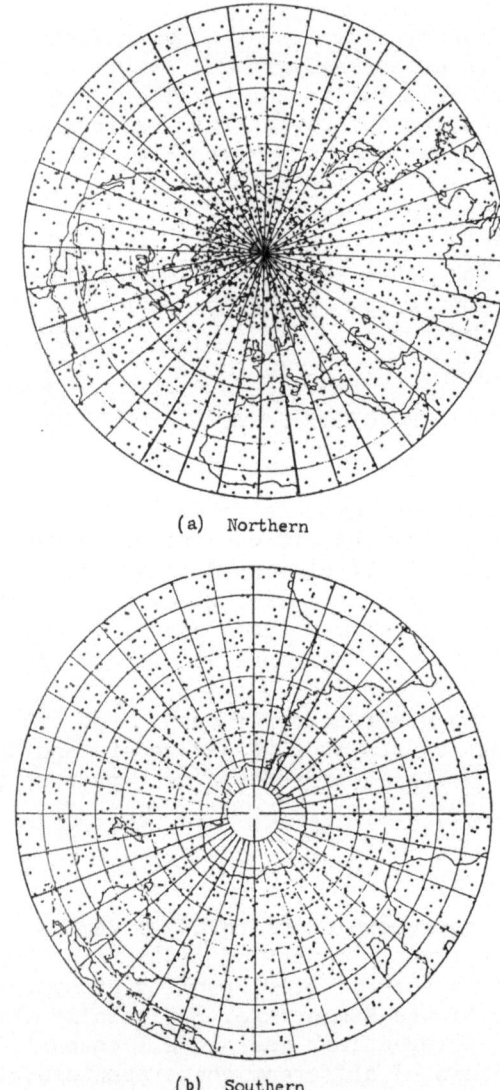

(a) Northern

(b) Southern

Figure 7   Taken from Roble (1975)
              Global data coverage possible for stellar occul-
              tation measurements. Only 5th magnitude stars in
              the Celestial Sphere ± 30° aft of the orbital plane
              in a polar orbiting satellite were considered.
              Forty-five seconds for occultation and slewing to
              the next star was allowed. Each point represents
              an ozone profile from cloud top to 100 km and the
              ensemble gives the 24-hour global data coverage
              that is possible.

and probably others giving even weaker features.

Finally an other improvement of such an instrument is to do the same type of observations, but using a star as a source and moving the pointing system from star to star during an orbit as evaluated by Roble (1975) an entire O_3 profile from cloud top height in the troposphere (ozone absorption in the Chappuis band near 6000 Å) to 100 km altitude giving the ozone distribution with altitude, along with the ones of other constituents listed above, could be made over the whole earth surface (see figure 7 taken from Roble, 1975). In other words, this observational technique, adapted to the observation of constituents in the earth's atmosphere, which is already giving very good observational data due to simultaneity in the case of solar occultations can be considerably improved in the future by giving a global coverage if stellar occultations are used.

REFERENCES

Gille J.C., Bailey P., House F.B., Craig R.A., and Thomas J.R.
The limb radiance inversion radiometer (LRIR) experiment (1975)

Ogawa M.
Absorption coefficient of O_2 at the Lyman alpha line and its vicinity - J.G.R., 73, 6759 (1968)

Weeks L.H.
Determination of O_2 density from Lyman alpha ion chambers - J.G.R., 80, 3655 (1975)

Roble R.G.
Atmospheric properties from solar and stellar occultations (1975)

Roble R.G. and Hays P.B.
A technique for recovering the vertical number density profile of atmospheric gases from planetary occultation data - Planet. Space Science. 20, 1727 (1972)

DISCUSSION

T.M. Donahue: Atroya, Reigler and I have succeeded in observing
a single UV absorption line produced by telluric H_2 using a
stellar source and the high-resolution scanning spectrometer
on OAO (Copernicus) two weeks ago. We think we can infer H_2
densities between 50 and 120 km from the data and plan to
try the same kind of experiment for NO.

A. Vidal-Madjar: As a matter of fact, the Copernicus Princeton
instrument is also very well adapted to this type of atmospheric study due to its high spectral resolution of 0.05 Å.
This result is very encouraging and shows how powerful this
observational technique is, even for low-altitude observations
(below 100 km).

HIGH RESOLUTION ATMOSPHERIC EMISSION SPECTROSCOPY AT 120 cm^{-1} AND 535 cm^{-1}, WITH A RAPID SCAN INTERFEROMETER AND 91 cm AIRBORNE TELESCOPE.

M. Anderegg, J.E. Beckman, A.F.M. Moorwood, H.H. Hippelein,

Astronomy Division, ESTEC, Noordwijk, Holland,

J.P. Baluteau, E. Bussoletti, A. Marten,

Groupe Infrarouge Spatial, Observatoire de Meudon, France,

N. Coron,

L.P.S.P., Verrières-le-Buisson, France.

1. INTRODUCTION

Instrumentation for observing infrared ionic lines from the interstellar medium has been developed by the present observers for use with the 91 cm telescope on NASA's G.P. Kuiper airborne observatory. Technical details have been described elsewhere[1] and the measurement of ionic lines from the Orion Nebula will be the subject of further published work. The system attains a spectral resolution of 0.02 cm^{-1} within a helium-cooled filter passband of $\Delta\nu/\nu \sim 0.1$, scanning rapidly to give fringe frequencies in an audio passband centred on 80 Hz, and fully resolved individual spectra in times ranging from 30 sec to 80 sec, for infrared bands centred on 535 cm^{-1} and 120 cm^{-1}. In a very preliminary account, presented at the CIRP[2], the capability of the instrument for atmospheric studies was indicated. Here the work is carried a step further, showing the results of using the interferometer from the ground (Pic du Midi, 3 km above sea level) and from the C141 aircraft at 13 km, in obtaining atmospheric emission spectra in the two passbands.

J. J. Burger et al. (eds.), Atmospheric Physics from Spacelab, 129–134. All Rights Reserved.
Copyright © 1976 by D. Reidel Publishing Company, Dordrecht-Holland.

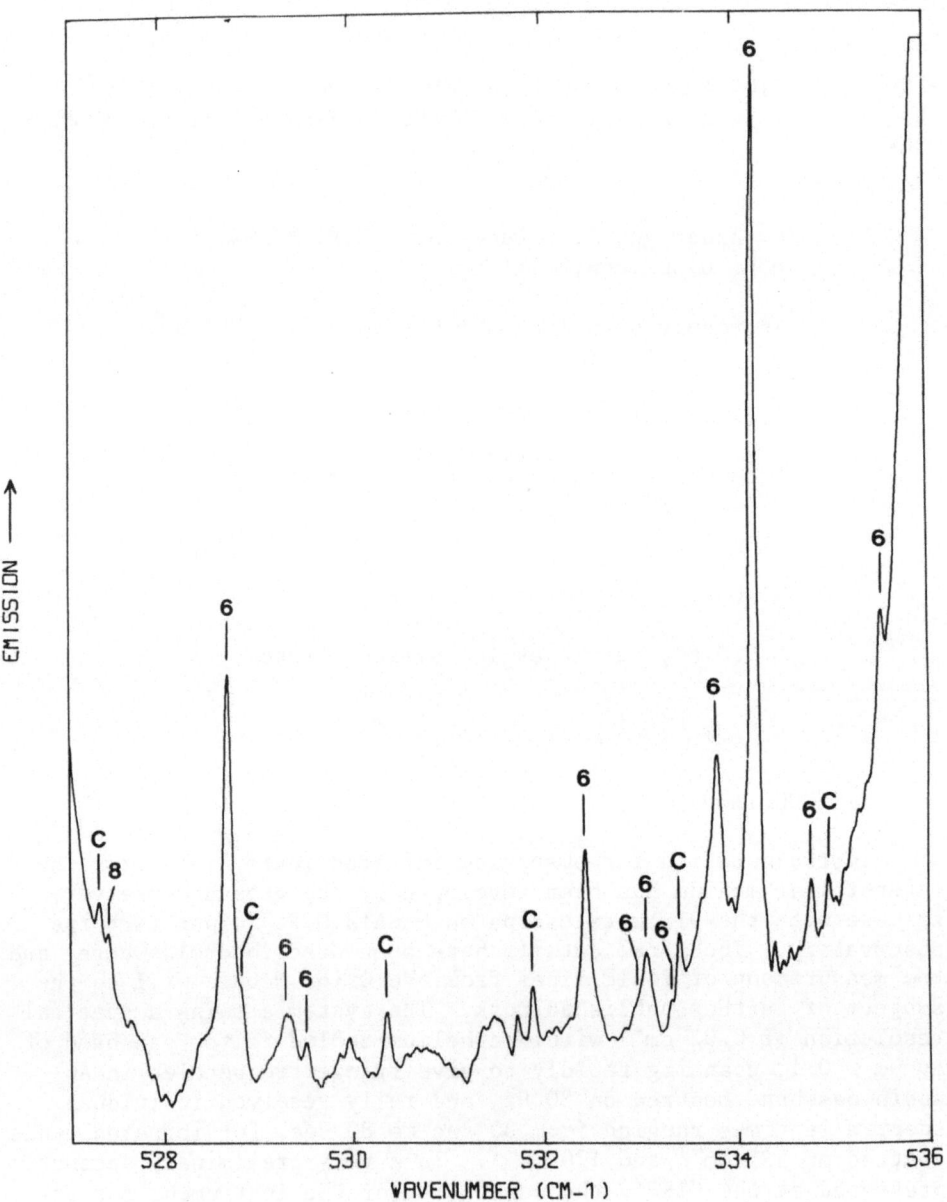

FIG. 1   Ground based (3 km altitude) atmospheric emission
          spectrum near 530 cm^{-1}.   C = CO$_2$;   6 = H$_2{}^{16}$O;
          8 = H$_2{}^{18}$O.

2. DATA

Fig. 1 shows a small portion of the spectrum of the atmosphere, obtained at an elevation angle of $65°$ from the Pic du Midi. Identifications are shown, as is the resolving capability of the instrument. The resolution enables line positions to be measured to within 0.005 cm^{-1}, provided that systematic errors due to imprecise dignment of the reference optical laser have been corrected. In checking these and line identifications, however, particular attention was paid to the intensities of the lines, as derived from their theoretical line strengths, and emitted from a standard model atmosphere. One can thus state with considerable confidence not only that many of the line positions agree well with those of McClatchey et. al.[3] but also that a number of lines are not in their predicted positions. To illustrate this we list a small sample of identifications for the isotopic species $H_2{}^{18}O$ and $H_2{}^{17}O$:

ν_{obs} (cm^{-1})	ν_{AFCRL}(cm^{-1})[4]	ν_{TM} (cm^{-1})[5]	Species	Transition
527.466	527.780	527.468	$H_2{}^{18}O$	6,6,1/5,3,2
524.029	524.070	524.045	$H_2{}^{18}O$	10,3,8/9,0,9
524.942	-	524.946	$H_2{}^{17}O$	10,3,8/9,0,9

TABLE 1

The work of Toth and Margolis[4] was on the vibration-rotation spectrum of H_2O^{18} at 3000 cm^{-1}, measured in absorption, from which the tabulated results are inferred. Our data gives an improvement in water ($H_2{}^{16}O$) rotational line positions; for example we measure the 14,5,9/13,4,10 transition predicted in the AFCRL listing[3] at 557.364 cm^{-1} to be at 557.208 cm^{-1}, a shift of 0.16 cm^{-1}. A transition of interest in making astronomical observations was the 13,9,4/12,8,5 transition, predicted[3] at 534.322 cm^{-1} but measured at 534.266 cm^{-1}. Although the shift here is only 0.06 cm^{-1}, the true position is important when designing an experiment to measure the SIII ionic transition at 534.39 cm^{-1}, which can be Doppler shifted in some astronomical sources by as much as 0.15 cm^{-1}. Our $H_2{}^{16}O$ frequencies are more precise within the range 500 cm^{-1} to 560 cm^{-1} than previous standard work, for example that of Izatt et. al.[5].

A second sequence of spectra, taken from 13 km aircraft altitude is of considerable interest. Fig. 2a shows a small portion of such a spectrum, obtained at a zenith angle of $40°$, with spectral resolution 0.02 cm^{-1}. A single such spectrum was obtained in

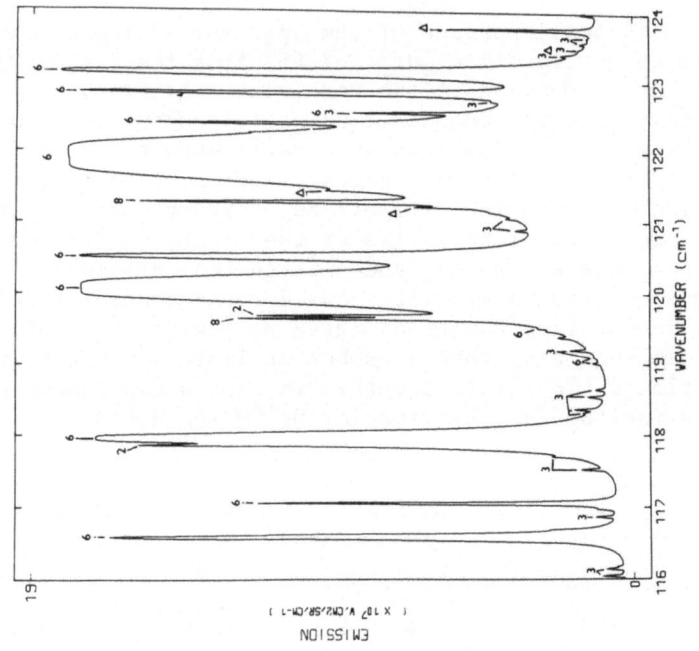

Fig. 2b.   Predicted emission spectrum
from 13 km altitude, based
on standard model.

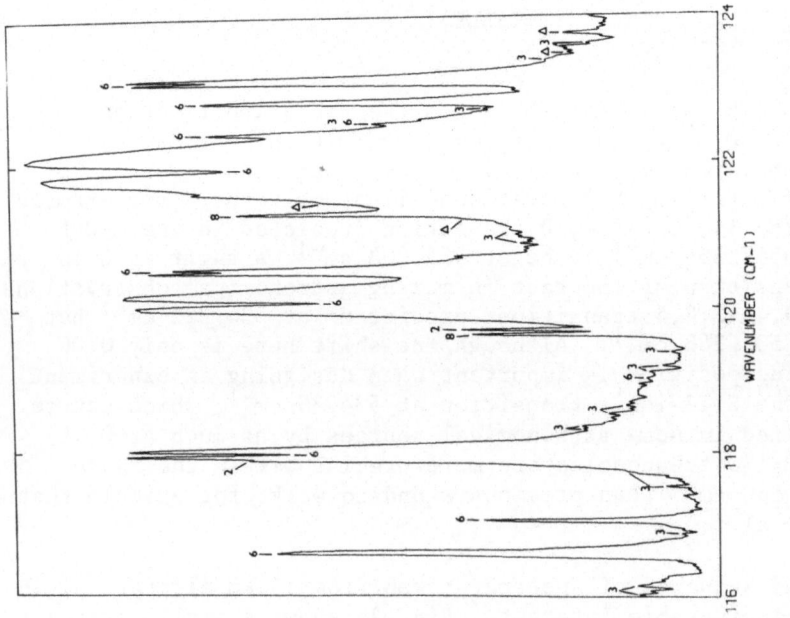

FIG. 2a.   Atmospheric emission spectrum
near 120 cm⁻¹ from G.P. Kuiper
airborne observatory at 13 km
altitude. $2 = O_2$; $3 = O_3$;
$6 = H_2^{16}O$; $8 = H_2^{18}O$; $\triangle$ = HDO.

30 seconds, and the average shown here was produced in 10 minutes.
A comparison spectrum produced using the AFCRL listing[3] and other
data is shown in Fig. 2b. Lines due to $O_3(3)$, $H_2^{16}O(6)$, $O_2(2)$
$HDO(\Delta)$ and $H_2^{18}O(8)$ are shown, and the signal to noise ratio in our
observations is sufficient to detect lines with intrinsic strenghts
as low as 0.5 x 10^{-22} cm^{-1}/molecule cm^{-2} in 10 minutes. As an
example of the comparison between these results and previous
estimates of $H_2^{18}O$ line frequencies, compare the line positions in
Fig. 2a and 2b, at $119.69cm^{-1}$ and $119.62cm^{-1}$ due to O_2 and $H_2^{18}O$
respectively. While the O_2 position agrees well with predictions
based on the compilation of Krupenie[6] the $H_2^{18}O$ predicted position
is incorrect. Using the total wavelength range from 100 cm^{-1} to
140 cm^{-1}, we have improved on the frequencies for $H_2^{16}O$ of Hall and
Dowling[7], whose laboratory spectra were at 0.07 cm^{-1} resolution,
and have identified a number of extra lines for this molecule.
In all 130 atmospheric emission lines have been detected in this
range, and close to 150 in the 500 cm^{-1} range.

A further measure of the use of our spectra is a determination of the ^{18}O : ^{16}O abundance ratio in the atmosphere, using the
$H_2^{18}O$ line at 109.970 cm^{-1}. A straight forward comparison with an
$H_2^{16}O$ line at 108.260 cm^{-1}, having similar strength, and assuming
a simple atmospheric model at temperature 250K, yields $|^{18}O|/|^{16}O|$
= 2.4 x 10^{-3} compared with 2.04 x 10^{-3} quoted in Goody[8]. Further
work on isotopic abundance ratios in the atmosphere based on our
data, but using more sophisticated models, is in progress.

3.   ATMOSPHERIC APPLICATIONS

The data presented here by no means represents the limiting
sensitivity available with our method. In particular if the astrophysical use, with high elevation pointing, were substituted by a
horizon-pointed experiment, the response to weak emission lines,
and hence to emission from minor constituents , could be improved
by a factor of over fifty. This rapid-scan method is also well
suited to obtain mixing ratio profiles with height, and a balloon-
borne system, for example, could give O_3 profiles to within a few
percent, with discrimination of a few hundred metres up to the
float altitude of 30 to 40 km.

REFERENCES

1. Anderegg, M., Moorwood, A.F.M., Hippelein, H.H., Baluteau,
    J.-P., Bussoletti, E. and Coron, N.; Far Infrared Astronomy
    (Ed. M. Rowan-Robinson) p. 171, Pergamon Press, 1976.

2. Anderegg, M., Beckman, J.E., Moorwood, A.F.M., Baluteau, J.-P.
    Bussoletti, E. and Coron, N.; Infrared Physics 16, 329, 1976.

3. McClatchey, R.A., Benedict, W.S., Clough, S.A., Bureh, D.E.,
    Calfee, R.F., Fox, K., Rothman, L.S. and Garing J.S.;
    AFCRL Atmospheric Absorption Line Parameters Compilation;
    AFCRL - TR - 73 - 0096.

4. Toth, K. and Margolis, J.S.;   J. Mol. Spec. 57, 236, 1975.

5. Izatt, J.R., Sakai, H. and Benedict, W.S.;   J.O.S.A. 59, 19,
    1969.

6. Krupenie, P.H.;   J. Phys. and Chem. Reference Data 1, 423,
    1972.

7. Hall, R.T. and Dowling, J.M.;   J. Chem. Phys. 47, 2454, 1967.

8. Goody, R.M.;   Atmospheric Radiation, p. 11, O.U.P. 1964.

MICROWAVE LIMB SOUNDING OF STRATO- AND MESOSPHERE

Erwin Schanda, Joachim Fulde, Klaus Künzi

Institute of Applied Physics, University
of Berne, Switzerland

1. INTRODUCTION

A considerable number of atmospheric constituents – as O_2, H_2O, O_3, CO, N_2O, NO, CO and others – exhibit a series of spectral lines in the millimeter wave region due to transitions between the rotational states of their electric or magnetic dipole moments. The maintenance of the thermodynamic equilibrium (collisional dominated) of this line spectrum up to mesopause altitudes is a specific advantage for the reliable determination of composition and temperature profiles throughout the strato- and mesosphere.

Space-borne experiments (e.g. Staelin et al., 1976 and Waters et al., 1975) have already demonstrated the feasibility of this method for global monitoring of the total water vapor and of the temperature profile in a downward-looking mode. The variable background radiation from the ground and the widths of the weighting functions are limiting the accuracy and the height resolution in this measuring situation. The weakness of the line radiations by the minor constituents would allow only very coarse determinations of the composition. The method of limb sounding as proposed by various groups for Spacelab missions (Waters et al., 1974; Schanda et al., 1974) allows for a height resolution of 3 kms or better throughout the range between the tropospause and the lower thermosphere. The long ray path at the tangential height and the absence of the more dense layers within the ray enhance the detectability of the weak emission by the less abundant molecules.

J. J. Burger et al. (eds.), Atmospheric Physics from Spacelab, 135–146. All Rights Reserved.
Copyright © 1976 by D. Reidel Publishing Company, Dordrecht-Holland.

2. ON THE DETECTABILITY

An extensive compilation of the resonance frequencies (up to
about 500 GHz) and of the proper absorption coefficients of important atmospheric constituents,computed from the molecular parameters as available in the literature, has been published recently (Fulde, 1976). A restrictive selection of lines, most promising for composition measurements, is presented in Table 1.
The center frequencies below 300 GHz and the absorption coefficients $\mathcal{H}_0$ (at the line centers ν_0) are given, the latter with
considerable uncertainty due to the lack of sufficiently reliable
spectroscopic parameters.

Table 1

Molecule	Center frequency ν_0 [GHz]	Absorption coefficient at line center, $\mathcal{H}_0$ [cm^{-1}]
O_2	~ 45 lines of reason- able strength between 48 GHz and 71 GHz	between 10^{-10} and $2 \cdot 10^{-5}$
	118.7503	$1.5 \cdot 10^{-5}$
H_2O	22.2351	$2.7 \; 10^{-5}$
	183.3101	$5.1 \; 10^{-3}$
O_3	101.7367	$6.8 \; 10^{-4}$
	110.8358	$1.1 \; 10^{-3}$
	184.3757	$2.5 \; 10^{-3}$
	231.2741	$4.2 \; 10^{-3}$
	235.7268	$6.0 \; 10^{-3}$
	237.1565	$5.9 \; 10^{-3}$
CO	115.2712	$3.3 \; 10^{-4}$
	230.5380	$2.5 \; 10^{-3}$
N_2O	appr.multiples of 25.1232, e.g.:	
	251.2115	$5.6 \; 10^{-3}$
HNO_3	106.4875 considered	$1.4 \; 10^{-3}$
	180.1309 only as a	$3.2 \; 10^{-3}$
	223.8372 rigid rotor	$1.2 \; 10^{-3}$
SO_2	129.5105	$1.0 \; 10^{-2}$
	221.9835	$3.0 \; 10^{-2}$
NO_2	255.5465 magn. fine-structure superimposed	$1.2 \; 10^{-2}$
H_2S	168.1836	$3.8 \; 10^{-2}$
	300.0726	$1.1 \; 10^{-1}$

For the calculation of the radiation transfer it is important to
note that for microwaves ($\lambda \gtrsim 1$ mm)
- the dominating line broadening mechanism up to mesopause altitudes is due to collisions, allowing the assumption of LTE,
- the spectral radiance is approximately given by the Rayleigh-
  Jeans formula $S \approx 2kT/\lambda^2$ (as long as $\nu/T \ll 2\cdot10^{10} Hz/K$) yielding a
  proportionality to the temperature,
- scattering of radiation may be neglected.
This yields an integrated equation of radiative transfer in an
arbitrary direction

$$T_B(h) = T_I \exp\left[-\frac{1}{\mu}\int_h^\infty \varkappa \, dz\right] + \frac{1}{\mu}\int_h^\infty \varkappa T_A(h) \exp\left[-\frac{1}{\mu}\int_h^z \varkappa \, dh'\right]dz \qquad (1)$$

where T_A, T_I, T_B are the temperature of the atmosphere, the background temperature incident on the atmospheric layer regarded
and the measured brightness temperature respectively, h is the
height measured in the vertical direction z and μ is the cosine
of the angle between viewing direction (S) and the zenith direction (Z).

   Assuming - for the sake of simplicity of this analysis - an
isothermal atmosphere, introducing $ds = -dz/\mu$ and using the optical depth $\tau = \int \varkappa \, dz$, a short-hand equation of radiative transfer
becomes

$$T_B(s) = T_I(0) \exp -\tau(0,s) + T_A[1 - \exp -\tau(0,s)] \qquad (2)$$

The width of the spectral lines $\Delta\nu$ is determined approximately
by $\Delta\nu^2 = \Delta\nu_c^2 + \Delta\nu_D^2$, with the collisional part $\Delta\nu_c = \Delta\nu_{co} \, p/p_o$
as a function of pressure p, where $\Delta\nu_{co}$ and p_o are the values
at a reference- (sea-) level; the constant Doppler width
$\Delta\nu_D = \delta \, \Delta\nu_{co}$ is only a small fraction of $\Delta\nu_{co}$ (approximately
$\delta \approx 3\cdot10^{-5}$ resulting from $\Delta\nu_{co} \approx 3$ GHz and $\Delta\nu_D \approx 100$ kHz). Therefore
$\Delta\nu^2 = \Delta\nu_{co}^2 [\exp(-2\beta z) + \delta^2]$ with $p/p_o = \exp(-\beta z)$, $\beta = Mg/kT \approx 0.11$ km^{-1}
at T = 300 K. The pressure dependent absorption coefficient as a
function of $(\nu-\nu_0)/\Delta\nu_{co} = \alpha$ (Lorentzshape) becomes

$$\varkappa = \varkappa_o \exp(-\beta z)[\exp(-\beta z) + \delta^2]^{-\frac{1}{2}} \cdot [1 + \alpha^2/(\delta^2 + \exp(-2\beta z)]^{-1} \qquad (3)$$

The transition between the regimes of Doppler broadening
($e^{-\beta z} \ll \delta$) and pressure broadening ($e^{-\beta z} \gg \delta$) respectively is
given by $z \approx -\frac{1}{\beta} \ln \delta$; this yields a transitional height of $z \approx 90$ km.
For the computation of the detectability of a certain constituent,
its effective absorption coefficient $\varkappa_e = \varkappa_o \cdot VMR$ has to be introduced into (3) where VMR is the volume mixing ratio and $\varkappa_o$ is
the absorption coefficient of the pure gas at line center.

The contributions of different layers to the measured (integral) brightness temperature can be taken into account by the concept of the weighting function. For looking tangentially through the atmosphere (limb sounding) into a cold background, the resulting brightness temperature becomes

$$T_B = \int_{h_T}^{\infty} g_T(h) \, T(h) \, dh$$

where $T(h)$ is the temperature at height h and $g_T(h)$ is the weighting function for a given tangential height h_T. Figure 1 presents the weighting functions for limb sounding at different tangential heights for an arbitrary constituent with an assumed constant volume mixing ratio in an isothermal atmosphere (Fulde, 1976). At the line center $(\alpha = 0)$ the division of the regimes of collisional and Doppler broadening as the dominant absorption effect at a height of about 90 kms can be recognized. The effect of collisional broadening in the lower-heights is demonstrated by the dotted lines for a frequency off-center by 10 times the Doppler width

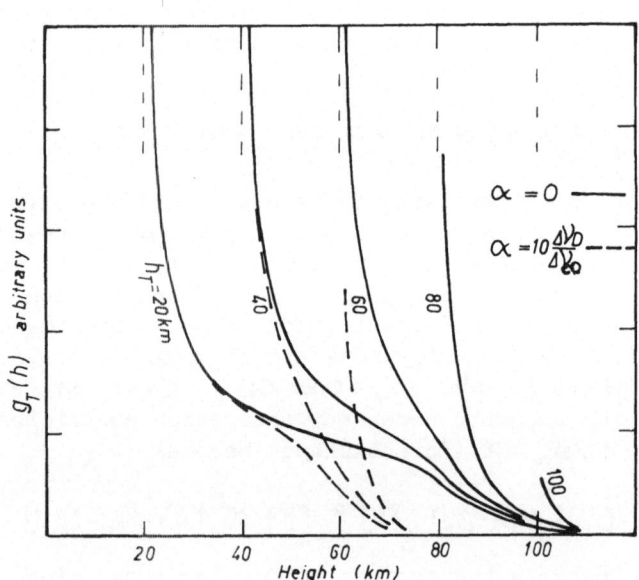

Fig. 1 Weighting functions for limb sounding (contributions by atmospheric layers above the tangential height h_T) at line center ——————— and on the wing — — — — — at $\nu = \nu_0 \pm 10\Delta\nu_D$

$(\alpha = 10 \delta)$. In order to achieve a narrow height resolution at h_T, i.e. to keep the contributions by upper layers to a minimum, one has to use a steep weighting function e.g. $\alpha = 0$ for $h_T = 80$ km but $\alpha = 10 \delta$ for $h_T = 60$ km etc.

Figure 2 shows the computed contributions to the brightness temperature by two different constituents at the respective line center frequencies against a cold background (3 K) as a function of the tangential height. The volume mixing ratios are crudely approximated for simplicity and a US standard atmosphere has been assumed. But these computed curves have to be used with caution: At the lower tangential heights the contributions by higher layers become relatively more important because the line center's absorption is effective at any height; in an experiment this would cause the loss of any useful height resolution. The necessary adjustment of the measuring frequency to a steeper weighting function however will result in a correct but lower brightness temperature. Nevertheless, taking an instrumental sensitivity of about 1 K the feasibility of measuring the height distribution is evident.

For the less abundant constituents we may follow an approximate treatment just to find estimates for their detectability. The absorption coefficient varies along the (tangential) ray path according to (3) as a function of the height at line center, taking into account the cuverature of the earth, as

$$\varkappa = \varkappa_e \left[1 + \delta^2 \exp 2 \beta h_T \cdot \exp(\beta s^2 / R) \right]^{-1}$$

where R is the earth's radius and s is some point on the ray path measured from the tangent point. The total optical depth is

$$\tau = \int_{-\infty}^{\infty} \varkappa \, ds = \varkappa_e \sqrt{2R/\beta} \cdot F(h_T) \tag{4}$$

where $F(h_T)$ assumes different shapes (taking now again an isothermal atmosphere and constant VMR's) for the regimes

$$h_T < -\tfrac{1}{\beta} \ln \delta : \quad F(h_T) = 2 \left(\ln \tfrac{1}{\delta} - \beta h_T \right)^{\frac{1}{2}} + 1/\sqrt{\pi}$$

$$\text{and} \quad h_T > -\tfrac{1}{\beta} \ln \delta : \quad F(h_T) = \exp(-\beta h_T) / \delta \sqrt{\pi}$$

The total optical depth of the less abundant constituents will certainly be $\tau \ll 1$, thus allowing (2) to be approximated by $T_B \approx T_A \tau$ for a cold sky background (emission measurement) and $T_B \approx T_I (1 - \tau)$ for an absorption measurement (solar occultation). Therefore we may substitute τ in (4) by T_B / T_A in a measurement with cold sky background, while for the occultation measurement $(T_I = T_{Sun} \approx 6000 K \gg T_A)$ we have $\tau = \Delta T / T_I$, where for $T_I - T_B = \Delta T$

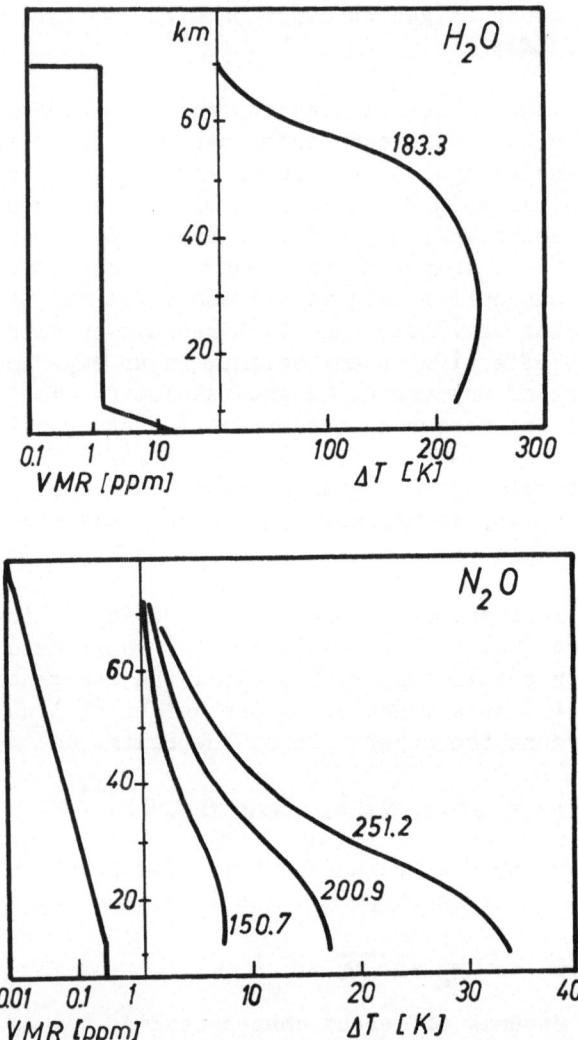

Fig.2 Brightness temperature caused by atmospheric H_2O and N_2O
as a function of the tangential height, computed from the
absorption coefficients at the respective line centers
(in GHz). The Volume Mixing Ratios (in parts per million)
assumed for the computation are indicated on the left parts
of the diagrams.

may be taken the smallest detectable temperature difference (3 K as a conservative estimate). Recalling $\varkappa_e = \varkappa_0 \cdot VMR$, the relation between the absorption coefficient of the pure gas and the smallest detectable volume mixing ratio is for the emission measurements

$$\varkappa_{o,em} \approx 3 \cdot 10^{-5} \; [\, VMR \cdot F(h_T) \,]^{-1} \tag{5}$$

and for the absorption measurement (solar occultation)

$$\varkappa_{o,abs} \approx 1.5 \cdot 10^{-6} \; [\, VMR \cdot F(h_T) \,]^{-1} \tag{6}$$

both in units of km^{-1}.

Table 2 gives the detectability of various minor constituents at the most favourable tangential heights, for which the approximate VMR's are taken from the current literature (see Fulde, 1976). According to formulae (5) and (6) the table can be used in both directions: at the given VMR's the values of $\varkappa_{o,em}$ and $\varkappa_{o,abs}$ are the minimum absorption coefficients of the pure gas needed at the used absorption line to be detectable; or vice versa. The lack of reliable spectroscopic data and the rather uncertain compositions make this approach necessary. NO seems to be detectable already in emission and at least NO_2 and HNO_3 in absorption.

Table 2

Molecule	h_T [km]	$F(h_T)$	$\varkappa_{o,em}$ [km^{-1}]	$\varkappa_{o,abs}$ [km^{-1}]	VMR
NO	100	0.056	50	2.5	10^{-5}
NO_2	20	1.5	20 000	1000	10^{-9}
HNO_3	10	1	3 000	150	10^{-8}
HO_2	30	5.5	18 000	900	$3 \cdot 10^{-10}$
H_2O_2	20	1.5	100 000	5000	$2 \cdot 10^{-9}$

3. INSTRUMENTAL CONSIDERATIONS

The specific design of the receivers and the antenna facility will be dependent on the scientific objectives: distributions of minor constituents, temperature or winds. The most important overall design criteria (see also: Schanda, 1976) are discussed in the following:

For achieving a height resolution of 3 kilometers in a limb mode
from a platform orbiting at 250 kms (maximum distance to the
tangent point about 1800 kms) the vertical dimension of the antenna has to be at least ~750 wavelenghts, yielding e.g. ~2 meters
for the 118 GHz O_2-line.
If it is demanded that in a 'side-looking' measurement (perpendicular to the Spacelab velocity vector) of the horizontal wind
component the Doppler broadening caused by the Spacelab velocity
(~7.5 km/s) is not more than the required accuracy of the wind
measurement (3m/s), the (horizontal) width of the antenna has
to be 3200 wavelenghts, yielding about 8 meters for the same
O_2-line. Measuring into other directions causes less Doppler broadening but a strong Doppler shift, which has to be compensated.
Figure 3 shows these effects as a function of the azimuthal angle
with zero elevation and an assumed flight velocity of 7.5 km/s.
Considering the possibility of computing the true wind velocity
from the Doppler broadened line by deconvolution, a modest reduction of antenna size (~1000 λ) may become permissible. A low

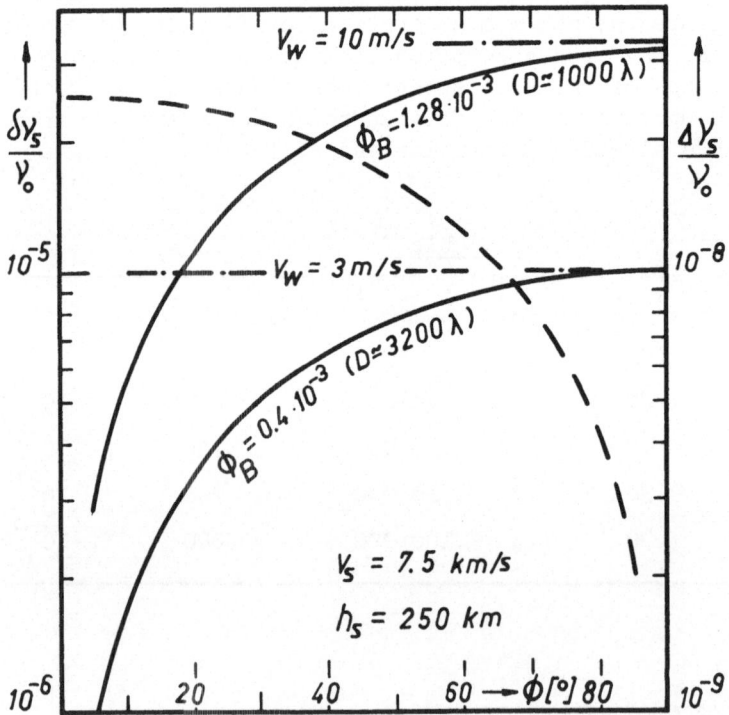

Fig. 3 The spacecraft velocity (7.5 km/s) causes a mean Doppler
        shift $\delta\nu_s$ — — — — and due to the final azimuthal width
        of the antenna beam Φ_B a Doppler broadening $\Delta\nu_s$ —————
        of a stationary air mass as a function of the azimuth
        angle ϕ ($\phi=0$ along the spacecraft velocity vector). Doppler
        shifts due to two radial wind velocities V_W — · — · —
        are compared with the Doppler broadenings

side lobe level of the antenna beam is required in azimuth (Doppler shift) and elevation (earth's radiation).

The specifications of the receiver sensitivity is dictated
primarily by the bandwidth B in order to resolve the narrowest
linewidths (due to Doppler broadening at high altitudes). If we
assume this 100 kHz and if we demand a resolution of the brightness temperature ΔT of 1 K, the required measuring time Σ according to

$$\Sigma \approx \frac{4(T_N + T_B)^2}{(\Delta T)^2 \cdot B} \qquad (7)$$

with the instrumental noise temperature T_N (single side band)
and the brightness temperature T_B can be calculated. Dependent
on the frequency and the state of the art of the receiver development, T_N may be taken between 1000 K and 3000 K. We assume the
atmospheric line radiation is entering only one side band, continuum radiation (sun and receiver noise) is entering both side
bands. Inserting $T_B = \frac{T}{2} \cdot T_A = 150\,K$ (worst case) into (7), the required measuring time assumes values between 70 and 440 seconds.
These long measuring times cause a spatial averaging over many
hundred to thousands of kilometers along the flight direction.
Due to the long transmission path (a few hundred kms) contribu-
ting to the measured brightness temperature in a limb sounding
mode, the 'side-looking' measurement will yield an averaging over
an area of at least 10^5 kms^2. In the lower (stratospheric) levels
the linewidths are much larger and the use of a e.g. 1 MHz receiver bandwidth would reduce the measuring time by a factor 10
according to (7).

In case of occultation measurements even these measuring
times may be too long. Sun-rise and sun-set composition measurements are of particular importance because of the rapidly changing intensity of the photo-chemical reactions during these short
periods; therefore we consider now the feasibility of these measurements separately. The detectability of a weak absorption coefficient of a constituent gas for a given volume mixing ratio
or vice versa has been considered above for cold backgrounds and
solar occultation measurements respectively (comparison of equations (5) and (6)). The above assumption of equal sensitivity of
the receiver in both measurements is arbitrary and not very realistic recalling (7), where instead of the atmospheric temperature the solar temperature has to be inserted for T_B . If we relax
the detectability of the absorption coefficient in an occultation
measurement to the same value as achieved by the emission measurement, we may gain considerably in measuring time. From (2) and
(7) and taking into account that the line radiation acts only on

one of two receiver side-bands, we find the ratio of the measuring times for absorption (solar occultation) and emission (cold
bakcground) measurements approximately as

$$\left[\frac{\Sigma_{abs}}{\Sigma_{em}}\right]^{\frac{1}{2}} = - \frac{T_A}{T_S - T_A} \cdot \frac{T_N + T_S - \frac{1}{2}(T_S - T_A)(1 - e^{-\tau})}{T_N + \frac{1}{2}T_A(1 - e^{-\tau})} \tag{8}$$

assuming $T_{sem} = 0$ and T_A constant throughout the atmosphere. In the
millimeter wave spectral range (50 GHz $< \nu <$ 300 GHz) the solar
temperature may be taken $T_S \approx 6000$ K and the atmospheric temperature is only a small fraction of this. Therefore the instrumental noise temperature T_N is determining the measuring times.
Figure 4 presents Σ_{abs}/Σ_{em} according to (8) as a function of T_N
for constant T_A and T_S and for two values of τ. For extreme low
values of the receiver noise temperature ($T_N \lesssim 300$ K, which seems
not feasible for operational use within the near future) the measuring time for the emission mode becomes even shorter than for
the absorption mode because of the much lower total system noise
temperature. The time available for occultation measurements is
determined by the orbit of the satellite related to the sun. The
worst case appears if the velocity vector is crossing the solar
disk: the tangential height (at tropospheric level) changes by

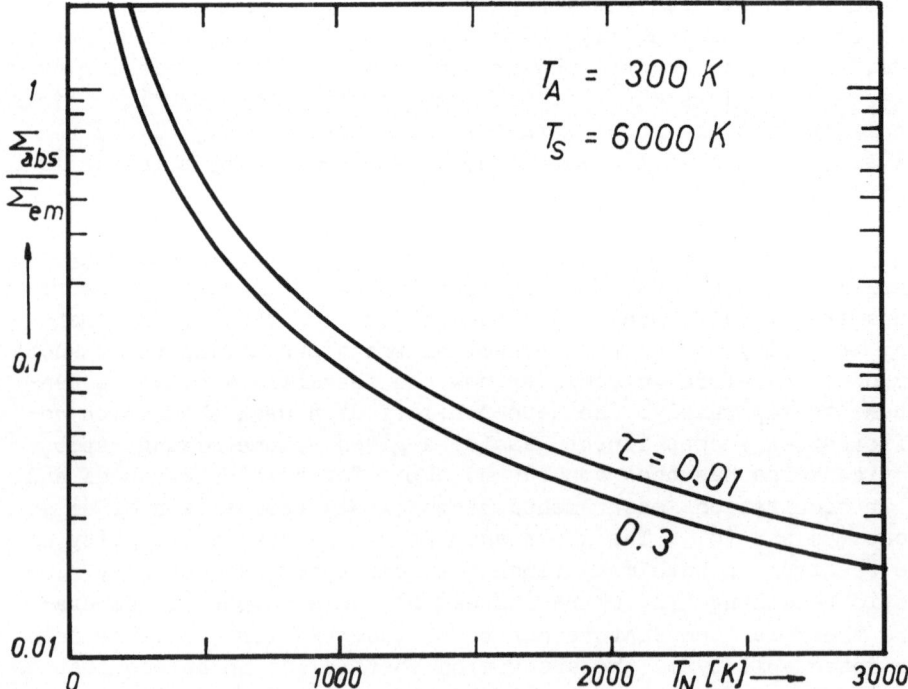

Fig. 4 The ratio of minimum measuring times needed in absorption
(solar occultation) and emission measurements respectively
for equal detectability of a given constituent as a function of the receiver noise temperature

3 kms already after about 1.5 seconds if the orbit level is
250 kms and the speed 7.5 km/s.

A Spacelab experiment on limb sounding by microwave and
infrared passive sensors is proposed in a cooperative effort
together with the Jet Propulsion Lab., Pasadena, USA, with the
Clarendon Lab., University of Oxford and the National Physical
Lab., Teddington, both U.K.

4. REFERENCES

Fulde J., Theoretische und numerische Grundlagen für die Mikro-
wellen-Spektro-Radiometrische Untersuchung der Atmosphäre, Ph. D.
Thesis, Univ. of Berne, Switzerland (1976)

Schanda E. et al., Millimeterwave Sounding of Trace Gases in the
Earth's Atmosphere, Proposal for the ESA definition of experimental objectives, Inst. Appl. Physics, Univ. of Berne (1974)

Schanda E., A millimeter-wave sensor for Spacelab astrophysical
and atmospheric investigations, Proc. 16th Convegno sullo Spazio,
Roma, 433-440 (1976)

Staelin D.H. et al., Remote Sensing of atmospheric water vapor
and liquid water contents with the Nimbus-5 microwave spectrometer, J. Atm. Sc., in print (1976)

Waters J.W. et al., Microwave Limb Sounder, AAFE Proposal, Jet
Propulsion Lab., Pasadena (1974)

Waters J.W. et al., Remote Sensing of atmospheric temperature
profiles with the Nimbus-5 microwave spectrometer, J. Atm. Sc.,
32, 1953-1969 (1975)

DISCUSSION

S.A. Bowhill: Have you examined the feasibility of using a microwave beacon transmitter on a subsatellite for occultation
measurements?

E. Schanda: Until now we have only considered the simpler
configurations discussed in this paper, but I agree that the
more sophisticated arrangements such as the inclusion of beacon
transmitters should be studied in detail because they offer
advantages, such as an important improvement in spatial
resolution compared with the purely passive method.

J.T. Houghton: Combined microwave and infrared experiments are
important for investigations of the mesosphere because
microwave instruments can measure kinetic temperature , which
is not possible with infrared instruments above about 80 km.
Measurements of rotational and vibrational temperature of
different molecules and measurements of minor constituents
are also largely complementary in the two spectral regions.

SESSION 3

METEOROLOGY IN THE IONOSPHERE AND STRATOSPHERE

CRITICAL CLIMATE PARAMETERS

AND THEIR MONITORING FROM SPACE

Hans-Jürgen Bolle[*]

Meteorologisches Institut
Universität München

ABSTRACT

Contemporary Research on the physical basis of climate
and its changes is based upon mathematical modelling
and the collection of carefully validated sets of measured data. A hierarchy of mathematical models with
increasing complexity is existent. These models can be
used to identify parameters which have a critical function in our climate system. "Critical" in this respect
means: if they are changing or are changed by man's
activity, then the climatological condition also changes if no negative feedback mechanism is activated. The
measured data sets are needed as input and check for
the numerical models as well as for the detection of
trends in the magnitude of climate parameters.

A survey is given of parameters which are presently regarded as critical, and to which our main interest
is aimed at with respect to global monitoring. Then the
monitoring needs for those quantities are especially
discussed which are presently still inadequately covered by existing observing systems such as the radiation balance, concentration of trace constituents, and
parameters which determine energy fluxes in the atmo-
sphere-ocean-system.

Three areas are identified in which SPACELAB can
assist meteorology in solving the climate problem. The
first is a contribution to the permanent observation

---

[*] Presented by E. Redemann.

J. J. Burger et al. (eds.), Atmospheric Physics from Spacelab, 149–169. All Rights Reserved.
Copyright © 1976 by D. Reidel Publishing Company, Dordrecht-Holland.

system e.g. by monitoring the solar flux. The second is
the use of SPACELAB for timely and geographically limited pilot studies in fields where basic problems have
to be solved. The third area is the test of new monitoring techniques.

1. INTRODUCTION: THE DEFINITION OF CLIMATE.

    Climate is an iridescent notion and even meteoro-
logists hesitate to give an ultimate definition. Climate phrases in its widest sense the comfort of mankind
in his natural environment. It is very difficult to sum
up all the parameters which contribute to this comfort
in order to describe a climate. We are furthermore talk-
ing of climates in different scales. There is the micro-
climate which is the climate of an environment in which
a single plant may grow, a corner near a stone wall
which protects the plants from wind, a small area under-
neath a tree. Then we are concerned about local climate
e.g. of a city or a valley in a mountainous region. The
next larger scale would be the regional climate of an
area in the order of an European country. From here we
can go to the climate of a latitude belt, and finally
of the whole planet.

    The wellknown climatologist H.H. Lamb (1972) has
defined climate as the sum of all weather situations
which are faced at one of the scales within one year
and over the years. Contemporary climate research uses
a more formal definition. It es considered that a number of parameters contribute to climate: The basic meteorological quantities such as temperature, pressure,
wind, humidity, then the solar irradiance, concentration of minor atmospheric constituents and many others,
in total say N. The instantaneous weather at one point
and at a specific time can be described by a vector in
the N-dimensional space. This vector oscillates in the
course of the year, and the space described by the
movement of its arrowhead is the envelope for the climate. Mathematically the climate can be described in
first approximation by the mean values $\overline{a_i}$ of all N

parameters $a_i = \overline{a_i} + a_i'$ within this "climate space". But

not only these mean values are of interest and importance for the descripti+n of climate conditions but

also the deviations from the means $\overline{(a_i')^2}$ or, as the

modellers use to say, higher moments $\overline{(a_i')^m}$ of the parameters, spectral distributions, and cross correlations.
These functions include transport processes and represent as well the bandwidthes of variability or the coincidences of events.

Another problem then remains to determine the time
over which the oszillations of the "weather vector"
have to be observed and the climate parameters have to
be integrated in order to derive meaningful climate
data. The integration time has of course to be one year
at least because otherwise the result would yield seasonal weather means instead of climate data. If the
mean planetary climate is to be determined for all ages,
then integration time must be about 4 billion of years.
In that case, however, no climate changes can be detected, all even strong variations are treated as if
they occur within a bandwidth of natural fluctuations.
The essential question is, where to place a boundary
between the admissable variability and essential changes of climate. Let us therefore take a look into the
history of climate variations.

From paleo-climatic investigations it is well known
that there have been dramatic variations of certain climate parameters such as temperature, ice cover or sea
level. There is strong evidence that at least the mid
latitudes have been warmer than today by about 10K for
a long period of about 50 million years. 20 - 30 million
years before present (BP) the temperature started to
decrease and about 5 million years BP oscillations
started which lead over into the ice ages 1 million
years BP. From the main features of climate periods the
time scales of the different events can be deduced which
are summarized in Table 1.

Table 1:   Time Scales of Observed Climatic Fluctuations
            (partly after Kutzbach 1974)

Inter annual variability	2 years
Recent warming period 0.5K/20 years (1920-1940)	20 years
Recent warming period 0.6K/60 years (1880 - 1940)	60 years
Little ice age (1350 - 1850) 1.5K/200 years	500 years

Major fluctuations within
    present interglacial
    2K/200 years
	1000 years

Duration of recent interglacials	10^4 years

Glacial - interglacial
    fluctuations
    10K/5000 years
	10^5 years

Duration of ice ages	$10^6 - 5.10^7$ years
Interval between ice ages	$1 - 2.10^8$ years

For comparison:	
Formation of rocks	$2.8 \cdot 10^9$ years BP
Formation of earth	$4.6 \cdot 10^9$ years BP

We would certainly like to detect early climatic trends
of smaller efficiency as those which lead to an ice age
and we see that these changes in environmental conditions
developed in time scales of hundredth of years. One now
believes that also within 100 years the climate can dra-
matically change. In order to study conditions which
may lead to such changes, time scales of some ten years
are therefore adequate. WMO has defined the minimum
length of climate series to be 30 years which was more
a practical decision since much longer observation
series were rare. It turns out, however, that such time
intervals are rather adequate to our problem. Within
the last 30 years the change of the CO_2 content of the
atmosphere has been about 6%, a very well established
number. The temperature anomaly of the northern hemisphere after a series of volcanic eruptions between
1880 and 1916 seems to have had a ascent period in the
order of 30 years. The recovery time of the stratosphere after the Agung eruption in 1963 was in the
order of 10 years. The period of 30 years also covers
roughly 3 solar cycles.

We have seen that climate research has to start
with three suppositions:

1. To specify the parameters which are relevant for
    climate and to point out those which are of pri-
    mary importance.

2. To specify critical numbers of changes which can
    be considered as beyond "normal" statistical fluc-
    tuations and which may indicate climatic trends
    or changes.

3. To specify the time base for climatological data
    assessment.

As has been explained before, we have to think of observation periods of a few decennia in the order of
30 years. One of the basic principles in meteorological
observations is the invariability of the observation
scheme. The introduction of innovations either in instrumentation or in location requires extensive comparisons and parallel measurements. One talks of the
"fit" between observation series. Sometimes it may even
be less important to improve the absolute measuring
accuracy than to destroy the relative relations between the data gathered at different times and at different locations.

The continuous improvement of measuring accuracy
and calibration is of course a major aim of meteorological instrument development but the new techniques
are implanted very carefully. This is an important
point also with respect to meteorological observations
from space. The progress has to be sought not so much
in the observation of spectacular events but in the
compatability and reliability of the system for long
term observations.

2. HOW IS METEOROLOGY APPROACHING THE CLIMATE PROBLEM?

There are basically two ways of investigating the
question whether or not climate may change in the
future. The empirical approach leads through the analysis of time series of observations and their comparison with past developments. Since instrumental observations are only available for some hundred years, the
studies of past climate events is mainly based upon
indirect observations such as analysis of sediments,
O^{18}/O^{16}-ratio in ice cores, tree rings and distribution
of certain fauna species. Reports from men about outstanding climate events including the archeologic
research on human settlements cover only 0.5%o of the
whole interesting period. From the old records gross
features of the structure of the ocean-atmosphere
system can be studied, and with long enough observation
series conclusions with respect to future developments
can be drawn by comparing the present dynamic system
with those of earlier periods.

During the last decennia, and especially because
of the critical food production, our economy got rather
sensitive to even slight climatic variations so that
we have to deal with events which are not well docu-
mented by paleo-climatology. Also a new component was
introduced, the changes in nature by the activity of
mankind which does not have a complement in former
times. Moreover we would like to understand the physical reasons of changes as they occured in the past
since we may need this knowledge very soon in order to
initiate counteractions to avoid possible and unwanted
climate changes in the future.

Meteorology is therefore developing another
powerful research tool: the mathematical modelling of
the atmosphere and its responses to changes in forcing
functions. By now a whole family of such models with

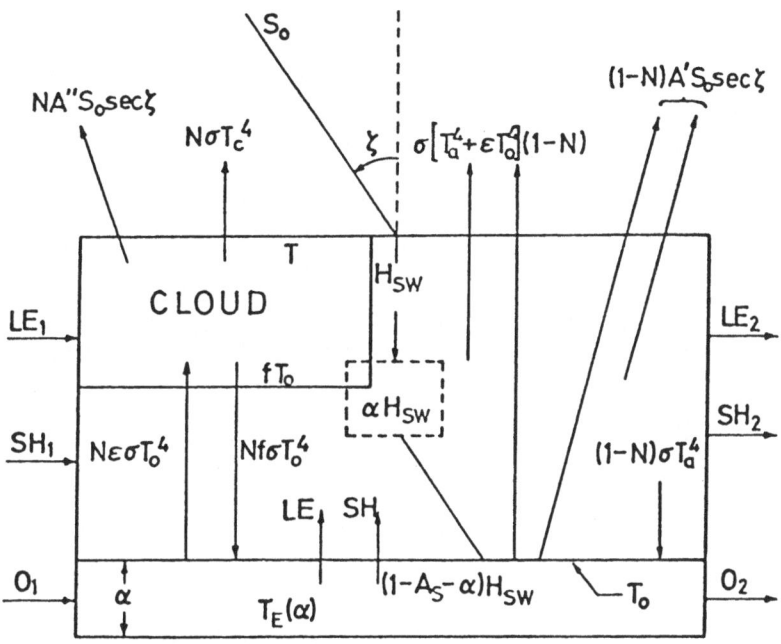

Fig.1. Energy Fluxes in an Atmospheric Box

A'   Albedo of cloudfree atmosphere-surface-system

A''   Cloud albedo

$A_s = NA'' + (1-N) \cdot A'$

T_a Effective atmospheric emission temperature

T_o Surface temperature

ε Surface emissivity

$T_e = (NT_c^4 + [1-N] \cdot [T_a^4 + \varepsilon T_o^4])^{1/4}$

f   Factor accounting for T(cloud base) $< T_o$

graduated complexity is available. In these models
certain boundary conditions are introduced and the
behaviour of the atmosphere can be studied in dependence of the boundary values as well as of the
composition and other physical properties of the
atmosphere itself. The development of such models
basically starts with a consideration of energy exchange in an atmospheric "box", which can be a cube
or a zonal ring around the earth.

The energy fluxes which have to be considered
in the energy budget are the following quantities
(compare Figure 1 and notation after equation 5):

Q_R radiant net flux density determined by

$H_{SW} = H_{SW}$ (S_o) solar radiant, flux density
or irradiance at the top of
the atmosphere (mainly short
wave flux)

S_o solar constant

H_{SW} $(1-A_S)$ outgoing shortwave flux density

$A_S = A_S$ (N) albedo

N   cloudiness

$Q_{LW} = \sigma T_e^4$ absorbed longwave net flux
density
$(T_e$ = equivalent temperature)

Q_C advective fluxes and stored energy consisting of

LE   flux of latent energy   Lq

SH   flux of sensible heat $c_p T$

O   oceanic heat flux

PE   flux of potential energy   gz

KE   flux of kinetic energy u^2/z

$G = C \int (dT_E/dt)$ dz stored energy

The different models which are available can shortly
be described by the following terms:

Model Type:	Main Objective:
Horizontally averaged, one-dimensional vertical coordinate models	Determination of planetary mean climate in dependence of composition changes
Horizontally varying energy banlance models a) function of latitude only b) function of latitude and longitude (land-ocean distribution)	Determination of surface temperature as function of energy balance with strongly simplified transport assumptions and mostly poor vertical resolution
Zonally symmetric models of the atmosphere-earth system	Models with detailed zonal dynamics as function of latitude. Zonal surface energy balance models
Zonal symmetric dynamic models of the atmosphere	Description of zonal structure and transports from observationally prescribed distribution of thermal and momentum sources
Atmospheric dynamics models emphasizing longitudinal variations	Constant zonal flow with superimposed perturbations. Parameterization of forcing by orographic and thermal features as well as by variations of latent and sensible heat
Stochastic models	Statistical treatment of climate variables as response to forcing by weather
Three dimensional circulation models of the atmosphere	General circulation models based on a set of time dependent primitive hydrodynamical equations. Simulation of large scale features with fixed boundary conditions.
Three dimensional circulation models of the earth-atmosphere system	Most advanced models including feedback between ocean, atmosphere and cryosphere

Because of restrictions in computer capacity
numerical models are normally not capable to treat
the atmospheric features and processes in great detail. Small scale processes must therefore be introduced in a simplified or "parameterized" form. Great
emphasis must therefore be laid upon the investigation
of such processes and their description by simplified
relations.

The primary driving force in all climate models
is the balance between incoming solar radiation and
outgoing infrared flux:

(1) $Q_R = H_{SW} (1 - A_S) - Q_{LW}$

or in an annual mean:

(2) $Q_R = \bar{\pi} R^2 S_O (1 - A) - 4\bar{\pi} R^2 \sigma T_{eff}^4$

where R is the earth radius, A the planetary albedo,
σ the Boltzman constant and T_{eff} the effective emission temperature.

The complete energy balance equation writes

(3) $Q_R - Q_C = G$

where G is the change of energy in the system equivalent to its net heating rate. If the whole energy
content of the system remains constant which can be
assumed over a period of a year, equation (4) reduces
to

(4) $\overline{Q_R} - \overline{Q_C} = 0$

From this conditions relations for the adjustment of
the internal parameters can be derived. Some of them
as the composition of the atmosphere, will be considered as fixed or as input parameters, others such
as the surface temperature will normally be treated
as dependent variable and will be computed.

In order to give an example the equation (4) is
written explicitly for a model which has been developed by Sellers (1973):

$$H_{SW} \left[1-A_S\right] - Q_{SW} =$$

(5)

$$\frac{1}{g} \int_{0}^{P_o} \overline{\frac{d}{dt} \left(Lq+c_p T+gh\right)} \, dp + C \int_{0}^{d} \frac{d \, \overline{T_E}}{dt} \, dz$$

Where g = acceleration due to gravity

L = latent heat of condensation

(590 cal g^{-1} = 2470 W s g^{-1})

c_p = spec. heat (1.005 W s g^{-1} K^{-1})

h = height above sea level

C = heat capacity

T_E = soil or water temperature

d = depth of ocean or land which still contributes to the vertical energy flux.

The last term at the r.h.s. represents the storage of heat in the surface layers of the soil respectively the oceans. The dashes stand for the mean over a certain period, say 1 month.

3. WHICH ARE THE CRITICAL CLIMATE PARAMETERS?

The significance of the parameters entering equations (4) can only be established by numerical experiments in which certain quantities are changed, and by which the effect of these changes on other parameters is studied. The parameter which most often is used as indicator for climate changes is the surface temperature. From numerical experiments with different models it turned out that the quantities listed in Table 2 affect the surface temperature in the given order of magnitude. This picture may quantitatively change if the ocean effects are more properly considered. It tells us, however, which atmospheric qualities have to be observed.

TABLE 2: CHANGES OF SURFACE TEMPERATURE (ΔT_o) IN
DEPENDENCE OF CHANGES OF OTHER ATMOSPHERIC
PARAMETERS AS COMPUTED BY DIFFERENT MODELS
(AFTER SELLERS), 1973, MANABE AND WETHERALD
1967, AND 1975)

Varied Parameter	Variation	ΔT_o	Remarks
CO_2	+100%	+ 2 K (0.1...3K)	Depends on Humidity. Effects strongest in polar regions.
	- 50%	- 1.5K (-1...-2.3K)	
H_2O	+ 10%	+ 1.5 K	
O_3	1mm STP	+ 1.4 K	
Albedo	+ 1%	- 1 K	
Cloudiness	+ 1%	- 8.2 K to + 0.17K	Strongly negative for dense low, slightly positive for thin cirrus clouds. Depends on optical cloud properties.
Optical depth (aerosols)	+ 7%	- 0.5 K	Corresponds to doubling of manmade turbidity.
Solar const.	+ 1%	+ 0.9 K	

For comparison: Mean temperature is now about 10 K
lower than 100 million years B.P.. Amplitude of temperature variations during ice ages was about 9 K.
N.H. temperature increase 1890 - 1950 was 0.6 K.

Since some of the important climate parameters (ozone, aerosols, carbon dioxide, water vapor) are integrated in a whole system of chemical and photochemical reactions, another cycle is attached here to the climate system: atmospheric chemistry and pollution. This relates climate research strongly to especially middle atmosphere research where most of the photochemical reactions take place which vice versa affect the energy budget. The reactions which have to be regarded in the stratosphere and mesosphere depend on the other hand on temperature and transport mechanisms in these layers or generally speaking on meteorology and climate of the middle atmosphere. The strong relation between e.g. the stratosphere and the troposphere can be demonstrated by the effect of the CO_2 increase. As shown in Table 2 an increase of CO_2 causes an increase of near surface temperature. The physical reason is that the direct energy exchange between surface and space is further reduced. As a consequence - since we have more emitting gas in the atmosphere - the equivalent emitting temperature of the planet must be reduced in order to maintain radiative equilibrium. This results in a cooling of the stratosphere in the order of 5K in 30 km altitude for a 100% CO_2 increase.

Also the exchange of gases through the lower atmosphere, the solid earth, vegetation and oceans is essential and not yet fully explored. We can therefore describe climate research as a system with a meteorological core which has strong interactions with the hydrosphere (oceans), the cryosphere (ice and snow cover), the solid earth surface (albedo), and the upper atmosphere.

In order to respond on the needs to investigate the climate problem, WMO together with ICSU have defined an observing program which consists of the following tasks:

- global observation of the radiation budget
- global observation of surface radiation parameters
- global observation of precipitation and hydrology
- global observation for validation of ocean models

- global observation of gases and particulate
  matter
- observation of secondary climate parameters
  such as distribution of snow and ice; mass
  of glaciers; biomass of vegetation; tree
  lines; land use; forest fires, etc.

As an example for the specifications given for the
observations Table 3 is synthesized from material
presented in GARP Publication Series Nr. 16 (1975).

4. HOW CAN OBSERVATIONS FROM SPACE AND ESPECIALLY SPACELAB CONTRIBUTE TO CLIMATE RESEARCH?

The contributions which meteorological satellites
already make to assess meteorological quantities are
referred to in the following paper (Houghton). We will
therefore concentrate here upon a few open questions
which are of basic interest with respect to SPACELAB.
For meteorological/climatological research in the
eighties we will have an observation system consisting
of 4 - 5 geosynchronous satellites, 1 - 2 polar
orbiters and probably a few small "climate satellites".
Also earth resources satellites contribute to meteorological data assessment.

A major problem is the accuracy, the long term stability and the "complete coverage", both in space and
time, with which meteorological data have to be
gathered for climate studies.

As an example we will consider the basic problem of
the radiation balance. What is needed here are the
energy fluxes entering and leaving a box like that one
demonstrated in Fig. 1. We may adequately cover this
problem if the box contains 99% of the total atmospheric mass or extends to about 50 km height. A monitoring of the flux entering or leaving a unit area of
this box is in principle only possible with a flat
plate radiometer suspended in the height of the upper
boundary of the box. Satellites measure either the flux
at satellite altitude or radiances under different
directions and from differently large areas.

Flat plate radiometers do not provide much detailed information in the regional scale since 50% of
the radiation (of an isotropic radiation field) comes
already from an area of 10^6 km^2 if the satellite is

TABLE 3: TENTATIVE SPECIFICATION OF CERTAIN OBSER-
VATIONAL REQUIREMENTS NEEDED FOR VALIDATION
OF CLIMATE MODELS IN ADDITION TO THOSE OF
THE FGGE.

Spatial resolution is 10^4 km^2 unless otherwise noted.

VARIABLE	ACCURACY (1σ)		TIME RESOLUTION
	Desired	Useful	
1. Solar irradiance at top of atmosphere (reproduction accuracy required)	2 Wm^{-2}	10 Wm^{-2}	$\frac{1}{4}$ - $\frac{1}{2}$ year
2. Net radiation budget at top of atmosphere (solar and terrestrial)	2 Wm^{-2}	15 Wm^{-2}	5 - 15 days
3. Clouds: horizontal distribution, cloud height and measure of diurnal variation	5% amount 1°C cloud top temp.		5-15 days
4. a) Sea surface temperature	0.5°C	1.5°C	5 days
b) Heat content of upper layer (200 m)	1 Kcal cm^{-2}	3 Kcal cm^{-2}	5 days
5. a) Snow (100 km resolution)	Presence/ Absence		5-15 days
b) Sea ice (50 km resolution)	Presence/ Absence		5-15 days
6. Surface albedo	0.01	0.03	5 days
7. a) Precipitation over land	1$\frac{mm}{day}$	3$\frac{mm}{day}$	5 days
b) Precipitation over sea	1$\frac{mm}{day}$	4 levels of discrimination	5 days

TABLE 3 CONTINUED:

VARIABLE	ACCURACY (1σ)		TIME RESOLUTION
	Desired	Useful	
8. Soil Moisture	10% of local field capacity	2 levels of dis- crimina- tion	5 days
9. Runoff (river basin)	10%		15-30 days
10. Land surface temperature	1 K		5 days
11. Relative humidity over land	10%		5 days
12. Ozone profile (2km vertical resolution)	0.5 ppm		5-30 days
13. Carbon dioxide (2-4 baseline stations, 10 re- gional stations)	0.1 ppm		15 days
14. Aerosols, number density	5%		daily
15. Wind stress over ocean	0.1 dyne cm^{-2}	0.4 dyne cm^{-2}	5 days

Frequent observations are needed for a limited time to determine the extent of solar irradiance variations (and their wavelength dependence) as a function of solar rotation period since such veriations could bias longer-term statistics based upon less frequent obser- vations.

in an 1000 km orbit. The total area which contributes to the measured signal is $9 \cdot 10^6$ km^2 (Europe has an area of 10^7 km^2). Such an area is not compatible with existing dynamical models which use a grid width of 300 - 500 km, neither is the desired flux measured which leaves the atmospheric box at 50 km altitude. We also do not get a representative time mean if only a polar orbiter is used for this type of measurements. It arose therefore already a controverse about the interpretation of existing measurements (Vonder Haar an Suomi, 1971, Winston, 1972). For an assessment of the fluxes leaving the earth it seems therefore to be advisable to use also radiance information such as obtained from geosynchronous satellites every half hour and from scanning radiometers on board of polar orbiters in order to establish the fluxes. This can only be done, if the following presuppositions are fulfilled:

a) the relation between the radiance leaving the earth under a specific angle and the flux into the hemisphere must be known

b) the relations of the spectral channels used in the different instruments to each other, and to the whole relevant spectrum must be established

c) the instruments must be calibrated and give constant or known response over long times.

SPACELAB can be used to carry out necessary pilot studies, instrumental comparisons, and complemental measurements over test areas. One application seems to be typical for SPACELAB: the determination of the solar irradiance outside the atmosphere (see e.g. Zirin and Walter, 1975). It is necessary to determine these quantities with high absolute accuracy in the order of 0.5% and to try to detect variations of the solar flux, at least over a period of one solar cycle with a frequency of one or two measurements per year. No continuous monitoring is necessary but either instruments with extremely high measuring accuracy and stability have to be flown, or a complete calibration in order to check the performance of the instruments during the mission.

It could be demonstrated (Table 2) that atmospheric aerosol is an important climate parameter

which has to be observed since it affects the optical
depth of the atmosphere and by this the energy budget.
Atmospheric turbidity can be monitored from meteorological satellites. However, it is still difficult to
determine the particle masses which are transported
through the atmosphere, and their vertical distribution. It is presently studied, whether Lidar experiments can yield particle size distribution and
densities of aerosols as needed for more detailed
radiative transfer studies. SPACELAB is a good vehicle
to test such techniques though a later application must
be planned on a more or less operational basis.

Another example is the assessment of precipitation.
NIMBUS 5 has proved that qualitative measurements of
shower activity are presently feasible with microwave
radiometers. In order to develop this powerful tool
into a quantitative technique SPACELAB can be uesed as
a test platform.

Great difficulties presently still exist in ex-
ploring the large scale photochemistry of the upper
atmosphere which includes precise measurement of minor
constituent concentrations as well as transports in
the stratosphere (compare Kunkel et. al., 1975). Also
here the global patterns of exchange phenoma, energy
budget and concentrations have to be established in
order to run climatological models for these atmospheric layers which later on have to be connected
to the tropospheric/oceanic meteorological models.
Here climate research joins the broad river of aero-
nomic research which is dealt with in other papers
of this symposium.

5. CONCLUSIONS

Climate research deals with small of changes in
the solar - atmosphere - earth - ocean system and with
the physical explanation of observed variations. It
needs high quality data for a number of parameters to
be observed over long periods in order to feed mathematical models. The main tool to obtain such data from
space will obviously be meteorological satellites since
only they can provide adequate observation continuity.
However, more quantitative methods are still desired,
with respect to atmospheric parameters such as the
aerosol content, precipitation, radiation balance,
winds, concentration of trace gases. SPACELAB will
play an important role in developing such techniques,

in crosschecking on the accuracy of different
types of instruments, and in carrying out case
studies. Solar radiation measurements will probably
be a preferred domain of SPACELAB, and in the exploration of the middle atmosphere SPACELAB is an
important platform for elaborate experiments though
the complete survey will also in this field require
a more operational approach.

Literature

Flohn, H., 1969. Ein geophysikalisches Eiszeit-Modell.
                  Eiszeitalter und Gegenwart 20, 204-231

Kulp, J.L., 1961. Geologie time scales
                  Science 133, 1105-1114

Kunkel, B., F. Wolz, H.-J. Bolle, and E. Redemann, 1975.
                  Air quality measurements from space
                  platforms. ESA CR-577 (ESRO-Contract
                  No. SC/70/HQ.

Kutzbach, J.E., 1974. Fluctuations of climate monitor-
                  ing and modelling. WMO Bulletin 23,
                  47-54

Lamb, H.A., 1972. Climate: Present, Past and Future.
                  Methuen, London.

Manabe, S. and R.T. Wetherald, 1967. Thermal equili-
                  brium of the atmosphere with a given
                  distribution of relative humidity.
                  J.A.S. $\underline{24}$, 241-259

Manabe, S. and R.T. Wetherald, 1975. The effect of
                  doubling the CO_2 concentration on
                  the climate of a general circulation
                  model.
                  J.A.S. $\underline{32}$, 3-15

Schwarzbach, M., 1963. Climates of the Past.
                  D. Van Nostrand Co., N.Y.

Sellers, W.D., 1973. A new global climate model
                  J. Atm. Sci. $\underline{12}$, 241-254.

Vonder Haar, T. and V. Suomi, 1971. Measurements of
              the earth's radiation budget from
              satellites during a five year period.
              J.A.S. <u>28</u>, 305-314.

Winston, J.S.:   Comments on Measurements of the
              earth's radiation budget from satel-
              lites during a five year period.
              J.A.S. <u>29</u>, 598-601.

WMO, GARP Publication Series No. 16, 1975.
              The physical basis of climate and
              climate modelling.

Zirin, H. and J. Walter (Eds.) The Solar Constant and
              the Earth's Atmosphere Workshop Pro-
              ceedings. California Inst. of Tech-
              nology. Big Bear City, Calif.
              (NRF grand DES 75-16101).

DISCUSSION

A.C. Durney: You mentioned that the solar constant would have to
   be monitored over a period of at least about 22 years.
   Presumably this is to try to detect the variation over the
   solar cycle. With what accuracy would this measurement have
   to be made? (Note: speaker said 1%, someone else said 0.1
   - 1%).

E. Redemann: If you want to detect the variation in solar
   radiation, the measuring accuracy has to be smaller
   than   this variation, which is about 1%.

G. Schmidtke: Because the variation in the solar constant is
   expected to be between 0.1 to 1%, an accuracy of   1% for
   its determination would not be sufficient. An accuracy of
   0.1 % might be appropriate to monitor variations in the
   solar spectral irradiance, e.g. with solar cycle.

A.C. Durney: Why is it necessary to take up calibration
   instrumentation in Spacelab to check the performance
   of the solar-radiation-measuring instruments? Calibration can presumably be done on the ground before
   and after flight to check whether there are drifts?

G. *Schmidtke*: Laboratory calibration has to be performed before
and after flights with Spacelab. However, the ambient
conditions on Spacelab are expected to be very different from
those found in the laboratory. Under these circumstances
different data are to be expected and it is necessary to
carry out an inflight calibration as well. This is especially
true for the EUV spectral region, where surface effects are
responsible for changes in the calibration parameters.

STRATOSPHERIC MODELS

R.J. Murgatroyd, A. O'Neill and S.A. Clough

Meteorological Office, Bracknell, UK

The processes that determine the composition, temperature and
atmospheric motions throughout the stratosphere are complex and
highly interactive. They include photochemistry and chemical
reactions between many minor constituents, absorption and scattering of solar radiation, emission and absorption of atmospheric
radiation, the energetics and transports associated with the
motions on all scales and the boundary transfers with the mesosphere above and especially with the troposphere below. It is
the aim of the modelling described in this paper to account for
the observed features of the atmosphere and particularly the
stratosphere in terms of the physical and chemical processes
considered to be important and the relevant governing equations.
If the agreement with the available observations is good there
can be some confidence in the value of the predictions of the
models of features not so far observed and also of the likely
effects of perturbing influences. In addition diagnoses of the
model results will be useful in examining the detailed nature of
the phenomena they represent and linking observations with theory.

The basis of current techniques for modelling the troposphere and
stratosphere and the processes considered to be important will be
outlined together with the limitations of the models, necessary
parameterisations and some of the present achievements. In
addition to discussing the problems associated with the full three-
dimensional global modelling (which is very demanding in computer
facilities), the nature of the approximations made in order to
produce useful two-dimensional (height-latitude) and one-dimensional
(height only and hemispherically averaged) models will be examined
and assessed.

J. J. Burger et al. (eds.), Atmospheric Physics from Spacelab, 171–197. All Rights Reserved.
Copyright © 1976 by D. Reidel Publishing Company, Dordrecht-Holland.

INTRODUCTION

One of the main outcomes of the great increase in stratospheric
research in the last few years is the general realisation that
many complex processes contribute interactively towards producing
its so-called 'natural' or average state. This natural state
is in fact widely variable in time and space as regards
composition, thermal conditions and motions and also to some
extent is changed sporadically and locally by both natural and
artificially produced perturbations. The meteorological elements
are determined by the types of processes illustrated in Fig. 1.
Absorption and scattering of solar radiation at any level (eg
by O_2, O_3, CO_2, H_2O, clouds, aerosols etc) depends on the
composition and for some gases leads, after dissociation
processes, to chains of chemical reactions which primarily
determine the nature and distribution of the minor constituents.
Some molecules such as CO_2, O_3 and H_2O also play a major role in
determining, by absorption and emission in the infra-red region
of the spectrum, the field of atmospheric radiation. The distribution of the net heating and cooling due to the
divergences of the solar and atmospheric radiation fields sets
up heat sources and sinks which provide drives for the
atmospheric motions and broadly establish the observed temperature
distribution. Transport of constituents by the air motions and
variations of the chemical reaction rates with temperature also
have major effects on the composition fields. These can in turn
lead to changes in heating rates and photochemical processes
thus establishing feedback mechanisms affecting both dynamics
and chemistry.

Even this broad picture of stratospheric mechanisms is much too
simplistic as boundary effects particularly those involving
transfer of constituent mass, energy and momentum to and from
the troposphere are of major importance in determining the
final state. Consequently a full model of the stratosphere
must have an adequate representation of tropospheric
conditions and is also likely to be inadequate at its higher
levels unless transfers at the stratopause can be parameterised
successfully.

It is at present impossible to obtain an overall description of
the composition, thermal and motion fields of the stratosphere
in space and time based on observations and there is little
prospect of doing this in the foreseeable future. Consequently
the method of obtaining improved understanding must be to make
as many measurements as possible on a planned basis in order to
develop and verify theoretical ideas.

In view of the interactions of the different processes, the
most comprehensive approach towards obtaining a better overall

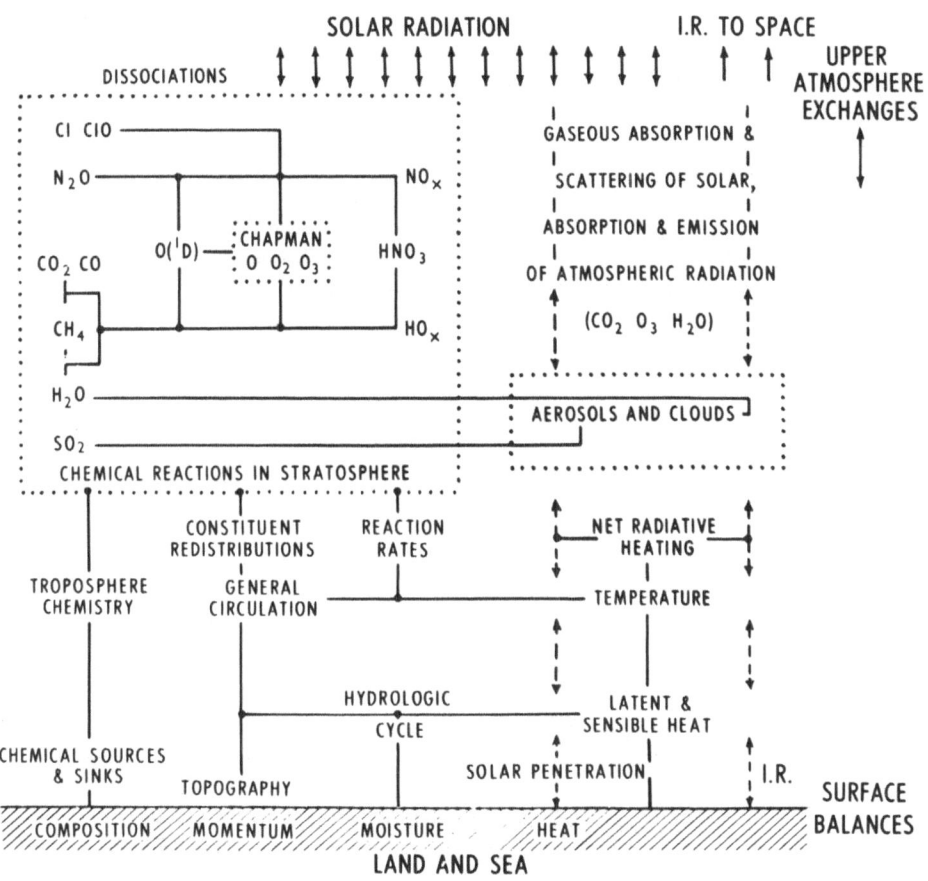

Figure 1 : Factors affecting the composition of the stratosphere

understanding would seem to be to devise a model of the
stratosphere which would fully represent the quantitative
evolution in time of the global fields of the different
meteorological parameters. Moreover such a model is required
for prediction of possible effects of stratospheric perturbations.
Its construction would require firstly a complete mathematical
framework to describe the processes which determine the
composition (dissociation, chemical kinetics and transport), net
radiative heating (absorption and scattering of the solar beam and
emission and absorption processes associated with infra-red
atmospheric radiative transfer), and temperature structure and
motions (thermodynamic equation, equations of motion and of
continuity and statistical treatment of small sub-grid scale
motions, turbulence etc). Secondly, a certain amount of basic
data is required of the relevant physical constants, solar
spectral intensities at the upper boundary of the model, major
constituents and probably other quantities depending on the basic
assumptions. In addition adequate boundary data are necessary.
If the lower boundary is selected to be the tropopause it is
necessary to specify values of many of the constituents there or
of their fluxes through this level as well as fluxes of momentum,
energy etc. Alternatively if the bottom boundary is at the
earth's surface (which is basically preferable) chemical, heat,
momentum etc transfers have to be specified or parameterised
and questions of how to deal with both transfers at land and
water surfaces (which may involve considerations of ocean
circulations) have to be considered.

It will be noted that the complications are such that various
assumptions and parameterisations have to be introduced in any
practical model and the possibility exists of adjusting or
'tuning' the model to seek improved agreement with observations.
This is of course a dubious procedure. Moreover although the
output of a large interactive three-dimensional global model is
easier to analyse than a very incomplete set of observations its
resolution particularly in the vertical may not be adequate for
many purposes eg studies of the effects of sub-grid scale
motions,stratophere-troposphere transfer etc. In addition since
such a model is so complicated the question arises as to the
extent to which this complexity is justified. It is necessary to
establish from detailed analyses what the basically important
processes are and whether acceptable approximations and
simplifications can be introduced in given problems. This
knowledge is required not only to understand and appreciate the
principal mechanisms but also to make it possible logistically
to carry out numbers of experiments for particular investigations,
many of which are concerned with variations over periods of
months or years. These variations are particularly important
for example in studies of the combined effects of a number of

possible perturbations. Before such studies can be undertaken
at all, however, it is necessary to establish that the model
used adequately accounts for the natural state as observed so
far.

It will only be   possible in this short review to mention a
few of the features of the stratosphere that have to be
modelled.

(i)   As regards composition,the central problem is to
account quantitatively for the ozone distribution in terms
of basic physical and chemical data and the solar
radiation and meteorological conditions in the
stratosphere. Studies of this type must also be able to
isolate the main processes so that others may safely be
rejected in series of experiments designed to study the
effects of perturbations. Amongst other features the
model should be able to account for the general vertical
profiles of the constituents (such as that of the ozone
layer with maximum concentrations 5-10 km above the
tropopause), latitudinal changes (eg maximum total ozone at
high latitudes), seasonal changes (eg maximum ozone in the
spring), and differences between hemispheres.

(ii) As regards radiation one of the main objectives after
a successful detailed model has been obtained must be to
seek acceptable simplifications. This is a necessary preliminary to comprehensive studies of the likely effects of
changes of stratospheric composition on climate for
example. The net radiative heating is expected to account
for major features in the temperature structure such as
the high temperatures in the upper stratosphere with maxima
near the summer pole. Other features such as the
minimum temperatures around the equatorial tropopause and
the maximum in middle latitudes in the lower stratosphere
in winter largely appear to be produced dynamically.

(iii) Finally the models have already been shown to be most
useful in studies of the dynamical meteorology of the
stratosphere. In the upper stratosphere the winds are
mainly westerly in winter and easterly in summer as a
consequence of the radiative fields. In the lower
stratosphere there are many aspects which follow from the
effects of the forcing of stratospheric air motions by
those in the troposphere below. These include such
aspects as equator to pole transfer of heat and momentum in
stantwise convection, sudden warmings, the quasi-biennial
oscillation and equatorial oscillations. In general the
upwards propagation of wave energy from troposphere to

stratosphere is inhibited in summer (Charney and Drazin,
1961) with the result that the stratospheric winds are
zonal and steady. In winter, however, upward transfer takes
place particularly at the longer wavelengths with the
result that the tropospheric systems (with wavenumbers of
about 5-10) die out in the lower stratosphere and are
replaced by very long wave systems mainly of wavenumbers 1-3.
For many purposes these are simpler to model than the
tropospheric synoptic scale eddies. In particular they can
often be treated as quasi non-divergent and their behaviour
approximated in a quasi-geostrophic system whilst
mechanistic models can achieve some success using linear
theory (see Holton,1975 for details).

PHYSICAL BASIS OF STRATOSPHERIC MODELS

(i)   Constituent distributions .

The local rate of change $\partial n_i / \partial t$ of the concentration n_i of
an atmospheric constituent is determined by its photochemical
production P_i and loss $Q_i n_i$ terms, the presence of any local
net source or sink S_i and the contribution due to transport
processes. The three-dimensional continuity equation for the
constituent will then be

$$\frac{\partial n_i}{\partial t} + \nabla . n_i \mathbf{v} + \frac{\partial}{\partial z} n_i w = (P_i - Q_i n_i) + S_i \qquad \ldots \ldots (1)$$

where ∇ is the horizontal del vector operator.

There will be a set of these equations, one for each constituent, to solve simultaneously. Starting from an
initial composition field the problem is to perform
numerical integrations with as few approximations as possible
to reach a quasi-steady state or alternatively to use this
equation to study equilibrium conditions. The two-
dimensional horizontal wind vector $\mathbf{v}$ and the vertical wind
component w will vary in space and time according to their
controlling equations. The terms P_i, Q_i are functions of
the concentrations of the constituents taking part in the
photochemical processes determining   and may contain
contributions both from photodissociation processes and
chemical reactions between different constituents.

The rates of the former have to be calculated from a
knowledge of the extra-terrestrial spectral intensities of
the dissociating solar radiation and the attenuation above the
point considered by the overlying atmosphere. Data on the
chemical reaction rates appropriate to stratospheric

conditions must be determined from laboratory studies.

There are several limitations to the numerical integration
procedures which can be used. In view of the large range of
rate coefficients the set of equations will be a 'stiff
system' and care will be needed to retain stability and con-
serve the total amounts of the basic species nitrogen,
oxygen, hydrogen, carbon, chlorine etc throughout. A
large stratospheric chemical kinetics scheme may contain
about 40 constituents and 100 different reactions.

(ii) Radiation

For modelling the temperature distribution and general
circulation of the stratosphere a first requirement is to
calculate the diabatic heating rate per unit mass Q where:-

$$\frac{Q}{c_p} = \left(\frac{\partial T}{\partial t}\right)_{SR} + \left(\frac{\partial T}{\partial t}\right)_{AR} \qquad \ldots \ldots (2)$$

and $(\partial T/\partial t)_{SR}$. $(\partial T/\partial t)_{AR}$ are the rates of
change of temperature due to solar and atmospheric
radiation processes respectively and c_p is the specific heat
at constant pressure. Q/c_p is usually a few $^\circ$C per day in
the upper stratosphere and fractions of a $^\circ$C per day in the
lower stratosphere.

Since the effects of the solar radiation are small at wavelengths above about $4\,\mu m$ and those of the atmospheric
radiation important only at greater wavelengths their
effects can be calculated separately. For solar radiation,
scattering and absorption by both gases and aerosols have to
be taken into account. The effects of the aerosols are
particularly difficult to estimate due to lack of data on
their distributions, sizes, composition, shapes and optical
properties. The major heating is due to absorption by ozone
in the ultra violet region of the solar spectrum (Hartley
and Huggins bands from about 220-340 nm) and it also absorbs
weakly in the Chappuis bands in the visible region.
Absorption by carbon dioxide, water vapour and other minor
constituents is also of importance in the solar near infra-red
region. In the upper stratosphere absorption by molecular
oxygen around 200 nm must also be calculated. For
atmospheric radiation the balance between emission and
absorption at each level must be found, the relevant constituents in order of importance being carbon dioxide
($15\,\mu m$ band), ozone ($9.6\,\mu m$ band) and water vapour

(rotation and $6.3\ \mu m$ bands). In the troposphere, cloud is
of major importance both as regards the solar and
atmospheric radiation but leads to many uncertainties due
to lack of knowledge of its distribution and optical properties.
The numerical treatment of even the effects of the various
gaseous constituents contains many complications and there
are considerable difficulties in devising suitable
formulations for use in models. It will only be possible
here to outline the main principles (see eg Goody, 1964,
Craig, 1965, Rodgers and Walshaw, 1966, and Hunt and
Mattingly, 1976 for further details).

Considering first the solar radiation, if z is the vertical
coordinate with $z = 0$ at the surface and $z = \infty$ corresponds
to the top of the atmosphere, where the specific intensity
of the solar beam at frequency ν is $I_{\infty\nu}$, then at a
selected level Z

$$I_{Z\nu} = I_{\infty\nu} \exp(-\int_{Z}^{\infty} \kappa_\nu\ \rho_a \sec\theta\ dz) \qquad \ldots\ldots (3)$$

where κ_ν is the absorption coefficient ρ_a the absorber
density and $\sec\theta$ the approximate magnification factor to
account for the slant path of the beam at zenith angle θ.
     If two constituents absorb at the same frequency their
absorptions are additive.   Integrating over all
frequencies, then at each level Z

$$\left(\frac{\partial T}{\partial t}\right)_{SR} = \frac{1}{\rho\,c_p} \frac{\partial}{\partial Z} \int_\nu I_{Z\nu} \cos\theta\ d\nu \qquad \ldots\ldots (4)$$

where ρ is atmospheric density.
In more detailed treatments absorption of radiation scattered
from other levels and also that reflected from the earth's
surface and clouds will have to be estimated.

The corresponding calculations of $(\partial T/\partial t)_{AR}$
have to take account of the flux F_ν from all directions
in a hemisphere about the vertical where

$$F_\nu = \int_0^{2\pi} \int_0^{\pi/2} I_\nu \sin\theta \cos\theta\ d\theta\ d\phi \qquad \ldots\ldots (5)$$

and θ here is the inclination of the radiation to the
vertical and ϕ is the azimuth. In this case I_ν is the
sum of three contributions $I_{G\nu}$, $I_{U\nu}$, and $I_{D\nu}$ originating
respectively from radiation from the earth's surface, the

atmospheric layers below Z and those above Z.

$$I_{G\nu} = B_{G\nu} \exp(-\sec\theta \int_0^Z \kappa_\nu \, \rho_a \, dz) \qquad \dots (6)$$

is the radiation intensity $B_{G\nu}$ from the earth's surface attenuated in its path up to the level Z.

$$I_{U\nu} = \int_0^Z B_\nu(z) \exp(-\sec\theta \int_0^Z \kappa_\nu \, \rho_a \, dz) \sec\theta \, \kappa_\nu \rho_a dz \qquad \dots (7)$$

is the integral of the emissions from all atmospheric layers below Z attenuated in their paths up to the level Z. $I_{D\nu}$ is formed by integrating similarly the contributions from layers above Z.

Then, forming F_G, F_U, and F_D as the fluxes from $I_{G\nu}$, $I_{U\nu}$ and $I_{D\nu}$ respectively

$$\left(\frac{\partial T}{\partial t}\right)_{AR} = -\frac{1}{c_p \rho} \frac{\partial}{\partial z}(F_G + F_U - F_D) \qquad \dots (8)$$

The principal difficulty in evaluating the I_ν terms for atmospheric radiation is that the κ_ν vary rapidly with ν due to the line character of the infra-red spectra and are also dependent on temperature and pressure. Assumptions must be made, for example the actual spectral distributions may be approximated by band models (see Goody, 1964), the absorber amounts scaled to weighted mean temperatures and pressures (the Curtis-Godson approximation) and the integrations over the hemisphere avoided by use of a suitable diffusivity factor to further scale the absorber amounts. In very simple treatments Newtonian cooling is often assumed, in order to avoid the complexities involved in repeated calculations on the lines above.

At present it does not appear possible to make adequate measurements to validate in any detail model calculations based on the above expressions. Any further progress in this area would be of great value. Meanwhile other measurements are urgently required of solar spectral intensities and atmospheric composition together with improvements in

laboratory data and techniques for obtaining
simpler radiation models of acceptable accuracy.

(iii) Temperature distribution and general
circulation.

The prediction of the distributions of the basic
variables pressure p, density ρ, temperature T and
the three wind components u (zonal), v (meridional)
and w ((vertical) involves the use of the conservation
equations of mass (continuity), energy (thermodynamic), momentum (motions in three dimensions) and
state, ie six equations in six unknowns. Individual
constituent continuity equations eg for water vapour,
ozone etc as treated in (i) above may be added as
required.

We define a latitude ϕ, longitude λ, height z
coordinate system with a spherical earth of radius a
and rotation rate Ω and put

$$\left.\begin{aligned}
dx &= a\cos\phi\, d\lambda \\
dy &= a\, d\phi \\
f &= 2\Omega\sin\phi \quad \text{(Coriolis parameter)} \\
\frac{d\chi}{dt} &= \frac{\partial\chi}{\partial t} + (\mathbf{v}.\nabla)\,\chi + w\,\frac{\partial\chi}{\partial z}
\end{aligned}\right\} \quad \ldots\ldots (9)$$

and

where χ is any scalar quantity. Then the diabatic heating
rate per unit mass, Q, produced by the radiative processes
in (ii) above is related to T, p and ρ by the
thermodynamic equation

$$c_p\,\frac{dT}{dt} = \frac{1}{\rho}\,\frac{dp}{dt} + Q \qquad \ldots\ldots (10)$$

where $(1/\rho)(dp/dt)$ is the adiabatic heating
rate. Also from the equation of state

$$p = R\rho T \qquad \ldots\ldots (11)$$

where R is the gas constant for air.

The equations of motion in Cartesian coordinates are

$$\frac{du}{dt} = -\frac{1}{\rho}\frac{\partial p}{\partial x} + fv + \frac{uv\tan\phi}{a} + F_\lambda$$

$$\frac{dv}{dt} = -\frac{1}{\rho}\frac{\partial p}{\partial y} - fu - \frac{u^2\tan\phi}{a} + F_\phi$$

$$\cdots\cdots (12)$$

(where F_λ and F_ϕ are frictional forces per unit mass)

$$\frac{dw}{dt} = -\frac{1}{\rho}\frac{\partial p}{\partial z} - g$$

For large scale motions $\frac{dw}{dt}$ may be neglected and the vertical motion equation reduces to the hydrostatic equation

$$\frac{\partial p}{\partial z} = -\rho g \qquad\qquad \cdots (13)$$

The set is completed by the mass continuity equation

$$\frac{d\rho}{dt} = -\rho(\nabla \cdot \mathbf{v} + \frac{\partial w}{\partial z}) \qquad\qquad \cdots (14)$$

With suitable boundary conditions these primitive equations after some manipulation can be integrated numerically from given initial fields to give the evolution of the p, ρ, T, u, v, w distributions. At the same time prediction of constituent fields (particularly water vapour) and interaction through radiation fields may be achieved using the equations in (i) and (ii) above.

Although there are some advantages in using height z as the vertical coordinate some models, making use of the hydrostatic approximation, use pressure p, $\log p$, or $\sigma = p/p_0$ where p_0 is surface pressure. In general the use of p leads to a simpler **mass** continuity equation but gives poor resolution in the stratosphere unless unequal intervals of p are used. With σ coordinates the effects of the earth's topography

are handled most effectively but detailed analysis
and zonal averaging of the model outputs present
greater problems.

The above continuous equations are non-linear and in
general they are able to produce solutions for all scales
of atmospheric motions irrespective of their immediate
application, although use of the hydrostatic equation
eliminates vertical propagation of sound waves. It is
also possible in models of the large scale systems of the
extra-tropical stratosphere to use quasi-geostrophic
equations in which gravity type oscillations are filtered
out. Models of this type have greater computational stability
for given horizontal resolution than those based directly on
the primitive   equations but are more difficult to
handle if detailed composition and radiation equations
are added (see eg Cunnold et al., 1975). In these models
the rotational component of the wind, which is dominant
in the large systems of the extra-tropical stratosphere,
is treated as being in geostrophic balance allowing use
of the thermal wind equation,while the vertical velocity
is determined from its divergent component. The first
prediction equation is the vorticity equation (which is
derived from the equations of motion and which provides
a relationship between the divergent component and the
evolution of the vorticity of the rotational component)
and a second prediction equation is provided by the
thermodynamic equation (see Lorenz, 1960 for basic details).
This system is particularly suitable for spectral modelling
techniques in which the number of harmonics can be
limited as desired to produce the required computational
efficiency.

The above outline has been restricted to the main processes
which determine the composition, radiation conditions,
temperatures and air motions in the stratosphere. For
tropospheric motions many other processes such as the
hydrological cycle, large-scale convection, topographical
effects, surface conditions and lower boundary transfers
have to be formulated (see eg Smagorinsky et al., 1965:
Kasahara and Washington, 1971: Corby et al., 1972 for
details). The friction terms F in the equations of **motion**
include the sub-grid scale effects by using an eddy
viscosity formulation, and surface drag is inserted at the
lower boundary. Land surface temperatures are computed
from energy balance considerations but, until ocean
circulations can be included in the models, surface
temperatures over the oceans have to be specified. All
models have to be designed to be energetically consistent

with their finite difference analogues adjusted to meet
the same constraints as the continuous equations.
Special measures are sometimes needed to avoid spurious
negative values of constituents. The nature and length
of the time steps must also satisfy conditions of
computational stability throughout, a problem which may be
severe in polar latitudes unless suitable precautions
are taken. Practical details of this type and
considerations of the size, resolution and possible running
time of a model with regard to the available computer
facilities are beyond the scope of this short paper
and reference should be made to the quoted papers and
textbooks for **further** details (eg Haltiner, 1971).

TYPES OF MODELS

In view of the variety and complexity of the processes
considered above a hierarchy of models has been produced
for different purposes. Three-dimensional models are
necessary to study the large scale motions and transports,
one-dimensional models are commonly used to study the
detailed photochemistry and two-dimensional models appear
attractive for combined transport and photochemical studies
provided that the various difficult averaging problems
can be solved. A number of mechanistic models have been
developed to study special phenomena eg the semi-annual and
quasi-biennial oscillations, troposphere - stratosphere
forcing and sudden warmings, stratosphere - troposphere
transfer and several others. No complete fully interactive
model of the troposphere and stratosphere has yet been
produced to the authors' knowledge. Several of these
types of models and the kinds of results they are
producing will be briefly outlined below:-

Three-dimensional models

Three-dimensional modelling of the troposphere and
stratosphere has made considerable progress in recent **years**
and has provided simulations of the different dynamical
processes which are taking place. For example, results of
several primitive equation model experiment s both for the
Northern Hemisphere and the globe have been reported (eg
Smagorinsky et al. 1965, Manabe and Hunt, 1968, Kasahara and
Washington, 1971 and Newson, 1974). These models in addition
to representing with some success the basic temperature
distribution and general circulation have also indicated
the nature of the transports of heat and momentum, the
basic energetics (the transfer between mean and eddy
forms of kinetic and available potential energy) and the

way tracers are transported by the air motions. They
also have demonstrated the importance of upward
propagation of energy from tropospheric systems in
driving stratospheric motions and establishing the
planetary wave systems of the stratosphere. Some success
has also been achieved in the simulation of stratospheric
sudden warmings (eg Newson, 1974). It has also been
demonstrated that simpler spectral quasi-geostrophic
models (eg Cunnold et al.,1975) are able to reproduce the
main features of the stratosphere and give valuable new
information on the dynamical processes which appear to
be most important. One of the major uses of all these
models is for diagnostic studies which hopefully can be
applied in the construction of simpler models. Due to
their complexity and computer requirements,
it has so far only been possible to use them in a
limited way for studies of the effects of perturbations
of composition, radiation etc. Therefore, simpler models have
been developed, which can at the same time incorporate the
results of these three-dimensional model studies by
using suitable parameterisations.

Two-dimensional models.

The solution of the complete three-dimensional equations
is computationally expensive and for many purposes a
two-dimensional (height-latitude) model may be adequate.
In this case there is a possibility of devoting more of
the computations to the physical and chemical aspects
while at the same time retaining essentially the
latitudinal and seasonal features due to the motions. We
define a zonal average $\overline{\chi}$ by

and write

$$\overline{\chi} = \frac{1}{2\pi} \int_0^{2\pi} \chi \, d\lambda$$

$$\chi = \overline{\chi} + \chi'$$

$$\qquad \qquad \dots (15)$$

$$\overline{v\chi} = \overline{v}\,\overline{\chi} + \overline{v'\chi'} \qquad \qquad \dots (16)$$

The equations for the zonal means of the fluxes of χ
expressed in this way are then the sum of the mean and eddy
terms of the form $\overline{v}\,\overline{\chi}$ and $\overline{v'\chi'}$ respectively. The two
dimensional constituent continuity equations will be of the
form

$$\frac{\partial \overline{\chi}}{\partial t} = -\left(\frac{\overline{v}}{a} \frac{\partial \overline{\chi}}{\partial \phi} + \overline{\omega} \frac{\partial \overline{\chi}}{\partial p} \right) - \left(\frac{1}{a\cos\phi} \frac{\partial}{\partial \phi} \overline{v'\chi'}\cos\phi + \frac{\partial}{\partial p} \overline{\omega'\chi'} \right) + \overline{C} + \overline{S}$$

$$(17)$$

mean transport   *eddy transports*

Here $\overline{\chi}$ is a mixing ratio and the equation is written
most conveniently in pressure coordinates with $\omega = dp/dt$.

$\overline{C}$ is the net photochemical rate of increase of $\overline{X}$, and $\overline{S}$ its external source, both being averaged round latitude circles. In order to integrate this equation it is clearly necessary to express the eddy flux terms as a function of the $\overline{X}$ field or introduce them as known quantities varying with time. Moreover the mean circulation components $\overline{v}$ and $\overline{\omega}$ as well as $\overline{C}$ and $\overline{S}$ have to be inserted as functions of time.

Considering these in turn:-

(i) C depends on the solar zenith angle and, provided that the model can be regarded as applying to a given longitude,this can be taken as $\overline{C}$ which will then vary with time diurnally and seasonally at each latitude. Any system involving averaging of C round latitude circles at a given time would give rise to considerable problems as regards the form of average to use for the dissociation terms in view of the varying solar illumination throughout the day at each longitude.

(ii)   To the extent that source terms in $\overline{S}$ are specified there are no special problems but empirical terms will be required for average rainout round latitude circles.

(iii) The field of the mean circulation term $\overline{v}$ to use is not well known observationally throughout the stratosphere particularly as regards its short term variability (eg in sudden warmings). One possible but somewhat limited approach is to use some average values for the solstices and assume sinusoidal variations with time. Another is to extract data from three-dimensional model outputs and a third is to attempt to derive $\overline{v}$ by solving two-dimensional equations derived from the equations of motion in a dynamical model. Once the $\overline{v}$ field is established the $\overline{\omega}$ values may be otained from the two-dimensional mass continuity equation.

$$\frac{1}{a\cos\phi} \; \frac{\partial}{\partial\phi} \; \overline{v}\cos\phi \;+\; \frac{\partial\overline{\omega}}{\partial p} \;=\; 0 \qquad \cdots\cdots (18)$$

The two-dimensional equations for $\overline{u}$ and $\overline{T}$ will be of similar general form, i.e.,

$$\frac{\partial\overline{u}}{\partial t} \;=\; -\left(\frac{\overline{v}}{a\cos\phi} \; \frac{\partial}{\partial\phi} \; \overline{u}\cos\phi \;+\; \overline{\omega}\frac{\partial\overline{u}}{\partial p}\right)-$$

$$-\left(\frac{1}{a\cos^2\phi} \; \frac{\partial}{\partial\phi} \; \overline{v^1u^1}\cos^2\phi \;+\; \frac{\partial}{\partial p} \; \overline{\omega^1u^1}\right) \;+\; f\overline{v} \;+\; \overline{F}_\lambda \;\cdots(19)$$

$$\frac{\partial \overline{T}}{\partial t} = -\left(\frac{\overline{v}}{a} \frac{\partial \overline{T}}{\partial \phi} + \overline{\omega} \frac{\partial \overline{T}}{\partial p} \right) -$$

$$-\left(\frac{1}{a \cos \phi} \frac{\partial}{\partial \phi} \overline{v^1 T^1} \cos \phi + \frac{\partial}{\partial p} \overline{\omega^1 T^1} \right) +$$

$$+ \frac{R}{p c_p} (\overline{\omega} \overline{T} + \overline{\omega^1 T^1}) + \frac{\overline{Q}}{c_p} \quad ..(20)$$

and it will be noted that since the eddy fluxes $\overline{v^1 u^1}$ and $\overline{v^1 T^1}$ at least are easier to observe than $\overline{v}$ there are possibilities also, by neglecting small terms, of using these equations diagnostically to determine approximately the mean circulation parameters $\overline{v}, \overline{w}$ (see eg Vincent, 1968, Louis et al., 1974).

Harwood and Pyle (1975) have developed a two-dimensional photochemical-dynamical model integrating with time using this approach and adding the thermal wind relation for the zonal wind component to obtain a closed set of equations. They attempted to express the eddy fluxes of angular momentum in terms of the fluxes of potential vorticity and potential temperature. These latter were considered to be quite well conserved and to this extent well treated by K theory (see below). Because of mathematical difficulties, however, observed horizontal momentum fluxes had to be used. Their treatment nevertheless produced satisfactory representations of the mean circulation.

(iv) The most difficult problem in the two-dimensional modelling is the formulation of the eddy terms $\overline{v^1 \chi^1}$, $\overline{w^1 \chi^1}$. If χ were a conservative quantity it might be thought that a possible approach would be to use a mixing-length theory in which

$$\overline{v^1 \chi^1} = -K_y \frac{\partial \overline{\chi}}{\partial y} \qquad \ldots \ldots (21)$$

where K_y is an eddy diffusion coefficient, with a corresponding expression for $\overline{w^1 \chi^1}$. However, this is not satisfactory in the stratosphere partly because observations have indicated that the transport of several quantities may be counter-gradient and there-fore requires a negative value of K_y or K_z and partly because the eddy transport is mainly by the very large scale systems and hence this relation is not physically plausible.

A commonly used but hardly satisfactory way of

circumventing the first objection is to use a tensor formulation for the eddy diffusion coefficients (Reed and German, 1965) where

$$
\left.
\begin{aligned}
\overline{v^1\chi^1} &= -\,(K_{yv}\frac{\partial \overline{X}}{\partial y} + K_{yz}\frac{\partial \overline{X}}{\partial z}) \\
\overline{w^1\chi^1} &= -\,(K_{zy}\frac{\partial \overline{X}}{\partial y} + K_{zz}\frac{\partial \overline{X}}{\partial z})
\end{aligned}
\right] \quad \ldots \ldots (22)
$$

The horizontal and vertical diffusion coefficients K_{yy}, K_{zz} are always positive and apparently have values of the order of 10^{10} and 10^4 cm^2 s^{-1} respectively while the cross coefficients $K_{yz} \approx K_{zy}$ may be either positive or negative and will be approximately 10^7 cm^2 s^{-1} (see Luther, 1974). In spite of its drawbacks, most current two-dimensional models make use of this scheme with some apparent success. For example, using estimated values of the mean wind components and Luther's values of the K's together with a basic photochemical scheme, some first experiments using a two-dimensional model in the British Meteorological Office have produced the type of results shown in Fig.2 (ozone and nitric acid cross-sections) and Fig. 3 (nitric oxide and nitrogen dioxide cross-sections).

One dimensional models

If the chemical and physical processes have to be modelled in great detail (as in current studies of potential stratospheric pollution by aircraft emissions and surface releases of chemicals) attempts may be made to model the whole of a hemisphere by taking a hemispheric average for each level in the vertical. The assumption is made that all the significant effects of the motions, both mean and eddy, can be described in this average over horizontal area ($\overline{}^A$) in terms of a vertical eddy diffusion formulation in which

$$
\overline{w\,n_i}^A = -\,K_z\left[\frac{\partial \overline{n_i}^A}{\partial z} + (\frac{1}{\overline{T}^A}\frac{\partial \overline{T}^A}{\partial z} + \frac{1}{H})\overline{n_i}^A\right] \,\ldots(23)
$$

The principal problem here is the specification of K_z and the common approach is to attempt to derive a suitable K_z profile from existing observations of tracers eg radio-active substances, ozone, water vapour, potential vorticity etc. Considerable variations with latitude (see eg Danielsen, 1968 for studies of the role of tropopause gaps), season, height, and constituent take place and there are

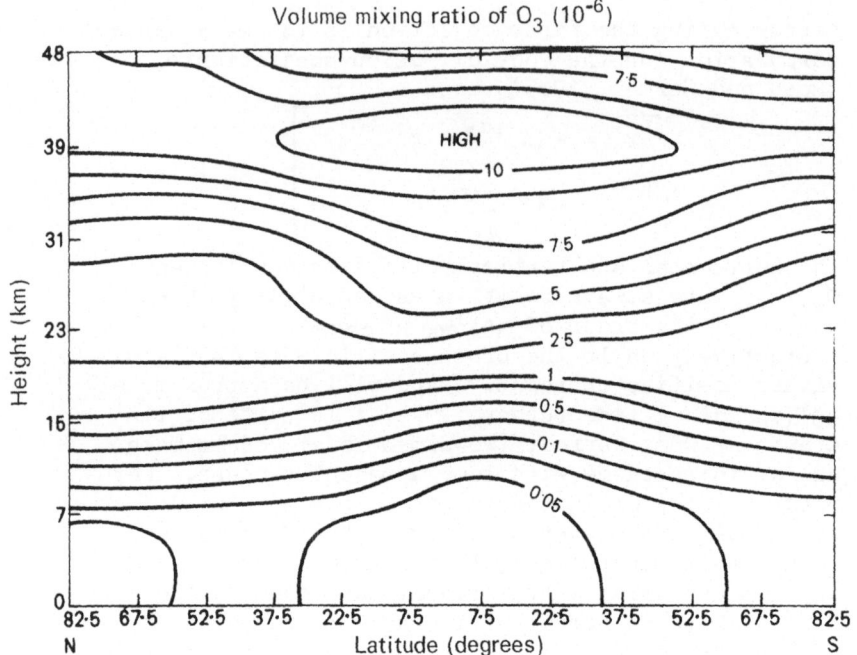

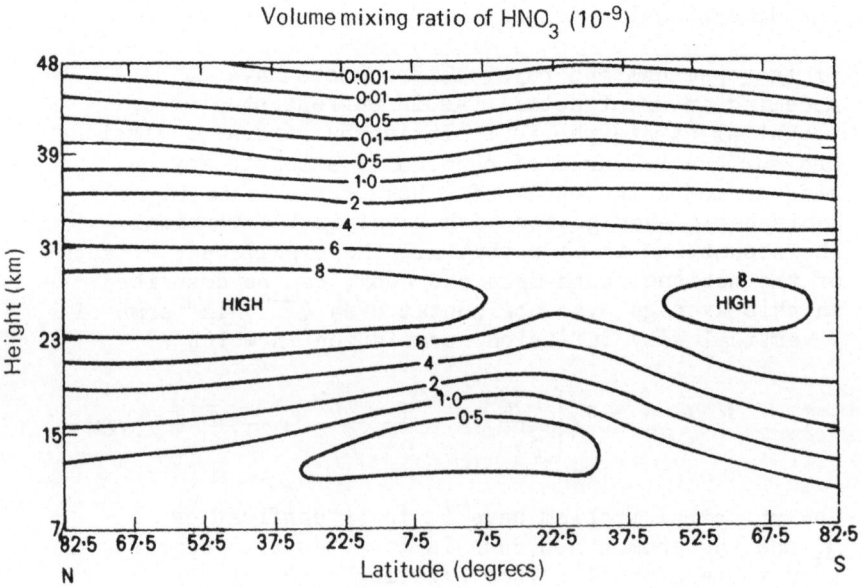

Figure 2 : Two-dimensional model cross-sections of ozone and nitric
acid mixing ratios at the northern hemisphere winter
solstice, midday.

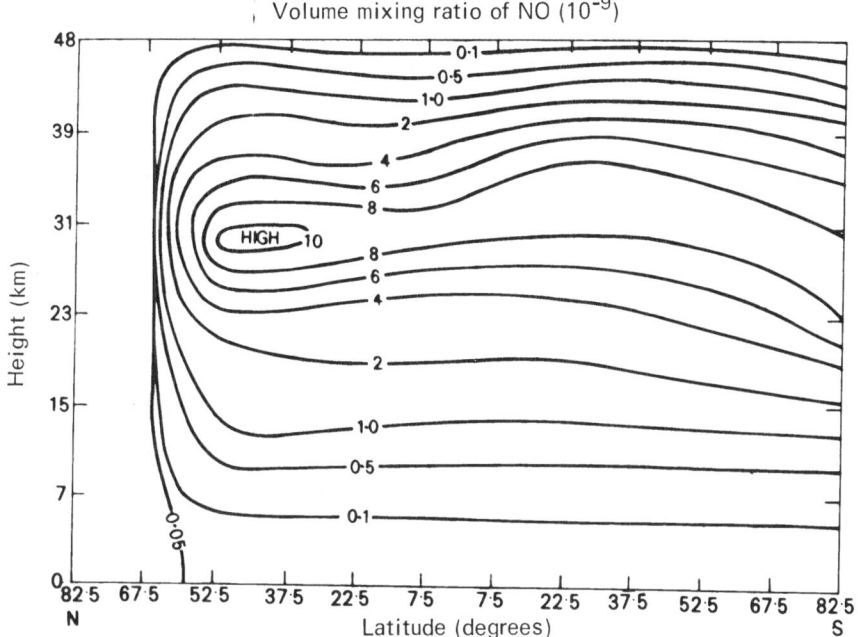

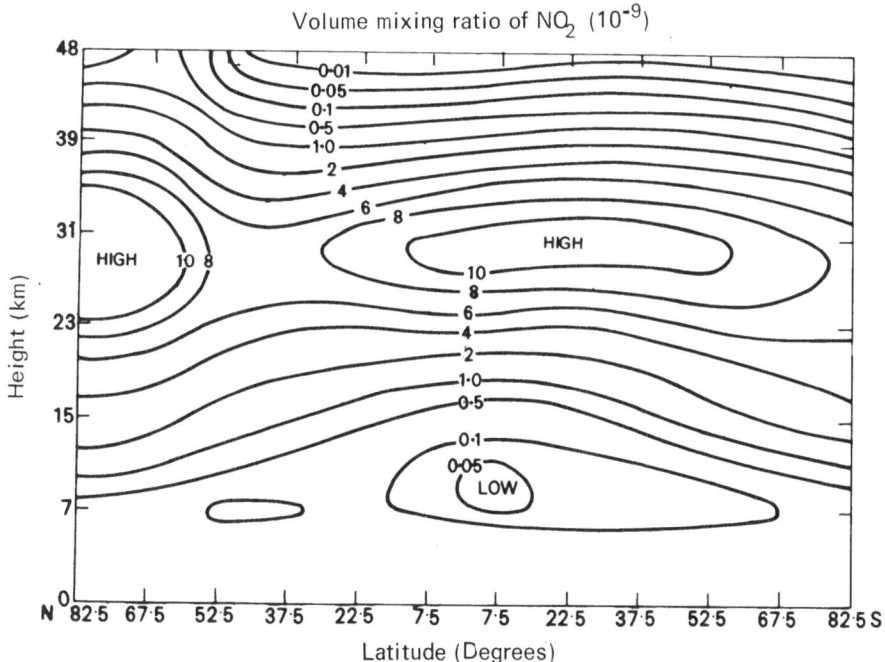

Figure 3 : Two-dimensional model cross-sections of nitric oxide and
nitrogen dioxide mixing ratios at the northern hemisphere
winter solstice, midday.

considerable problems regarding the K_z profile to select
and whether it should vary with time. It has for
example been shown from the results of integrations
with the British Meteorological Office's three-
dimensional model that although average values of global
profiles can be extracted from the results of a year's
integration, the concept breaks down if applied only
to a hemisphere or for shorter periods (in the sense
of apparently producing negative K_z's due to the effects
of the mean circulation terms). Clearly the problem
requires further study and quantitative results
obtained from one-dimensional models should be treated
with appropriate reservations.

However, primarily because of limited computer facilities,
one-dimensional models have been used extensively and have
greatly added to understanding of the photochemistry of
the natural stratosphere as well as providing some first
estimates of the possible effect of perturbations. Figs. 4
and 5 illustrate the type of results that can be obtained
and are examples kindly provided from one of his experiments
by Dr A.F Tuck of the British Meteorological Office. It will
be noted that these show considerable differences from
the magnitudes of the mixing ratios given in Figs 2 and
3. Both, however, are broadly within the range of
existing measurements. Clearly many more measurements are
needed at different times and locations in the stratosphere
and preferably of several species simultaneously to
provide validation and guidance for future model
developments.

Most of the general considerations discussed above as
regards model complexity and the type of model to use
apply also to the modelling of possible changes of climate
due to stratospheric perturbations. In some respects,
these problems are even more complicated than the
estimation of changes in stratospheric composition. The
information that can be obtained from simplified climate
models is very limited and their results should again only be
accepted with the appropriate reservations.

CONCLUDING REMARKS

Modelling of the stratosphere is still a relatively new
and rapidly changing field developing both in response
to its great potential for improving overall understanding
of the basic processes and its use in investigating
possibilities of environmental changes arising from

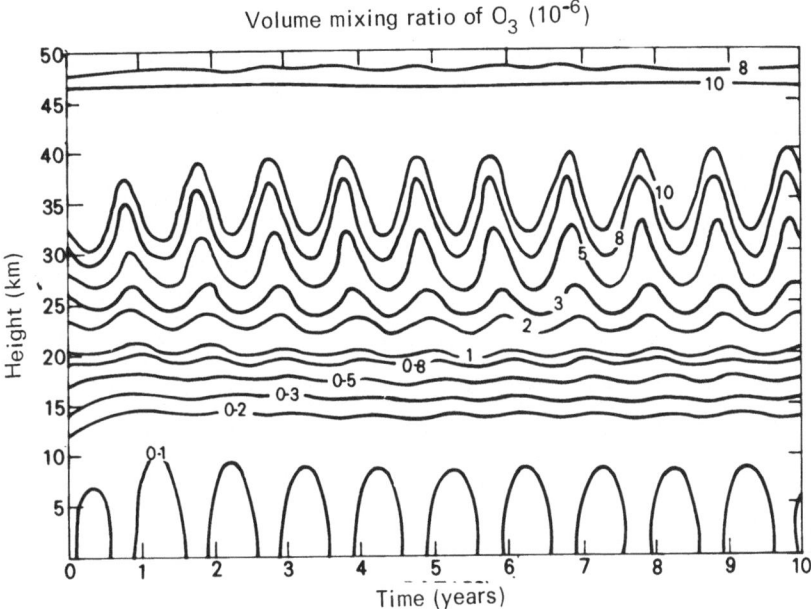

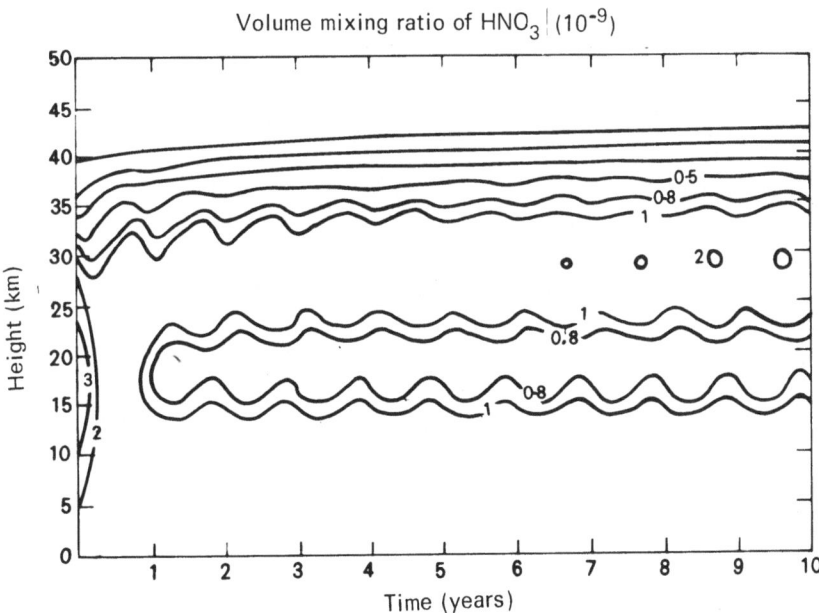

Figure 4 : One-dimensional model time series of ozone and nitric acid mixing ratio profiles (midday values starting at the spring equinox), 45° latitude.

Volume mixing ratio of NO (10^{-9})

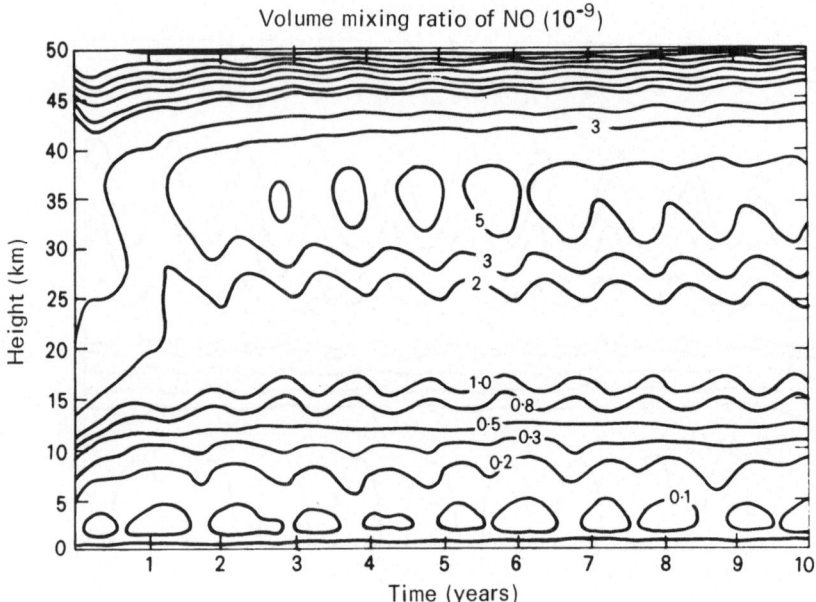

Volume mixing ratio of NO_2 (10^{-9})

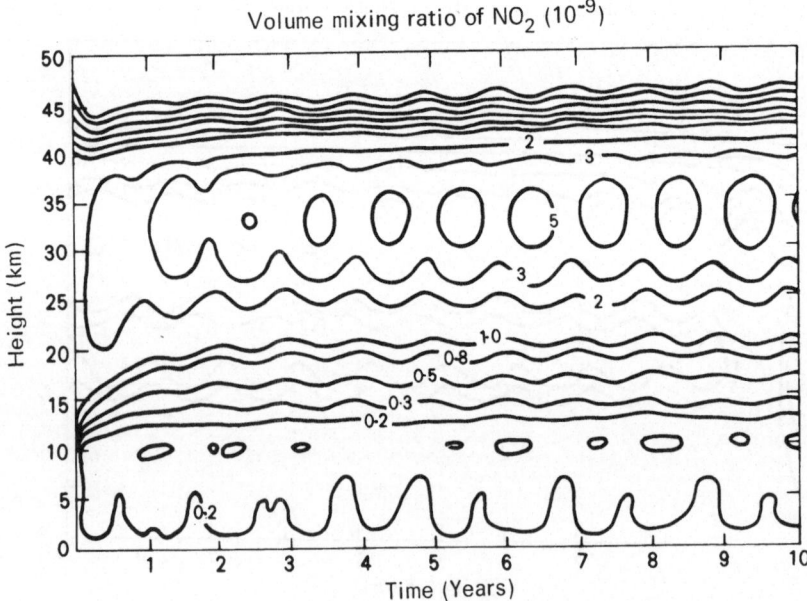

Figure 5 : One-dimensional model time series of nitric oxide and nitrogen dioxide mixing ratio profiles (midday values starting at the spring equinox), 45° latitude.

stratospheric perturbations. As regards composition, since it has been realised in the last few years that many nitrogen-oxygen-hydrogen-carbon and halogen species could affect the ozone layer, a great deal has been learned about the processes determining the detailed distributions of the various constituents. The modelling has,however,not yet advanced sufficiently to give reliable quantitative simulations and in particular much more must be known about natural sources and sinks before this can be achieved.

There is also a major **requirement** for further progress in modelling of radiation,particularly to incorporate more realistically the effects of clouds and aerosols for use in climate studies. The large scale modelling of atmospheric motions and temperatures particularly in the upper stratosphere could be improved greatly if there were more measurements available. Another aspect of great importance is the study of stratosphere -troposphere exchange which is not satisfactorily understood, at least in a quantitative sense. This is very necessary in particular for the formulation of transport mechanisms in the simplified two- and one- dimensional models. Finally there is a continued need for the development of simple 'mechanistic'models designed to improve understanding of special features including sudden warmings, the quasi-biennial oscillation, equatorial waves etc and the basic dynamics of the stratosphere in general (see Holton,1975, for a detailed discussion of current achievements).

We are indebted to several of our colleagues ät Bracknell for examples of their work illustrated in this paper. It is hoped that detailed accounts of these modelling studies will be published in due course.

REFERENCES

Charney,J.G. and Drazin,P.G.	1961	J. Geophys.Res., 66, 83-109.
Corby,G.A.,Gilchrist,A. and Newson,R.L.	1972	Q.J.Roy. Met. Soc., 98, 809-832.
Craig,R.A.	1965	The Upper Atmosphere - Meteorology and Physics. Academic Press, New York.

Cunnold, D.M., Alyea, F.N., 1975   J.Atmos.Sci., **32, 170–194.**
  Phillips,N.A., and
  Prinn, R.G.

Danielsen,E.F.   1968   J.Atmos.Sci., 25, 502-518.

Goody,R.M.   1964   Atmospheric Radiation,**I,**
                                    Oxford University Press,
                                    Oxford.

Haltiner,G.J.   1971   Numerical Weather
                                    Prediction, Wiley, New York

Harwood,R.S. and   1975   Q.J.Roy. Met. Soc., 101,
  Pyle,J.A.   723-747.

Holton,J.R.   1975   Dynamic Meteorology of the
                                    Stratosphere and Mesosphere
                                    Met. Monographs, 15, no. 37,
                                    Amer.Met.Soc.

Hunt,G.E. and   1976   J.Quant.Spec.and Rad.
  Mattingly,S.R.   Transfer **16, 505–520.**

Kasahara,A. and   1971   J.Atmos.Sci., 28, 657-701.
  Washington, W.M.

Lorenz,E.N.   1960   Tellus, 12, 364-373.

Louis,J.-F.   1975   C.I.A.P(1974) Monograph 1,
                                    Report No.**DOT-TST-75-51**
                                    Section 6–23, U.S.

Luther, F.M.   1975   Dept. of Transportation
                                    ibid, **6–31.**

Manabe,S. and   1968   Mon.Wea.Rev., **96,** 477-502
  Hunt, B.G.

Newson,R.L.   1974   Proceedings of the Anglo-
                                    French Symposium on the
                                    Effects of Stratospheric
                                    Aircraft,**2, XXVII-1-22,**
                                    Bracknell, U.K.

Reed,R.J. and   **1965**   Mon. Wea. Rev., 93,
  German,K.E.   313-321.

Rodgers,C.D.   1966   Q.J. Roy. Met. Soc.,
  and Walshaw C.D.   92, 67-92.

Smagorinsky, J., Manabe, S., and Holloway, J.L.	1965	Mon. Wea. Rev., 93, 727-768.
Vincent, D.G.	1968	Q.J.Roy. Met. Soc., 94, 333-349.

DISCUSSION

J. Gregory: With respect to topography, is it yet possible to
account for sea-surface temperature changes?

R.J. Murgatroyd: No. Sea-surface temperatures are at present
represented as means for chosen months, and these are long-
term.

J.T. Houghton: I would like to make a comment about the transports
in the stratosphere by mean motions on the one hand and eddies
on the other. One of the results which has come from the two-
dimensional model built at Oxford by Harwood and Pyle (and
which incorporates momentum transports obtained from satellite
data) is that transport by the eddies is very nearly compensated by transport by the mean motions, the net transport
being the difference between two comparatively large quantities.
Considering the crude way in which the transports are included,
it is therefore surprising that models can produce anything
like the right result. Perhaps somewhere in the dynamics
the correct compensation is automatically arranged? Perhaps
Dr. Murgatroyd would like to comment on this?

R.J. Murgatroyd: The compensation you have noted has also been
found in several other more detailed studies, e.g. by
S. Manabe and B.G. Hunt Mon. Wea. Rev. 96, 1, p. 477 (1968)
in their 18 level three-dimensional model experiments and
also by R.L. Newson in the Meteorological Office 13 level
three-dimensional model results. It seems to be a general
feature (except perhaps at low latitudes) but disappears in
periods of rapid change when one of the terms becomes
dominant and there is then a large enhancement of the
resultant transport. Effects of this type are lost in simple
models which assume constant or smoothly varying changes with
time in their parameterisations of the transport mechanisms.

V. Domingo: As I showed yesterday, the ion density in the low
stratosphere changes drastically during the solar cycle,
particularly   at   high latitudes, due to the change in
cosmic-ray flux. As a consequence, of course, the
composition will change; in particular the NO and therefore
the ozone. Have you introduced those changes in any model?

R.J. Murgatroyd: No, we have not attempted this particular
experiment. At first sight this could be a possible link
between the sunspot cycle and climatic statistics showing
changes with its period. However, other calculations (e.g.
those of Prof. M. Nicolet reported in the Proceedings of the
Joint COMESA-COVOS Symposium, Oxford, 1974) suggest that
cosmic-ray production of NO in the stratosphere is only a
fraction of that due to the N_2O reaction with $O('D)$, and other
studies by Dr. P. Crutzen and his colleagues at NCAR suggest
that NO production by solar flares can also be more
important. Moreover, the limited three-dimensional model
calculations we have made at the British Meteorological
Office in which large changes were made in the mean ozone
amounts in the stratosphere only indicated small consequent
changes in global surface temperatures. Consequently, it
appears that the demonstration of the effect you propose by
our currently available modelling techniques would be very
difficult.

REVIEW OF OBSERVATIONS FROM METEOROLOGICAL SATELLITES

J.T. Houghton
Department of Atmospheric Physics,
University of Oxford, UK

The advantage of satellites for the collection of meteorological information is mainly that very much better coverage in both space and time can be realised compared with conventional observation. A satellite in geostationary orbit can make continual observations of about a quarter of the atmosphere; a polar orbiting satellite can observe all parts of the atmosphere twice per day. From such satellites the observation of radiation reflected, scattered or emitted in various parts of the electromagnetic spectrum can yield information about the atmospheric structure.

A brief review was given of the instrumentation that is to be used for imaging in both the visible and infrared and for the measurement of temperature, density and composition. Results from recent satellite missions were presented and the potential for the future was discussed.

J. J. Burger et al. (eds.), Atmospheric Physics from Spacelab, 199–200. All Rights Reserved.
Copyright © 1976 by D. Reidel Publishing Company, Dordrecht-Holland.

DISCUSSION

G. *Hunt:* How accurate are the cloud amounts obtained in temperature
    retrievals, and can they be used to obtain a three-dimensional
    cloud climatology?

J.T. *Houghton:* Cloud data from temperature sounding channels
    refer to fractional cloud cover and height of the cloud top.
    No information on cloud base is, of course, available.

W. *Renger:* You had about 1 weighting function between 1000 mb
    and 900 mb. How could you find a temperature inversion in
    this region?

J.T. *Houghton:* From the satellite radiance measurements alone, it
    is not possible to infer atmospheric structure of smaller
    scale than a few kilometres in the vertical. However,
    statistical information about the atmospheric temperature
    profile is available which if utilised in the retrieval
    process enables a solution to be obtained for the most likely
    temperature structure consistent with the observations,
    which includes such smaller scale structure.

SESSION 4

SPACELAB FACILITIES FOR ATMOSPHERIC RESEARCH

SPACELAB AS AN ORBITING ATMOSPHERIC LABORATORY

J.J. Burger

Space Science Department, ESA
ESTEC, Noordwijk, The Netherlands

Spacelab is designed as a re-usable, general purpose laboratory, which will be flown to and from space in the cargo bay of
the Space Shuttle's orbiter. Once in orbit it will offer its
support to the many space experiments proposed already in a variety
of different disciplines.

The above concept makes Spacelab also attractive for investigations of the earth atmosphere. This paper discusses the resources which are available to this end, as well as some constraints
which have to be observed in using Spacelab.

Introduction

Past experience has clearly demonstrated the benefits and
scientific returns from space research. The view on the earth
with a global coverage, unobscured viewing to the sun and space,
access to the space plasmas and finally the cancellation of the
effects of gravity represent features of space which are unique
for research in many disciplines. To provide a more cost-effective
access to space for exploiting these features with the increasingly
more ambitious equipment foreseen, NASA is developing the Space
Transportation System.

At the same time it is well recognised that on the ground
one of the basic drives for a cost-effective progress in science
and technology is provided by laboratory research. Here, the
competent investigators are provided with adequate equipment to
carry out their experiments in a favourable environment, operating
with their instruments and observational data in a close feed-

back loop. To allow this laboratory approach to take place in a
cost effective way in space research as well is the prime aim of
Spacelab, which ESA is developing for augmenting the capabilities
of the Space Transportation System.

The Spacelab Concept

    The large variety of experiments foreseen in several completely
different disciplines has led to the concept of a general purpose,
modular and re-usable space laboratory where the following two
basic conditions are met:
(1) provision of adequate working conditions for experimenters
     (men and women), which naturally implies a habitable atmo-
     sphere and environment with easy access to the experiment
     equipment and
(2) provision of adequate support to the experiments equipment,
which is based on an efficient usage of general supplies, and
which permits an easy integration of this equipment into Spacelab.
Also the need for special experiment equipment designs is minimised
as much as practically possible, by standardizing the interfaces
to the resources which are quite ample for space research and
provided in a benign environment.

    The approach selected for satisfying the two above conditions
best can be compared with the approach applicable in general on
the ground. First, as ground-based laboratories draw resources,
like basic power and communication links, from general public
utility services, so Spacelab uses the several general utilities
provided by the orbiter. Then the Spacelab offers all general
type support, like power conditioning and distribution, air conditioning, structural support, etc., which is usually available in
ground laboratories. In addition Spacelab provides several commonly
needed special support items. Finally, the experiment specific
equipment in Spacelab has to be provided by the experimenters.

    Operationally, the Spacelab concept is depicted in Figure 1.
Pre-integrated equipment of Spacelab users is integrated with the
Spacelab, being assembled from the several modular elements which
constitute an optimum for that specific mission. Subsequently
the fully integrated Spacelab is loaded in the Orbiter, launched,
flown and after fulfillment of the mission, landed again. The
Spacelab and payload are then unloaded, and, as necessary, returned
to the experimenters, refurbished and/or prepared for a next flight.

Shuttle Support to Spacelab

    The prime service from the Shuttle to Spacelab is the trans-
port to and from space. The actual orbits which can be achieved

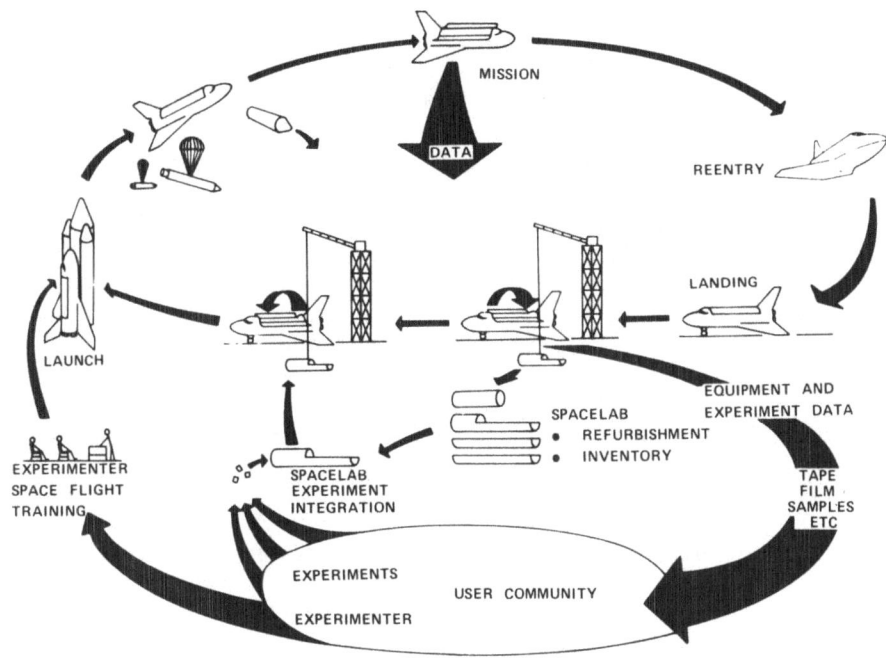

Figure 1 : Shuttle-Spacelab Operational Profile

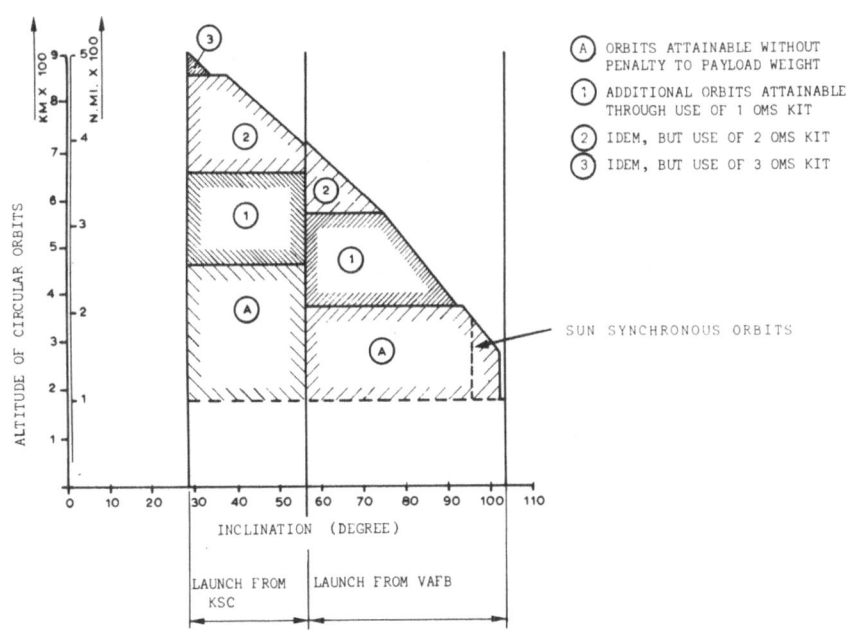

Figure 2: Typical Ranges of Circular Orbits
For Spacelab Missions

are of course, a multi-parameter function, the details of which
can be found in references (1) and (2). For Spacelab missions,
with a maximum landed weight of 14.515 kg, Figure 2 shows a typical
range of circular orbits attainable. The altitudes are given as a
function of inclination, where it is assumed that launches take
place from Kennedy Space Centre for inclinations between 28° and
57°, and from Vandenburg Air Force Base for inclinations between
56° and 104°. The effect of adding Orbiter Manoeuvering Subsystem
(OMS) kits, which reduce the allowable payload launch weights, but
not the basic 14.515 kg land weight, can be read from the Figure.
The orbit ranges achievable appear quite reasonable for atmospheric
investigations.

The services provided by the Orbiter to Spacelab in orbit are
schematically illustrated in Figure 3.

First, the basic pointing is provided by the Orbiter, with an
accuracy of roughly half a degree. Any pointing direction can
be achieved, but a restriction may in particular cases arise from
the thermal conditioning of the system which may require that a
pointing direction cannot be maintained continuously.

Secondly, the primary power for Spacelab is delivered by the
orbiter, together with the associated heat rejection capability.

Also, the communication system of the orbiter is used for
communications between Spacelab and the ground or subsatellites.

The living quarters for the crew, with all accommodation
features required, are located in the orbiter. In addition, the
orbiter provides some resources (volume, power, etc.), for usage
by payload equipment in the so-called payload specialist station
located in the orbiter aft flight deck.

Finally, the Spacelab experimenters may use a remote manipulator
system which can be operated from the orbiter cabin, for remote
access to experiment hardware outside the Spacelab module, thus
avoiding, when possible, the need for the crew to go for extra-
vehicular activities.

Further details on these orbiter provided resources can be
found in (1) and (2).

General Laboratory Support by Spacelab

Spacelab consists of two basic elements, the habitable
pressurized compartment (module) and an unpressurized equipment
mounting platform (pallet), (Figure 4). Module and pallet elements

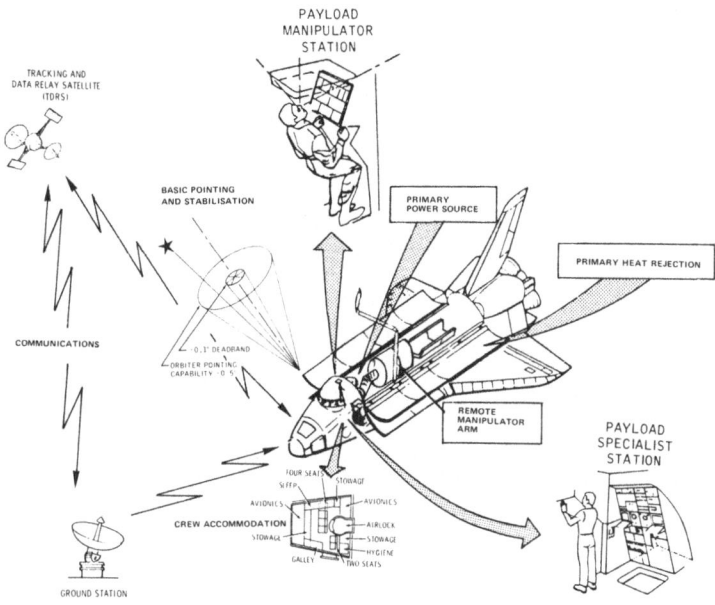

Figure 3 : Shuttle-Support to Spacelab

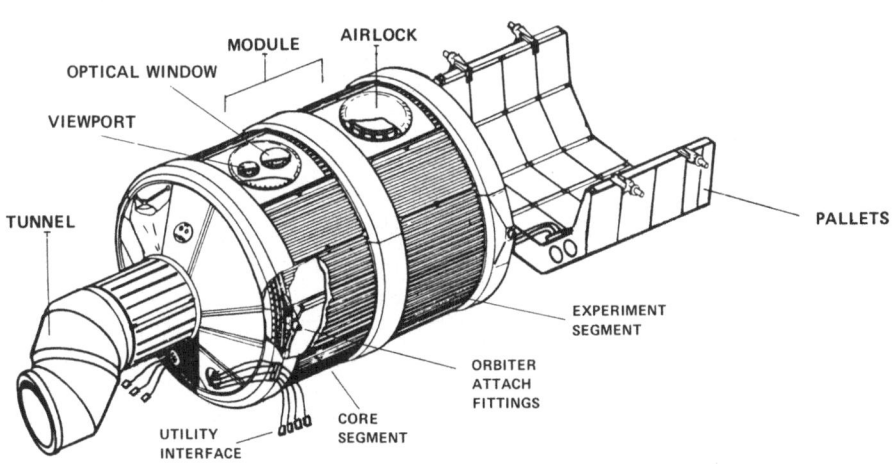

Figure 4 : Spacelab External Configuration

can be flown separately or in various combinations.

The module consists of cylindrical pressure shell segments
and cone-shaped end closures. Inside the module an oxygen/nitrogen
atmosphere is maintained at 1 bar pressure and within 1°C from
any present value between 18°C and 27°C. The air system provides
humidity and contamination control, thus reducing the material
control requirements to be imposed on experiment equipment. The
life support system in the module is sized to accommodate up to
four payload specialists per mission.

In addition to this, the Spacelab environment control system
comprises cooling capabilities for module and pallet equipment.
Inside the module the instruments will be placed inside racks
(Figure 5), in which the instruments will be cooled by forced air.
Equipment with large localised heat production can be interfaced
with an experiment dedicated liquid/liquid heat exchanger. On
the pallets experiments can be cooled either passively or actively
by interfacing to cold plates, which are part of a freon coolant
loop. The total cooling capability available to Spacelab and its
payload amounts to 8.5 k watts continuously and 12 k watt peaks
during 15 minutes every 3 hours.

Spacelab's volume is limited by the orbiter cargo bay. The
diameter of the module measures 4 metres and the length can be
chosen to be 4.2 or 6.9 metres. The volume available for experiment equipment amounts to 7.6 m³ in the short module and to 22.2 m³
for the long module configurations. This volume is available
inside standard 19" wide racks, on the centre aisle, under the
module floor and inside storage containers at the ceiling.

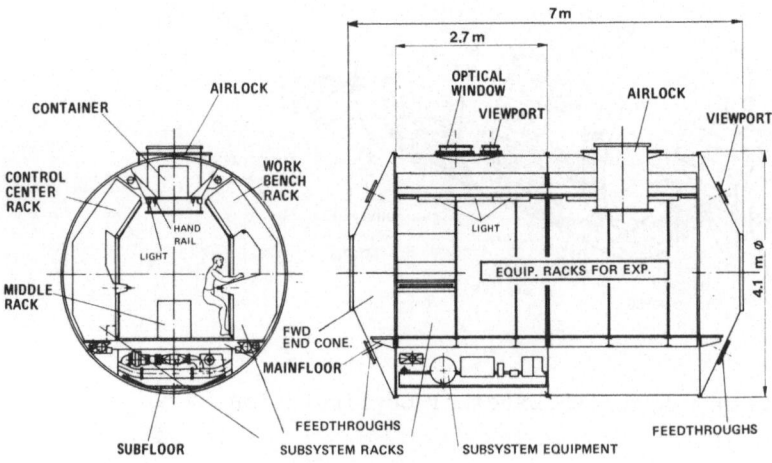

Fig. 5 Sectional views of Spacelab Module

Each pallet element is 3 meters long, offers a payload volume which can be approximated by a 4 meter diameter payload envelope over the full pallet length, and a surface mounting area of roughly 17 m^2. The total length for mounting payload equipment runs up to 15 meters for the pallet-only mode.

Depending on the Spacelab configuration a total payload mass between 5.500 kg and 9.100 kg can be carried and landed. The payload mass can be distributed as in nominal ground-based laboratories. For instance the module ground floor can carry loads of 500 kg/m^2. In the racks equipment can be installed up to 300 kg/m^3. The total load carrying capability of the pallets and module exceeds 1000 kg/m length.

The electrical power and distribution system conditions the primary power from the Orbiter (7 k watt maximum continuous and 12 k watt peak (up to 15 minutes per 3 hours)) for distribution to Spacelab subsystems and experiments at a nominal voltage of 28 volts DC and 115 volts AC at 400 Hz. The power available on-orbit for experiments and some mission dependent Spacelab subsystems (incorporation of which is at the discretion of the experimenters) amounts to 3.7 - 5 k watts continuously and 7.7 - 9.5 k watt peak, dependent on the Spacelab configuration. The power buses are routed to all experiment racks and pallets.

The standard electrical energy available to Spacelab is 890 KWh. For 7-day missions this implies that, dependent on the Spacelab configuration, 420 - 590 KWh is available to experiments and mission dependent Spacelab subsystems.

Naturally, the Spacelab design includes like nominal ground laboratories facilities for storage of loose equipment (film, tapes, tools, etc.) and a working area for the crew.

Special Laboratory Support by Spacelab

Apart from the general support to experiments described in the preceding chapter, Spacelab will offer some additional services which are not quite common to nominal ground laboratories. These include the command and data management system (CDMS) and some special common payload support equipment.

The Spacelab CDMS has for experiments three main functions. First, the CDMS will collect, format and merge the data from all simultaneously operating experiments into a single data stream, with a data rate capability up to 30 megabits/sec. These data will be transferred to the orbiter communication system. A second task of the CDMS is to command, control and monitor Spacelab subsystems and experiments, which includes "caution and warning" for

non-nominal situations. Thirdly, the CDMS offers data processing
and compression capabilities.

A block diagram of the Spacelab CDMS is shown in figure 6.
Experiments will interface with the CDMS through "Remote Acquisition Units" (RAU's) which are available in all racks and on the
pallets. They offer a bi-directional link between CDMS and experiments for acquisition of digital and analog data and for com-
mands. They also provide accurate time-signals. The RAU is of a
modular design, so that hte number of channels can be adapted to
the experiment needs. The RAU's are connected through a 1 mega-
bit/sec data bus to the CDMS computer system, which controls the
CDMS operations. Three identical computers are available, one for
experiment operations, one for Spacelab subsystems control and
one as a back-up for either of the two computers in case of fail-
ure. All computers have a 64 K core memory of 16-bit words and
can handle 350.000 operations/sec of the so-called Gibson-mix
instruction set. A mass memory unit, CRT's and keyboards are
standard peripherals to the computers.

Acquisition and merging of data from experiments with data
rates between 64 k bits/sec and 16 Mbit/sec can be achieved by
means of a high rate digital multiplexer. The output data stream
(up to 32 Mbits/sec) will be channelled either directly in to the
orbiter ku-band communication system or, into a high rate digital
data recorder used as an intermediate buffer. The recorder speed
can be selected at 6 different rates between 1 and 32 Mbits/sec,
according to mission needs. Tape exchange will be possible as well.

The CDMS system will be programmable in high order languages.

In summary, the Spacelab data handling can be distinghuised
in three regions. At experiment data rates below say 50 kbits/sec
the CDMS will be capable to continuously process the data, provided
that the actual processing time matches this rate. At higher experiment data rates and up to roughly 300 k bits/sec the data
can be acquired and routed to the ground via the CDMS RAU's and
experiment data bus. At still higher rates and up to 30 Mbits/sec
the CDMS multiplexer shall be used for the data acquisition.

Another special facility incorporated in the Spacelab development is a one meter diameter, one meter long airlock, which can
be accommodated in a top opening in the module. The airlock is
equipped with an extendable experiment platform to permit attached experiment to be exposed into space. Also viewports and an optical quality window are provided as mission dependant equipment.

Those experiments on the pallet which require better pointing
accuracy and stability than is provided by the orbiter, can use

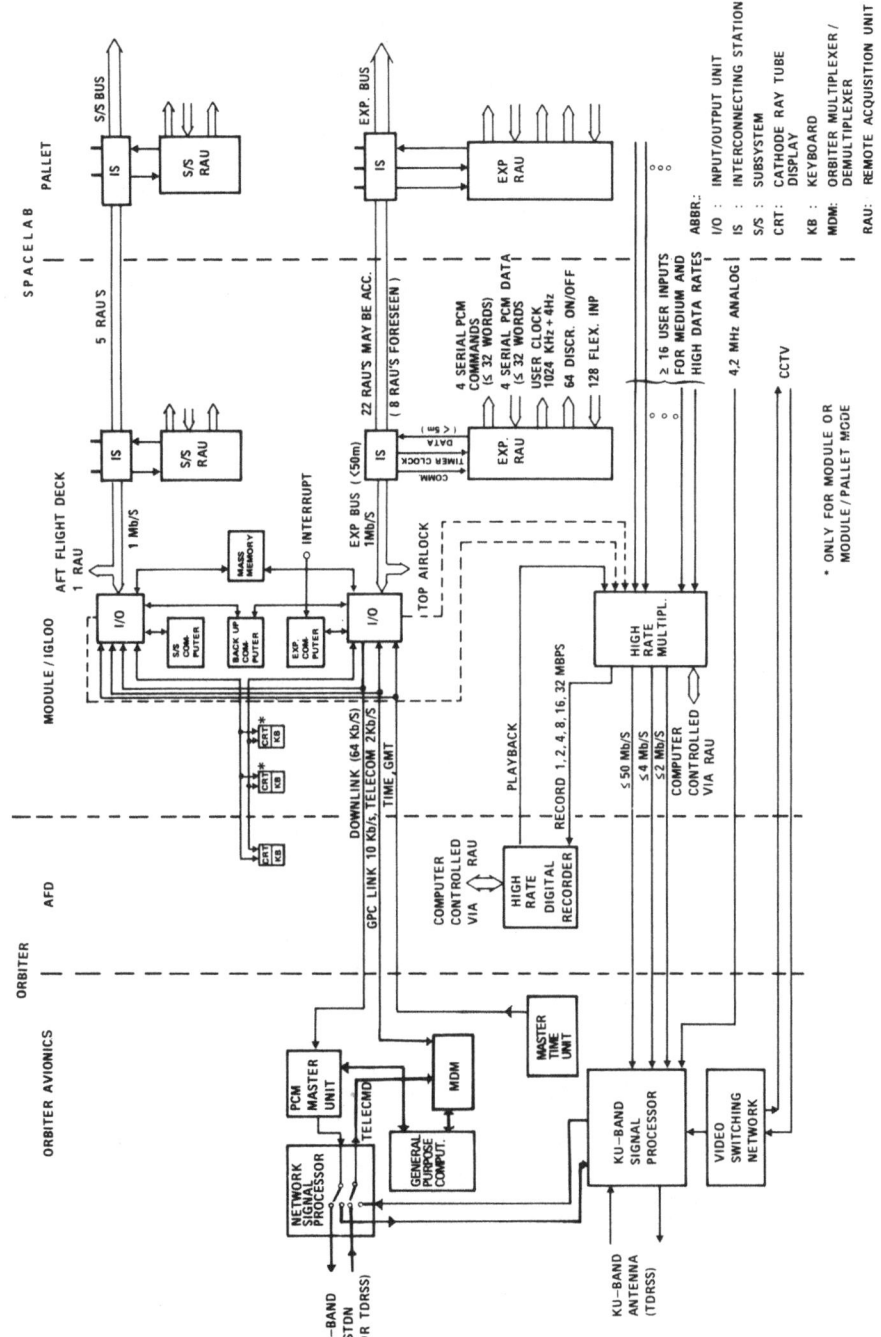

Figure 6 : CDMS Block Diagram

the Spacelab Instrument Pointing System (IPS). The IPS contains
a three-axis gimbal system (azimuth-, cross-elevation and elevation-
axis) as illustrated in Figure 7. It is mounted to the floor of
a pallet by means of a soft mount consisting of springs and dampers,
which reduce attitude disturbances caused by orbiter or crew move-
metns. Different sizes and types of payloads can be easily accommodated and attached in-orbit to the payload integration ring. Interfaces to the SL CDMS and electrical power system are available for
payload usage.

During launch and return, the IPS soft mount is inoperative
and locked by a soft mount clamp. During this phase the payload
is physically separated from the IPS by a payload/gimbal separator
mechanism and the payload is supported by a payload clamp assembly,
which distributes the loads to the pallet.
On-orbit the 700 Kg heavy IPS will be able to point payloads up to
more than 3000 kg and a diameter up to 3 meters with arc second
pointing accuracy and stability.

Spacelab Constraints

All Spacelab provided resources have their limitations, which
have to be observed in planning Spacelab missions. However, the
indications are that the available support will generally be quite
sufficient for experiment accommodation, especially when compared
with present-day space systems. In this situation obviously

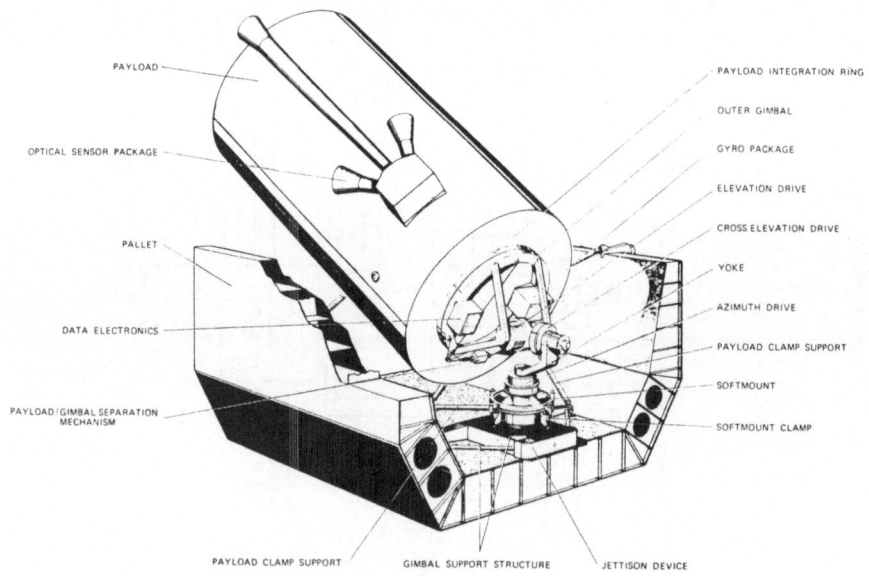

Figure 7 : Instrument Pointing System

the most significant constraint for atmospheric research in Spacelab will come from the limited duration of Spacelab flights.

The nominal on-orbit stay time of the Orbiter/Spacelab is
seven days. Extended missions can be achieved by adding consum-
ables and their containers, primarily needed for the crew life
support system and for electrical energy. The overall impact on
Spacelab payloads for extending Spacelab flights has been analysed under the assumptions of a total crew of four, and a power
consumption of 12.5 kwatts by the Orbiter and 7 kwatts by Spacelab and its payload. The most important results are given in Table
1, which shows the relation between maximum landed payload weight
and mission duration for three Spacelab configurations. Although
these data are preliminary, it is obvious from the table that the
reduction in payload mass becomes quite significant when the
Flights extend beyond 12 days. This is primary due to the roughly
400 kg landed weight added by each energy kit providing 840 kwh
of energy. Although several possibilities exist for improving
this situation (e.g. jettison of empty fuel cells) it appears
probable that the limited flight duration will remain Spacelab's
most significant constraint in the early 1980's.

SPACELAB CONFIGURATION	LONG MODULE	SHORT MODULE + 9 MTR PALLET	PALLET ONLY (9 MTR PALLET)
nominal payload mass =	5500 kg	5500 kg	9000 kg
mission duration =	13 days	9 days	12 days
5000 kg payload mission duration =	15 days	12 days	26 days
30 day mission payload mass =	400 kg	--	3600 kg

TABLE 1: The Relation of Payload Mass and Mission Duration for:
- 4 Crew Members
- 7 kwatts Spacelab Power and 12.5 kwatt Orbiter Power

Concluding Remarks.

The concept and design of Spacelab offers an approach to
Space investigations which is in many aspects similar to the
approach which is common in ground based research laboratories.
This will lead to a considerably more cost-effective and less
complex exploration of space in many disciplines.

Also for atmospheric physics Spacelab appears to offer
good prospects. The achievable orbits are quite suitable and the
general support to experiment equipment is quite adequate.
Although free flying spacecrafts will remain needed for monitoring and long-duration observations it is envisaged that the
use of Spacelab can significantly contribute to a cost-effective
progress in our understanding of earth atmosphere.

REFERENCES

(1) Spacelab Payload Accommodation Handbook, issued by ESA.
    Ref. No. SLP/2104, May 1976.

(2) Space Shuttle System Payload Accommodation, issued by NASA,
    Ref. No. JSC-07700, Vol. XIV, November 1975.

DISCUSSION

A. Nagy: What is the size of the planned pointing system?

J.J. Burger: The interface ring to experiments has a 1 metre
    diameter. The weight of the system is 750 kg. The
    Instrument Pointing Subsystem (IPS) is a piece of mission-
    dependent equipment, which can be flown as and when
    required.

THE DESIGN OF A SPACELAB LIDAR FACILITY

D. Dale[x]

Systems Studies Office,
European Space Research and Technology Centre,
Noordwijk, The Netherlands.

ABSTRACT

The LIDAR instrument for active atmospheric top-side sounding will
be a facility offering a basic capability over a range of wavelengths (2000 A to 10.6 µm). The core of the facility is a 1 m
class telescope (10-3/10-4 rad FOV) which in early missions will
be hard-mounted to a Pallet segment; later missions (1983/84)
will see the inclusion of a telescope rocking system for rocking
in a plane perpendicular to the velocity vector. Accommodation
of a variety of lasers (up to two per flight), with operating
wavelengths anywhere in the indicated range, and input power
levels up to about 2 kW average will be possible. For the early
missions, there may be the possibility of mounting the laser(s)
inside the module, with the laser being Pallet-mounted with the
telescope for later missions. Various detector systems can be
used with the LIDAR, with up to three different units on any one
flight: these will be Pallet-mounted with the telescope.

Other capabilities offered by the facility include automatic
optical alignment (transmitter/receiver/detectors), thermal
control of telescope lasers and detectors, contamination control
and control electronics.

---

[x] Presented by G. Haskell

J. J. Burger et al. (eds.), Atmospheric Physics from Spacelab, 215–235. All Rights Reserved.
Copyright © 1976 by D. Reidel Publishing Company, Dordrecht-Holland.

1.1   General

       A SPACELAB-borne (Light Detection And Ranging) has
been under phase A study for about 7 mths. The study, under ESA
sponsorship, was performed by CNES as prime contractor with MATRA
and CNRS as sub-contractors. The LIDAR technique involves stimula-
tion of atmospheric phenomena by use of a relatively high powered
laser and opticaly observing the causal effects via a receiving
telescope and optical detectors (see fig 1). The baseline study
philosophy was design and procurement of a facility instrument
which could be orbited on many SPACELAB missions; different
scientific aims could be achieved by accommodation of a variety of
laser stimuli and corresponding detector systems. Thereby the
basic instrument would offer both flexibility and growth potential.
Model scientific objectives assumed for study purposes are contained
in table 1, and serve to illustrate the variety of laser/detector
systems to be accommodated.

       Listed below are the major requirements placed on the
LIDAR facility.
       (A)   TELESCOPE   - 2000A to 10,6 μm.
                           - Diffraction limited at 10,6 μm.
                           - Dia 1m class.
                           - Compatible with detector F.o.v 10^{-3}/
                              10^{-4} rads.
                           - For later flights telescopes to be
                              rocked perpendicular to vel. vector.

       (B)   LASER SUPPORT SUB-SYSTEMS
                           - Capable of accommodating Dye, Nd-Yag,
                              Ruby and Gas lasers.
                           - Capable of accommodating lasers with
                              input powers fo up to 2KW.
                           - Provision of all laser management
                              and safety control functions.
                           - Provision of standard Spacelab/alser
                              interfaces.
                           - Accommodations of lasers either
                              inside the module or pallet mounted.

       (C)   DETECTOR SUPPORT SUB-SYSTEMS
                           - Accommodation of a variety of
                              detector packages up to 50kg mass.
                           - Provision of standard detectors/tele-
                              scope interface.

2.2   Configuration Concepts

       SPACELAB system offers the possibility of configuration
flexibility for payload instruments. In the case of the LIDAR

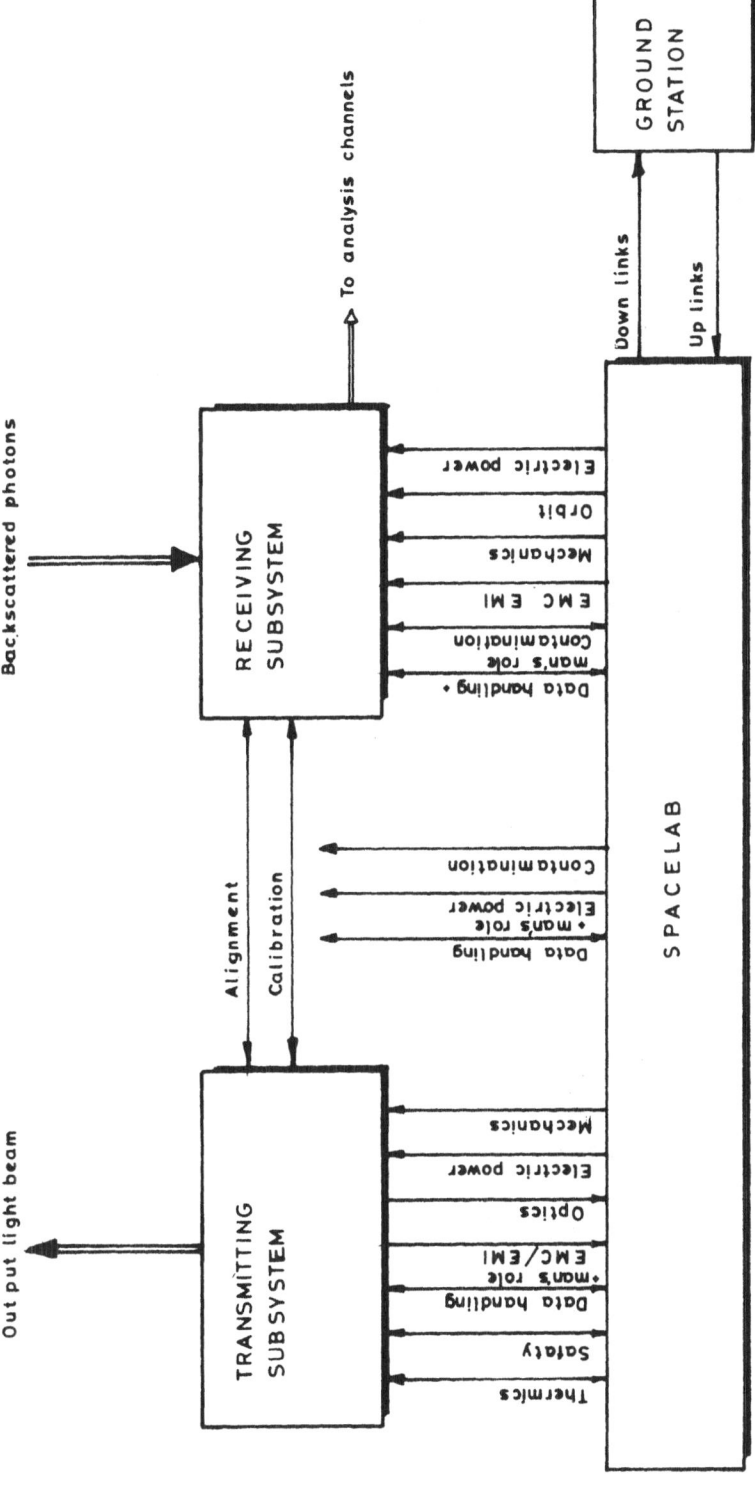

Figure 1. LIDAR functional scheme

Table 1

SCIENTIFIC OBJECTIVES AND ASSOCIATED LASERS

Flights	Scientific objectives	Laser
F.S.L.P.	a) Aerosols concentrations	Nd-Yag (F + SH) or Ruby (F + SH)
	b) Sodium concentrations and Mesopheric temperatures	R.6.G)F)
Subsequent Flights	a) Consolidation of results from first flight	Same as F.S.L.P.
	b) Improved measurements of aerosols	Dye pumped by Nd-Yag (or two lasers)
	c) Measurement of winds	TBD
	d) Measurement of minor gaseous constituents	
	α) resonance fluorescence methods	
	. K	DOTC (F)
	. Li$_+$	Cresyl violet (F)
	. Ca	DOTC (SH)
	. Fe	DOTC (SH)
	. OH	R6G (SH)
	β) differential absortpion methods	
	. O_3, Cl O, SO_2	R6G (SH)
	. H_2O	Ruby or DOTC (F)
	λ Infra red range	CO2 - CO

N.B.: F = fundamental
       SH = second harmonic

having recognised that the receiving telescope must be pallet
mounted, there remains the question of mounting lasers and detectors either inside or outside the module. A comprehensive study
of all options has shown that the most promising configurations
are as illustrated in Figs 2a and 2b. Because of alignment
problems with the Receiving Telescope output it has been demonstrated that detectors should be co-located with the receiving
telescope whilst laser(s) could be either inside or outside the
module. Potentially having the laser inside the module offers
the capability for crew access which could be attractive for
more advanced laser types where reliability is uncertain. However
this configuration, apart from engineering penalties of severe
safety control and provision of a dedicated aft-bulkhead optical
window, also reduces the LIDAR performance capability. Specifically, because of relative pallet (receiver telescope)/module in-
flight motions of up to 2 degrees, the minimum design LIDAR
detector Fov is 10^{-3} rads for this configuration. Essentially
this relaxed Fov compared to 10^{-4} rads for the all-pallet
configuration, increases background noise and if acceptable signal
to noise ratios and corresponding measurement operation is viable.
For these reasons the "all-pallet" configuration (Fig 2b) which
allows both night and day operations is considered to be the most
attractive at this time.

It should be noted that conceptually both configurations
are First Spacelab Payload (FSLP) compatible, providing accommodation incompatibilities with other payload elements do not exist.

For later missions beyond 1982, scientific objectives require
both maximum sensitivity and rocking of the transmitted and
received optic axis in a plane perpendicular to the orbital
velocity vector. For these reasons it was judged that the "all-
pallet" configurations would be required in this time scale. (see
fig 2c). Therefore included in the LIDAR design is the capability
to grow from a module laser configuration to an "all-oallet"
configuration without major modification.

For early flights (no LIDAR rocking) the system is hard
mounted to the pallet and can overhang to minimise pallet area
resource requirements (approximately 22% of 1 plalet area). For
the later flights (with LIDAR rocking) centre pallet mounting is
essential utilising and about 28% of the one pallet is required;
this is increased to 50% when a 45deg swept rocking is included.

2.3 LIDAR Facility Design

Telescope

A comparative analysis between Newtonian and Cassegrain

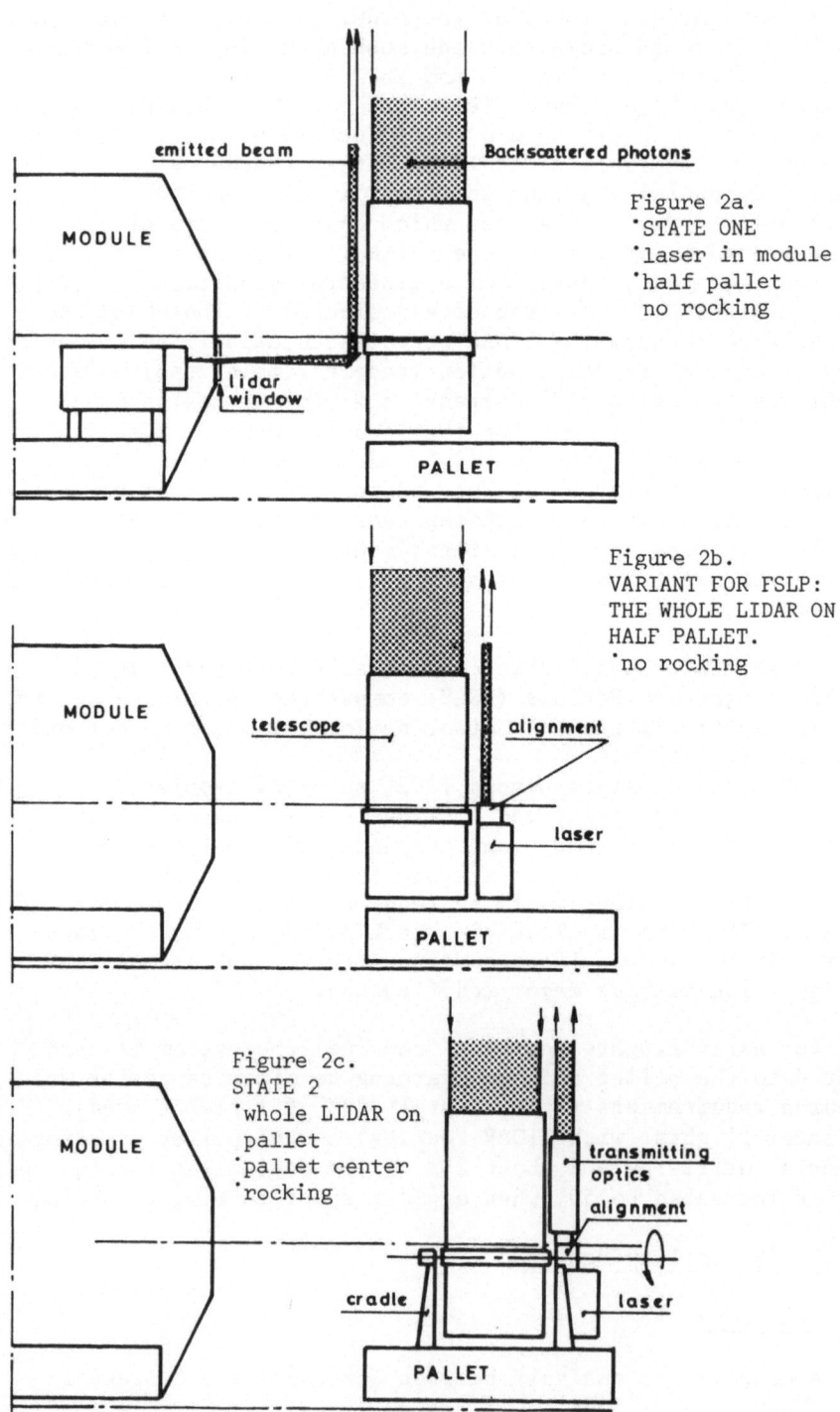

Figure 2a.
·STATE ONE
·laser in module
·half pallet
 no rocking

Figure 2b.
VARIANT FOR FSLP:
THE WHOLE LIDAR ON
HALF PALLET.
·no rocking

Figure 2c.
STATE 2
·whole LIDAR on
 pallet
·pallet center
·rocking

telescopes has been made and summary conclusions are included in table 2. For Space-borne LIDAR applications it was a conclusion that a Cassegrain (Dall-Kirkham) configuration was the most attractive; main factors were lower number of anciliary optical surfaces, less sensitivity to thermal distortions, easier integration of experiment packages and cof g. close to mounting axis.

A schematic dimensioned illustration of the telescope configuration (using Alshell + Invar rods structure) is included in fig 3. The primary mirror (of 1m dia class) may be either lightened Zerodur (50% lightening factor) or U.L.E. construction with F-number of F/2 and surface tolerance is compatible with diffraction limited operation at 10,6μm. The secondary mirror can be driven along the optical axis to compensate for in-flight thermal defocusing. For the telescope structure a number of options have been pursued:

(i)   Invar tube shell with aluminium rings

(ii) Aluminium tube shell with invar secondary mirror motering rods.

(iii) A carbon fibre reinforced plastic (CFRP) shell

Summary mass budgets for the three configuration options are given in table 3, and as expected option (iii) involves least mass (305 kg) with the more traditional structure of option (i) consumming the highest (345 kg). Choice between the various options will be based on EMC considerations (because of Invar magnetic properties) and cost (higher development cost will CFRP material): this is considered a follow-on phase task.

Thermal design includes a completely insulated concept using multilayer insulation, and utilises aperture doors when the telescope is not Earth pointed: these doors also minimise contamination particle deposition during non-operational phases.

Transmitting System

The transmitting system essentially consists of a LASER and transmitting optics. Requirement for the latter depends on the laser divergence which should be controlled to $<5 \times 10^{-5}$ rad (Det Fov 10^{-4} rads). With laser divergence better than 5×10^{-4} rads a suitable magnification factor of 10 is considered appropriate. To cover the total wavelength range of 2000 A to 10,6μm it is necessary to consider tho transmission optical systems:

Table 2   <u>TELESCOPE OPTICAL COMBINATION CHOICE</u>

Optical combination class	NEWTON		CASSEGRAIN	
Detector F.O.V.	10^{-3}rd	10^{-4}rd	10^{-3}rd	10^{-4}rd
Optical considerations				
Primary F number	F/3	F/3	F/2	F/2
Relay-Optics need	Yes (4 mirrors)	Yes (4 mirrors)	Yes (2 mirrors)	No
Total mirrors number (with telescope ones)	11 (	12 or (14) (depends on mirrors number of the collo-mating optics)	8	7 or (9) (depends on mirrors number of the collo-mating optics)
Global optical transmission (Out of occultation and window)	~0.63	~0.61 (0.56)	~0.72	~0.75 (0.69)
Linear global occultation	~0.4	~0.4	~0.45	~0.25 to 0.33
Primary mirror manufacturing	Easy	Easy	More difficult	More difficult

Table 2 (Continued)   TELESCOPE OPTICAL COMBINATION CHOICE

Mechanical quantitative considerations (Primary ring referenced)				
Focus depth at telescope focus	±0.9 mm	±0.09 mm	±5.7 mm	±0.57 mm (mirror spacing 40 μm)
Refocusing system need	Yes	Yes	No	Yes
Lateral tolerance at telescope focus	± 0.9 mm	±0.9 mm	±2 mm	±2 mm
Mechanical qualitative considerations				
Refocusing actuation	At relay-optics level	At relay-optics level	N.A.	At secondary mirror level
D.P. set-up and compatibility with rocking	Can be difficult		No problem	
Ground integration	Difficult (accessibility to SO and DP)		Easier	
Mechanical reference and tolerance trade-off	Difficult		Easier	
Centering	Cd G far from rocking axis		Cd G near to the rocking axis	
Load resistence Thermal control	Structure with appendagues		Simpler structure	

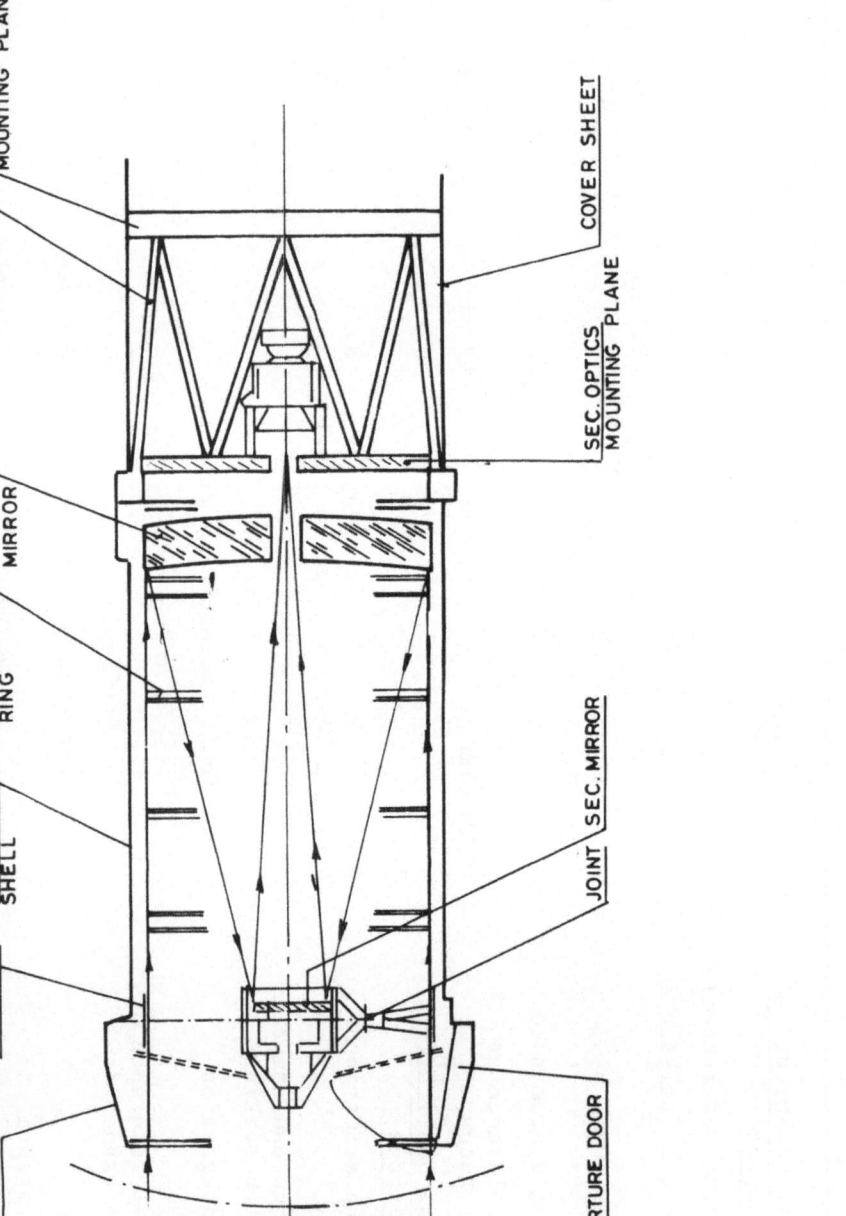

Figure 3. Overall telescope design

Table 3

Telescope weight in KG

	INVAR shell + Alu rings	Alu shell+ INVAR rods	Carbon fiber
– Structure	155	138.5	118
– Primary mirror	150	150	150
– Secondary mirror	2	2	2
– Additional optical units (corner reflector)	2	2	2
– Harness	5	5	5
– 10 % development margin	32	29.5	28
TELESCOPE WEIGHT	346 KG	327 KG	305 KG

System (1) (fig 5a) a confocal mirror system provides
wavelength coverage of 2000 A to 1.06 μm.

and System (2) (fig 5b), lens systems providing coverage in
the IR range ⁻1.06 μm to 10.6 μm.

The optical diameter is sized at about 300 mm. Concerning
the design of the laser support sub-system facilities, the
intention was to maximise the number of common laser support
equipments (thermal, alignment, management etc) that could be
provided by a central Agency, and at the same time would be
compatible with a variety of laser types, having electrical
input powers up to about 2kw. Table 1 summarises a sample
of typical lasers candidate for the LIDAR, and illustrates the
range of parameters likely to be experienced.

Schematically represented in Fig 4, is the conceptual laser
support facility design. The exlposive proof canister is pressur-
ised to 0.5 atmos with dry N_2 and internal separation compartments
preclude man exposure to high voltages or toxic elements, which
is an essential safety design feature. The canister is manufac-
tured from Aluminium primarily for purposes of shielding and
EMC/EMI control, and can be used either in the module or in the
pallet. An attractive feature of the design is that for all
LIDAR flights, standard Spacelab interfaces are maintained with
variation of laser head interfaces (because of the desire for
laser change) taking place inside the canister unit.

Laser thermal control (fig 6) includes:

- Cold plate for P.S.U. cooling

- Pump-filter

- Interface with S.L. Expt heat exchanger

- PCM (Phase Change Material) and by-pass valves.

- Cooling circuit to cool laser flash head.

The unit can operate using either water (module) or freon
(pallet) cooling fluid, and is designed to provide laser head
temperatures of 40^0C. Unfortunately Ruby lasers require temperature of 19^0C which therefore requires an additional refrigeration
machine interface: the space application feasibility of this machine has yet to be demonstrated.

Although the optical platform has been sized to accommodate anticipated laser types, in the case of the dye laser large

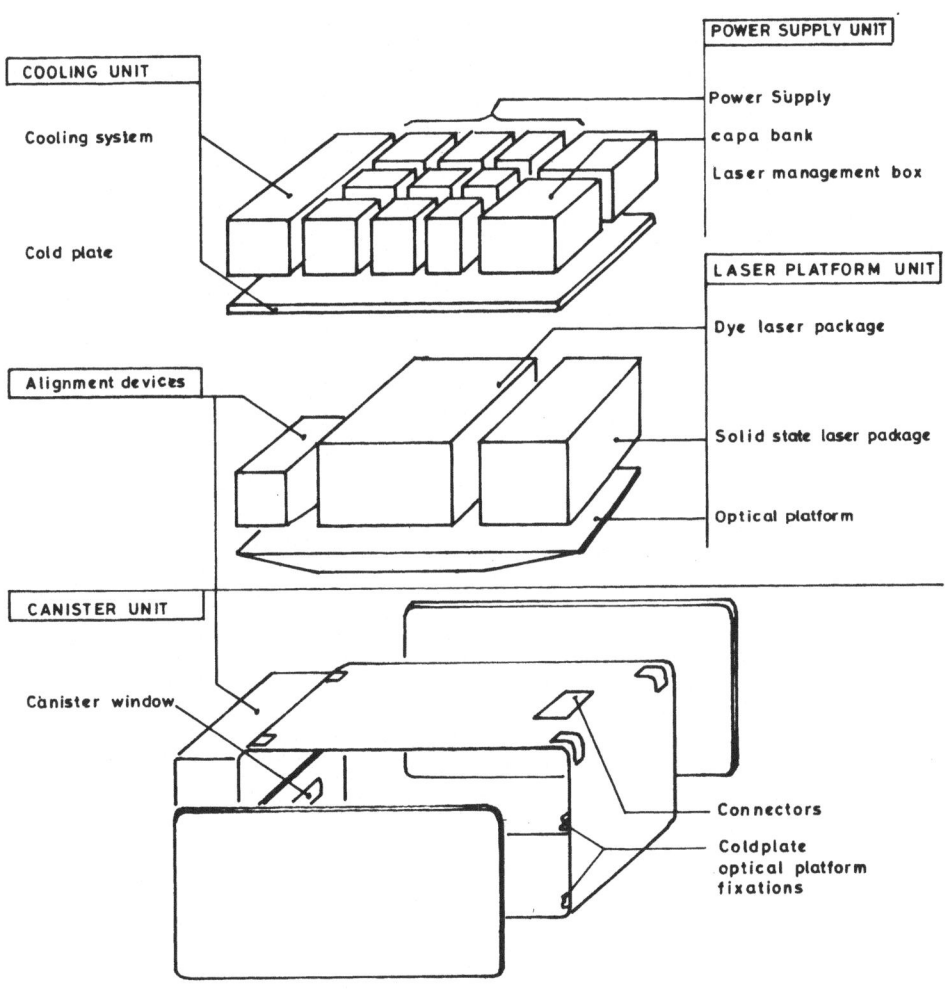

Figure 4. Typical design of the laser subsystem

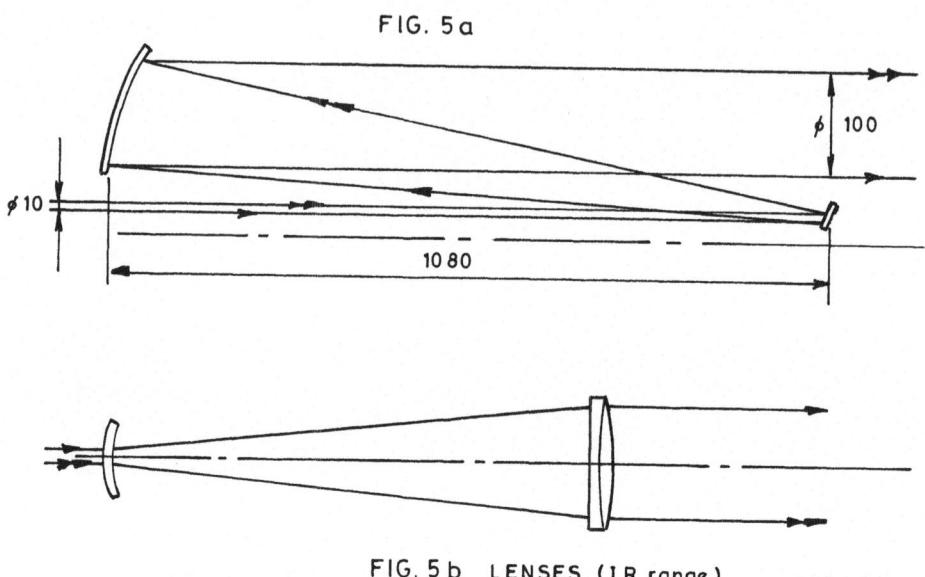

FIG. 5a

FIG. 5b   LENSES (IR range)

	Mirrors optics	Lenses optics
Spectral range	200 nm $< \lambda <$ 1060 nm	Associated to each IR wavelength
Magnification	10	10
Angular admittance	±20'	±20'
Diameter	~300 mm*	<300 mm
Overall height	~1200 mm	T.B.D.
Weight	<15 Kg	T.B.D.

TRANSMITTING OPTICS CHARACTERISTICS

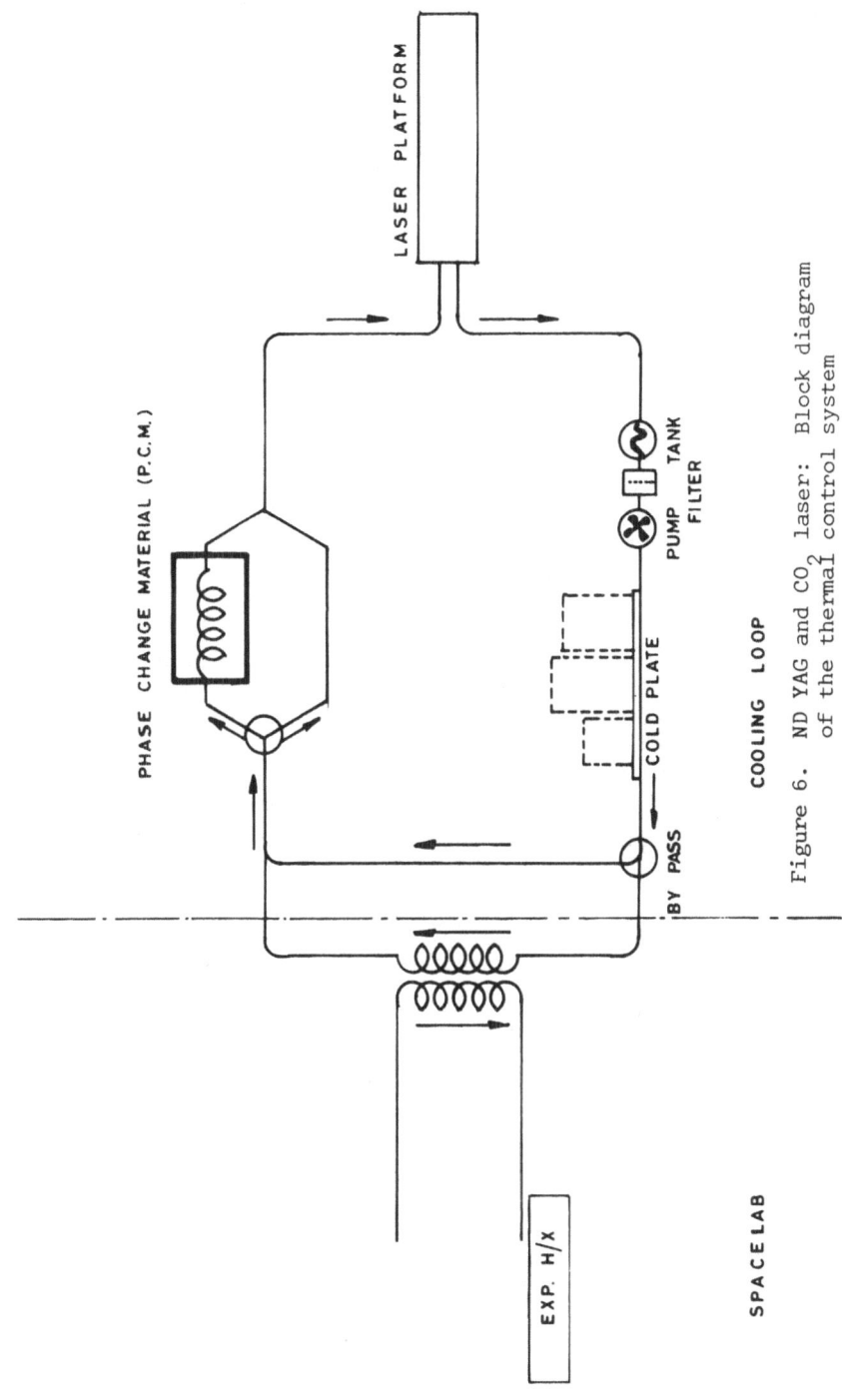

Figure 6. ND YAG and CO_2 laser: Block diagram of the thermal control system

dye volumes in the range of 100 - 200 litres for 10^{-6} shots, would
have to be accommodated outside the canister unit. Because of
the toxic nature of the dye, connections between the canister
and dye tank are double walled hoses with double sealed connectors. It could be possible however for the module laser configuration for the crew to interchange dye tanks in-flight with a
view to changing laser emitted frequency.

Basic Laser support facility items are considered to be:

- canister unit (with optical window and connectors)

- optical platform

- alignment device

- thermal control sub-system

- laser management/control

- safety engineering.

The estimated mass of these facility items is approximately 70 kg.

Detector Package Accommodation

Location of detector package compartment is below the
receiving telescope primary mirror (fig 3) and a volume of 0.6 m^3
is available. The detector compartment is maintained at a
standard temperature of 20^0C, and capability for Peltier or
cyogenic cooling of elements is available. The structure of the
compartment is sized for detector package masses up to 50 kg, and
the possibility for optically switching between a variety of
detector units exists.

Detector package interface with the receiving telescoep is
via a standard secondary optics systems, which fulfills the
following functions:

- provides a collimated beam 2.5 cm dia (standard interface)

- provides a mechanical shutter to inhibit reception
  by the detector unit of both the ground echo and echo
  from the gaseous cloud surrounding Spacelab.

- provides fine alignment sensing (using ground echo).

- provides sensing for telescope refocusing system.

THE DESIGN OF A SPACELAB LIDAR FACILITY

Transmitter/Receiver Alignment

Automatic alignment will be provided using an additional
collimating light source for coarse adjustment and taking advantage of the ground echo signal for fine alignment For the case
of detector F.O.V. 10^{-3} rads (laser in module) a maximum operational misalignment error of $\pm$ 40 arc secs is acceptable whilst
with an equivalent FOV of 10^{-4} rads the misalignment error
requirement is $\pm$ 2.5 arc sec" this can only be achieved in the
all pallet configuration). Alignment actuators consisting of 2
angled mirrors are situated before the transmitting optics
whilst quadrant sensors are located in the secondary relay optics
system.

2.4   Interfaces with Spacelab

These interfaces are summarised in the following table.

ELECTRICAL	550 - 1470 W FSLP 2500 W Later Flights	EPDS boxes Module 2 max Pallet 1	Depends on selected objectives. Max capability
THERMAL	As above	Experiment Heat Exchanger Pallet cold plates low power FSLP	Primary thermal control req, is for laser.

MECHANICAL	Module Floor-FSLP Option	Module Floor + Pallet Hardpts	Laser Module option
	Half pallet-FSLP otpion	Pallet Hardpoints	requires aftcone optical window which would not be Spacelab Project provided.
	Centre pallet-Later Flights.	Pallet Hardpoints	
MASS	In the range 550 - 800 kg FSLP		Includes typical lasers, but not detector systems
CDMS	Max Expt Computer Core Memory requirement 34k. for LIDAR Control and management. Science Data Dumped on mag. tape.	RAI pallet or module	Possibility to provided limited scientific data processing in real-time for display purposes.
EMC/EMI	Below safety limits for S.L. sub-systems. Min. interference with other Expts.	Radiated + Conducted.	EMC/EMI to be controlled by special LIDAR design features.
SAFETY	No hazard to S.L. or Crew.	Mechanical Integrity + Laser Features	Safety control throughout project + special tests.

Mission Dependant Equipment requirements are:

Digital recorder

Expt. Computer

Expt. input/output unit

Expt RAU (1 RAU for all pallet configuration
         2 RAU's for module/pallet configuration)

Keyboard

Display Unit

Note that in addition to the above equipment the module/pallet configuration will require an aft and cone optical window. This equipment would not be Spacelab provided and therefore chargeable to the exepriment.

2.5 Spacelab Operations.

The 1IDAR will be only operated in Earth pointing mode and the operational programme from an instrument standpoint can be flexible providing the timeline constraints of Spacelab are not violated. Note that full flexibility in terms of both day/ night oeprations can be achieved by the "all pallet" configuration. Bearing in mind that the LIDAR turn-on sequence and de-activate sequence take about 12 mins and 6 mins crew time respectively, maximum time operational sequences should be planned whenever possible. Fig 7 is a general illustration of the operation time/ orbit to accomplish 10^5 or 10^6 laser shots. The total number of shots required is deepndant on scientific objectives see section (1).

Activation/control and de-activation of the LIDAR is exercised by the crew (payload specialist) via the Spacelab Expt Computer. All scientific data will be sotred on magnetic tape for dumping to the ground via the Shuttle Communications link. In addition there may be the capability to provide some limited scientific data processing and display on-board Spacelab if this is a strong Experiment requirement.

1.6 Conclusions

The phase A study has demonstrated the engineering feasibility of a Spacelab-borne Lidar. It has been shown that a suitable instrument can be built which will provide scientific flexibility and growth potential. Within the Lidar it is expected that the majority of equipment would be provided as facility items. This facility would be compatible with accommodation of a variety of laser heads and detector systems, which in turn provide flexibility of scientific objectives.

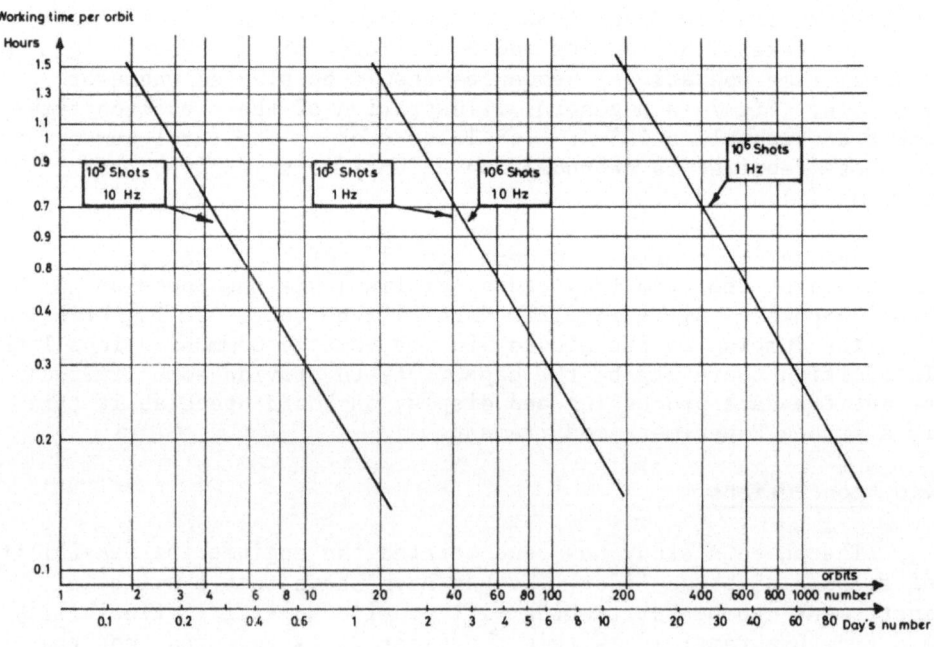

Figure 7. Next flights: relation between the working time
per orbit and the number of orbits for two values
of the laser lifetime (10^5 and 10^6 shots) and two
values of the repetition rate

DISCUSSION

S.A. Bowhill: How will the alignment be achieved to 10^{-4} rad?

G. Haskell: In two stages. Coarse alignment between the emitted
   and received beam is achieved by means of a supplementary
   low-power laser beam which is deflected from the transmitting
   telescope into the receiving telescope using a system of
   prisms. Fine alignment is performed by exploiting the strong
   return signal reflected from the Earth's surface or clouds.
   Alignment sensors are located at the secondary optics of the
   receiving telescope. Alignment actuators are located in the
   laser output beam.

J.T. Houghton: What is the status of the pointing facility?
   Will it be built for the first flight?

G. Haskell: The Instrument Pointing Subsystem (IPS) is being
   developed as part of the Spacelab Programme. It will not be
   available for the First Spacelab Mission, but is expected to
   fly on the Second Spacelab Mission.

SPACELAB-BORNE LIDAR - SCIENTIFIC OBJECTIVES
Volume I of ESA phase "A" study DP/PS(76)4

Lidar Consultant Group:

G. Fiocco, CNR, Frascati, Italy (1)
W. Renger, DFVLR, Oberpfaffenhofen, W. Germany
L. Thomas, Appleton Lab., Slough, U.K.

1. INTRODUCTION

This study follows in time two previous investigations related to a Spacelab-borne laser.

As in the previous reports, the study has considered a lidar device, to be used in the Earth-looking mode, mainly for measurements of the atmospheric properties.

The principal techniques under consideration are those of backscattering, by resonant or non-resonant processes, and those of differential absorption.

The main aims are to improve our knowledge of the structure, composition and dynamics of the atmosphere.

It is recognised, however, that the device will have applications to other geophysical areas such as geodesy, gravimetry and oceanography, and some applications are mentioned.

2. THE NEEDS OF ATMOSPHERIC RESEARCH IN THE 1980's

The future interest in the thermosphere will probably be in its interaction with the F region and magnetosphere. It seems likely that studies in these areas during the early 1980's will be well served by the Electrodynamic Explorer satellites, the EISCAT and other incoherent scatter facilities, and the use of interactive and chemical release experiments carried out with Spacelab, together with sub-satellites.

(1) Presentation of study at the Symposium.

J. J. Burger et al. (eds.), Atmospheric Physics from Spacelab, 237–254. All Rights Reserved.
Copyright © 1976 by D. Reidel Publishing Company, Dordrecht-Holland.

Although the region below 150 km will have been explored with the
Nimbus series of satellites by 1980, supplemented by rocket and
balloon - borne experiments, it will still represent the least
studied and understood part of the earth's neutral and ionized
atmosphere.

Perhaps the primary justification for an atmospheric research
programme is its relevance to environmental problems. The requirements for improved weather forecasting have been specified
within the Global Atmosphere Research Programme (GARP) and the
corresponding needs for safeguarding against atmospheric changes
which could lead to biological and climatological effects have
been identified. At the same time, a very active interest in the
basic science of the region below 150 km exists, and plans for a
coordinated approach are being formulated within the Middle
Atmosphere Programme, an international enterprise initialed by
the ICSU Special Committee on Solar Terrestrial Physics (SCOSTEP)
and endorsed by COSPAR, URSI and IAGA.

In order to consider the likely aims of atmospheric research
in the 1980's, it is convenient to consider in turn the areas of
Atmospheric Structure, Dynamics and Composition. Of course, much
of the interest and complication in atmospheric studies arises
from the interactions between these areas.

The most obvious need in the area of atmospheric structure
is for world-wide tropospheric measurements required in connection with numerical models being developed for meteorological
purposes. More specifically, the data is required for providing
initial conditions for such models and for testing their predictions. The accuracy requirements specified by GARP, namely $+ 1°$
in temperature and $+ 3\%$ in pressure are very demanding and are
unlikely to be met by 1980 or by experiments on the early flights
of Spacelab. For greater heights, interest will be shown in the
world-wide atmospheric structure and its relationship to sources
and sinks of energy; as an example, the relative importance of
the troposphere and the auroral region as source of gravity
waves needs to be established.

Furthermore, an examination of the transfer of energy and
momentum from one level to another will need information about
motions on a scale depending on the type of disturbance or wave
motion under consideration.

In this connection, the information obtained from the Nimbus
G and earlier satellites in this series will probably be adequate to study effects associated with planetary-scale waves, but
a better height resolution, about 2 km, and rather better horizontal resolution will be required for investigations of gravity
waves.

An improved understanding of the atmospheric energy balance awaits a better knowledge of atmospheric motions on all scale sizes and at all atmospheric heights. In addition, information on transport systems at specific height levels is required for two and three-dimensional circulation models which are being applied to composition studies.

A complete description of the wind field requires both the variation with height and position, and also the changes with time. The available information indicates that for the middle part of the atmosphere between about 10 and 70 km a height resolution of about 10 km for measurements of mean wind over horizontal distances of a few thousand km will be adequate. However, for measurements capable of resolving the marked variability at tropospheric heights and reproducing the marked wind structure indicated at mesospheric and thermospheric heights (gravity waves), a height resolution of 2 - 3 km and spatial resolution of 500 km or better will be needed. One aspect of atmospheric dynamics which will deserve special attention is the magnitude of vertical movements and, particularly, the relative importance of turbulent transport and that arising from convective motions. This is of interest at all atmospheric levels but particularly at the mesosphere/thermosphere boundary (the turbopause) and in the troposphere and stratosphere. There is considerable doubt about the validity of the concept of eddy diffusion coefficients normally invoked to represent turbulent transport processes, and about the values adopted in atmospheric models.

A knowledge of atmospheric composition, and particularly of minor species, will continue to be required for an understanding of the thermal balance of the atmosphere, the neutral and ion chemistry, the airglow, the absorption of solar radiations and other aspects of aeronomy. In addition to the present concern about the effects of man-made gases on the ozone layer, a very large number of pollutant gases existing at both stratospheric and tropospheric heights will need to be monitored on a worldwide scale. Furthermore, for tropospheric heights, the possibility of localized regions will need to be kept in mind, and the observation of these will demand very good spatial resolution. For climatological models information will be required on a limited number of constituents, namely water vapour, carbon dioxide, and ozone.

It seems likely that during the 1980's the aim will be to make coordinated measurements of constituents involved in particular chemical or physical problems. It seems certain that many of the gases presently recognised as important will have been monitored extensively but, in view of the complex interactions expected, there will be a continuing need to observe these simultaneously with additional species. One type of species which will require particular attention will be the free radicals such as OH and HO_2. Because of their reactivity, they play a central role in the chemistry at all levels up to about 80 km but are only present in concentrations smaller than about 1 part in 10^8. In addition, the measurement of the spatial distributions of aerosols and of related gaseous species such as SO_2, H_2O, NH_3, will continue to be of major atmospheric interest. The importance of backscattering and absorption of solar radiations by aerosols to climatology is well known but the chemistry of aerosols themselves and their possible roles as catalytic surfaces will probably attract increased attention.

3. IMPACT OF LIDAR

The particular feature of lidar experiments is their capacity for providing good spatial resolution. In principle, the range gates and pulsewidths can be chosen to provide information on the height variation of the quantity or parameter to be measured at intervals considerably less than 1 km, and the horizontal structure measurable depends largely on the available transmitter power. In providing good height resolution the lidar measurements are not very dependent on an accurate control or knowledge of the Spacelab attitude. A deviation of the laser beam of 1° from the vertical corresponds to a relative Doppler shift of less than 10^{-6}, arising from the spacecraft velocity, which is considerably less than the linewidth of atoms and molecules being measured. The height resolution capability permits measurements down to tropospheric heights for which the available passive sounding techniques are not very effective at present.

The horizontal resolution possible with lidar experiments will permit certain measurements to be made on a regional scale (< 100 km), for instance, the distributions of aerosols in the lower troposphere and of certain gases. The horizontal resolution available with limb-scanning passive sounding experiments, if capable of being operated at tropospheric heights, is limited to about 400 km.

Laser sources provide greater spectral power densities than those provided by the natural sources : 10^5 times more than those from the sun, moon and atmospheric thermal emissions. Consequently, they offer the possibility of observing those trace constituents present at concentrations below the detectable thresholds of passive techniques.

The derivation of the neutral distribution of constituents from lidar observations is, in general, unique and does not require the use of inversion techniques.

This presents a particular advantage when a marked structure is present, as in the case of the sodium and hydroxyl layers.

For wind measurements, the Doppler shift imposed on the signals returned from aerosols, and constituents such as Na, can be used. The analysis of the simultaneous return from the Earth's surface or clouds, will enable the contribution due to the spacecraft velocity to be assessed. The use of aerosols as the atmospheric tracer has the advantage that the movement of these particles imposes a small spectral width on the returned signal. This is particularly important in the measurement of small wind velocities in the lower atmosphere, for which a high resolution will be required. At upper stratospheric and meso-spheric heights, only Doppler broadening contributes to the scattering linewidths and the Doppler shift imposed by the rela-tively large wind velocities should be resolved with a reasonable accuracy.

4. APPLICATIONS TO ATMOSPHERIC STUDIES

4.1 Introduction

Laser probing experiments will find applications in the three areas of investigation: Structure, Motions and Composition. Concerning structure, observations of the Rayleigh signal, when differentiated from the returns from aerosols, are capable of providing information on total molecular densities with good height and horizontal resolution at heights below about 60 km. In addition, the frequency analysis of the Rayleigh scattered signal will produce the corresponding information on molecular temperatures. The combination of these data provide the atmos-pheric pressure which is the required parameter for many purposes.

For wind measurements it will be necessary to make use of offvertical transmissions, and therefore a manoeuvrable telescope system. As already mentioned, the measurements will make use of the Doppler shift imposed on the backscattering species and will require stable narrow-band lasers and receivers.

In measurements of composition, use can be made of the resonance scattering or fluorescence of radiations of wavelengths chosen in relation to the constituents being considered, in both the ultra-violet/ visible spectral regions and in the near infra-red part of the spectrum. Examples of constituents measurable in these regions are OH, NO_2 and K. However, in the ultra-violet and visible regions difficulties can arise from the competing Rayleigh scattered signal and also collisional deactivation of the electro-nically excited states. The measurement of the differential ab-sorption of two adjacent wavelengths for which the absorption cross-sections of a particular constituent are significantly

different promises to be the most powerful approach for composition studies. For ultra-violet and visible parts of the
spectrum the observation of signals derived from tunable dye
lasers and elastically scattered from molecular nitrogen and
oxygen and from aerosols will be used. Such measurements will
lend themselves to studies of constituents such as ozone in the
ultra-violet, NO_2 in the visible and water vapour in the near-
infra-red.

An extension of the wavelength region into the middle infrared part of the spectrum will benefit from the large number of
absorption features present and from the availability of high
energy tunable gas lasers. The most direct application of the
method in this part of the spectrum is to observe the total absorption of signals returned from the Earth's surface or from
clouds, thereby providing information on total column density of
the constituent under consideration; O_3, NO_2, H_2O, SO_2 are
examples of likely constituents to be measured.
This would represent a useful approach to monitoring regional
distributions of atmospheric pollutants. For certain constituents,
information on height distributions might also be possible provided the laser pulse energy and aerosol concentrations are high
enough to provide sufficient atmospheric returns at infra-red
wavelengths. It is evident that the incorporation of heterodyne
detection systems will increase the sensitivity of these infrared measurements.

It will be useful to briefly examine the application of the
different techniques by considering in turn specific types of
investigations.

4.2 Typical Investigations

4.2.1 Aerosols

Although there is considerable scientific interest in the
possible existence of aerosol particles at all atmospheric heights,
it is for the tropospheric and stratospheric regions that information is required most urgently. It is realised that the backscatter and absorption of solar radiations by tropospheric and
stratospheric aerosols is of major importance to the Earth's
radiation balance and climate. A continuous monitor of aerosol
particles is desirable, both for effects associated with natural
sources, such as volcanoes, and man-made agencies. The initial
experiments on Spacelab will be directed towards the measurement
of aerosol backscattering coefficient.
Ideally, information on concentrations, size distributions and
refractive index properties will be required for quantitative
treatments in climatology, and the future use of more than

one probing wavelength and polarization measurements will help in deriving such information.

The spatial distribution of tropospheric aerosols is related to their origin. In connection with the dynamical models of climate referred to above, it will be important to establish the relative importance of the different sources. For instance, the contributions of wind-borne dust from desert areas, maritime aerosols, and dust produced in industrial or densely populated areas should be identified separately.

A study of the mode of formation of aerosols at stratospheric heights will also benefit from world-wide measurements of particles and of relevant minor constituents gases. It is known from balloon and aircraft measurements that the major element present in stratospheric aerosols is sulphur, in the sulphate form. It is believed that much of this sulphate is produced at stratospheric heights by the oxidation of SO_2 followed by reactions involving water vapour and ammonia.

Principle and feasibility of measurement of aerosols

The principle of the measurement is to record the backscattered signal for a Neodymium-Yag laser; operation at 1.06 µm serves to enhance the contribution of Mie scattering relative to Rayleigh scattering compared with measurements in the visible and ultra-violet part of the spectrum.

The feasibility calculations are based on a Spacelab height of 250 km, and 1 m diameter mirror. Table 1 indicates the laser characteristics, the electron counts in 1 km height channels and the horizontal resolution presently available; this will be improved with increased photomultiplier efficiency at 1.06 µm. For conditions of low aerosol concentration at stratospheric heights, the signal received is only about one-third that expected from Rayleigh scattering. Thus the interpretation of the measurements will require the adoption of an atmospheric model for allowing for the Rayleigh signal. Alternatively, simultaneous measurements could be carried out with the 5300 Å emission generated by second harmonic generation; the difference in wavelength dependence of Mie and Rayleigh scattering would permit the aerosol contribution to be deduced. For conditions of large stratospheric-aerosol concentration, and for tropospheric levels at all times, the Mie scattered signal will dominate the returns. Attention is drawn to the very good horizontal resolution that should be possible at trospheric heights.

Table 1 Aerosol Measurements – System Performance

Height	Backscatter Coefficient	Pulse Energy	Pulse Repetition Rate	Power Input	Receiver Efficiency				Electron Counts Per Second	Resolution for 10% Accuracy
					Photo-multiplier	Filter	Other	Overall		
Troposphere (5 km)	8.0×10^{-9} (cm ster)$^{-1}$	0.3 J	20	600 W	0.02	0.5	0.5	0.005	2900	300 m
Stratosphere—large aerosol concentration (15 km)	2.4×10^{-10} (cm ster)$^{-1}$	0.3 J	20	600 W	0.02	0.5	0.5	0.005	86	10 km
Stratosphere—small aerosol concentration (15 km)	2.4×10^{-11} (cm ster)$^{-1}$	0.3 J	20	600 W	0.02	0.5	0.5	0.005	9	100 km

4.2.2 Clouds

Within the heat budget of the earth-atmosphere system about 70% of the solar electromagnetic wave energy is absorbed the remainder being reflected back to space, chiefly by clouds and also by atmospheric aerosols, the Earth's surface and molecules. Energy is also returned to space by infra-red radiation, mainly from the Earth's surface and by atmospheric gases such as H_2O, CO_2 and O_3. The heights of emission of the latter gases are determined in part by clouds, and increasing the proportion from higher and, thereby, colder regions will change the balance between absorbed and emitted energy. Numerical studies have demonstrated these effects of changing cloud cover and height on surface temperature and have shown that such changes are equivalent to variations in solar constant.

In addition to the cloud cover and cloud height, the thickness and optical depth are quantities which need to be measured. Information on cloud cover will be provided by operational satellites in the late 1970's but the resolving power offered by the laser probing experiment is particularly well suited to the measurement of the other three quantities. This application represents an extension of the aerosol measurements described previously.

Principle and feasibility of cloud measurements

The height resolution made possible with narrow laser pulses will provide an accurate measure of the height of the top of clouds from the sudden change in the count vs range relationship expected for scattering from molecules and background aerosols. Depending on the particle density within the cloud, and therefore the type of cloud, simultaneous measurements on the signals returned from the upper and lower sides will yield information on the geometrical thickness and optical depth.

Based on the system parameters described previously and the 0.3J, 20pps Neodymium-Yag laser, calculations have been carried out to illustrate the effects of cumulus clouds located near 2 km altitude and of cirrus clouds situated between 10 and 11 km. Use has been made of scattering models of Diermendjian for these clouds. It has been found that for cumulus clouds, the attenuation of the laser beam is so severe that no measurement of geometrical thickness or optical depth would be possible. For the cirrus clouds, the attenuation is again severe, but the large scattering coefficient of the background aerosol near the ground gives rise to returned signals. It is evident that in an experiment dedicated to cloud studies the widths of the laser pulse and of the photon counting channels will need to be smaller than 1 km considered in this study.

4.2.3 Sodium

(i) Spatial distribution

Although measurements of the sodium layer based on observations of resonance scattering of solar or laser radiations have
been carried out at a number of sites, little information is
available about the morphology of the layer. The origin of the
sodium atoms is not established and, in particular, the relative
importance of meteors and terrestrial particles transported up to
the mesosphere. Observations of the spatial distribution prior
to and following meteor showers will bear on this question. In
addition, the importance of sodium sublimation from particles,
of either meteoric or terrestrial origin, could be examined by
searching for systematic differences in the daytime layer for
conditions of varying earth albedo e.g. over oceans, land, and
cloud covered regions. Finally, the observations of the sodium
atoms as tracers can provide valuable information on transport
processes. The horizontal distribution should be relevant to
studies of atmospheric waves, the resolution being sufficient to
study gravity waves; the height distributions will also be required for such wave studies and, more importantly, could be
interpreted to yield information on the eddy diffusion coefficient
just below the turbopause height.

Principle and feasibility of the experiment

The principle of the experiment is to observe the resonantly
scattered radiation for a laser tuned to one of the sodium D lines
at 589.0 and 589.6 nm. A dye laser, operating with Rhodamine 6G,
is to be used with Fabry Perot etalons included within the cavity
to limit the spectral width of the pulse. It is assumed that the
laser linewidth is comparable with the Doppler width of the sodium line, in which case the differential backscattering coefficient
is taken as 3×10^{-13} cm^2 sterad^{-1}. A sodium peak concentration
of 10^4 cm^3 has been adopted and it is assumed that there is no
attenuation between the spacecraft height and the sodium layer.
A 1 km vertical resolution is again considered.

It is useful to consider two forms of dye laser incorporating
different methods for exciting the dye.

(a) Flash-lamp pump

(b) Frequency-doubled Neodymium Yag laser pump

The performance of arrangement (b) is assessed on the basis
of a 0.25 efficiency in the frequency doubling 0.3 efficiency in
illuminating the dye and 0.5 efficiency for the Fabry Perot etalons. For arrangement (a) only the latter figure is relevant.

The figure shown in Table 2 indicate a slightly greater efficiency in (a) but, more importantly, the higher pulse rate in (b) limits its use to nighttime conditions since the background radiation arising from solar radiation will be too large, even with a receiver bandwidth comparable with that of the laser transmissions and allowance for Fraunhofer absorption. However, the need to develop long-lifetime flashlamps and to provide large quantities of dye solution represent two problem areas in arrangement (a). Arrangement (b) should offer a greater reliability and require only about one-tenth of the quantity of dye solution.

(ii) Temperature

The temperature of the neutral atmosphere between 80 and 100 km can be deduced from the neutral sodium Doppler line measurement. Several methods have been studied. The absorption technique which has been proved successful on the ground will only give one measurement per orbit if an accuracy of 2.5% is needed (for a 1 Joule, 1 Hz laser). Using a laser emitter with a spectral width of 0.1 pm successively centered on the maximum and the minimum of the line shape, the integration period will be reduced by more than 2 orders of magnitude and with the same accuracy ($\pm$ 5° K), a resolution of 200 km will be attainable.

4.2.4 Minor Constituents

As mentioned in Section 2 the measurement of the height distributions of a wide variety of minor neutral constituents represents a major need in atmospheric science. In addition to the requirements for studies in various aspects of aeronomy, increasing concern is being shown in the possible impact of certain constituents on the Earth's environment.

Perhaps improved information of the gases H_2O, O_3 together with CO_2 is the most fundamental need since they play a major role in the atmospheric thermal balance. However, an improved understanding of the observed variations of these gases will need associated measurements of those constituents involved in their chemistry, and also of tranport processes. The possible biological and climatological effects arising from reductions of the stratospheric ozone layer caused by man-made agencies has highlighted the need for information on critical constituents such as chlorine and nitric oxide, and their compounds. At lower heights, considerable attention is being paid to the chemical cycles of constituents such as N_2O, CO, and CH_4, and specific needs have been identified in relation to the monitoring of pollutant gases such as SO_2, NO_2, and NH_3. It is evident that the identification and study of sources and sinks of particular gases could benefit substantially from information on the world-wide distributions. The measurement of minor constituents represents

Table 2   Sodium Measurements - System Performance

Laser					Receiver					
Primary Pulse Energy	Efficiency	Pulse Repetition Rate	Power Input	Transmitted Pulse Energy	Photo multiplier Efficiency	Filter Efficiency	Other Losses	Overall Optical Efficiency	Electron Counts Per Second for Layer Peak	Latitude Resolution (10% Accuracy)
Case (a)										
1J	0.002	1 s^{-1}	500 W	0.5 J	0.2	0.2	0.5	0.02	230	< 1°
Case (b)										
0.3J	0.01	20 s^{-1}	600 W	0.011 J	0.2	0.2	0.5	0.02	92	< 1°

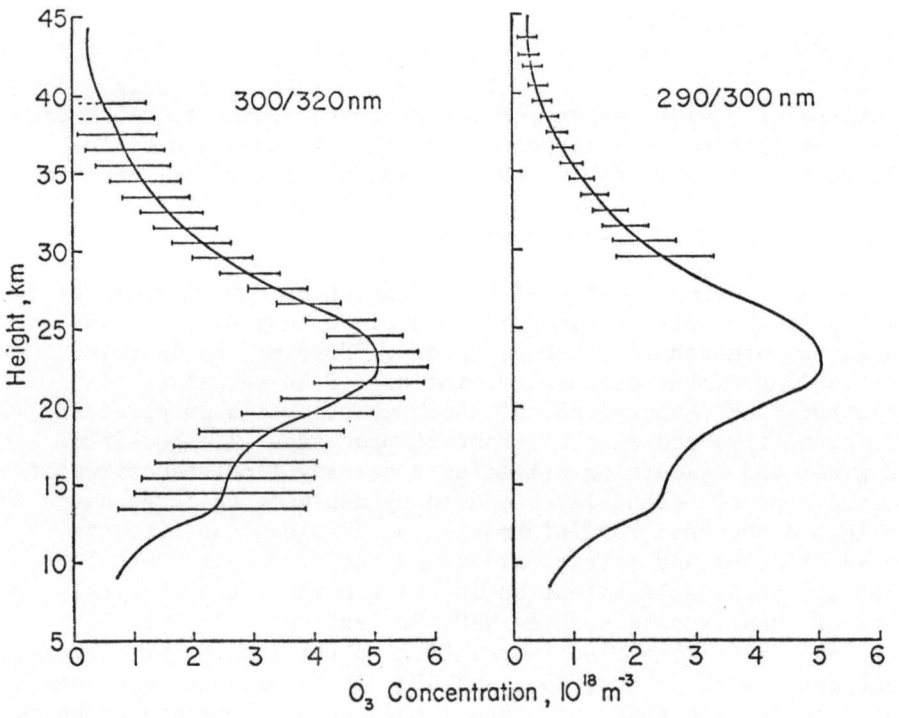

Figure 1

one of the most promising future applications of Spacelab-borne
laser experiments.

Two basic spectral ranges can be used in sensing atmospheric
constituents; the visible or ultra-violet radiations interacting
with electronic transitions, and infra-red radiations interacting
with the vibration-rotation transitions. In each of these spectral
ranges, resonant scattering, fluorescence, or selective absorption,
in which the frequency of the laser line transmitted is tuned to
coincide with that of a spectral line of a particular constituent,
enables the identification and measurement of that constituent at
considerable distances below the spacecraft. The application of
measurements of resonance scattering at visible wavelengths to
the study of the distributions of metal atoms, and also of certain
metal ions, has been illustrated in relation of sodium in Section
4.2.3. The use of ultra-violet wavelengths should enable the observation of constituents such as nitric oxide and hydroxyl but
will require the generation of wavelengths down to about 210 nm
and pulse energies in excess of what is presently feasible. Ozone
does not have a convenient resonance transition but information on
this constituent can be obtained from the differential absorption
of Rayleigh and Mie scattered signals at two laser wavelengths
near 300 nm; the absorption cross section of ozone varies rapidly
with wavelength in this spectral region. For illustrative purposes,
the results derived for pulses of energy 100 mJ, corresponding to
a 0.1 efficiency in frequency doubling of a visible beam from a
1 J tuned dye laser, and an overall optical efficiency of 0.03
are shown in figure 1. These standard errors shown refer to 1000
pulses, which at 5 s^{-1}, correspond to a latitude range of about
11°. It is to be noted, however, that the power input required to
obtain this spatial resolution is rather high, about 2.5 KW.
This differential absorption technique can also be applied to the
measurement of H_2O in the troposphere using radiations near 832
nm and to NO_2 in the stratosphere using radiations near 450 nm.

The future application of this technique in the infra-red
region probably provides the most promising approach to the
observation of neutral, minor constituents. The high efficiency
offered by gas lasers, and the relative richness of the near and
middle infra-red region in absorption features are two advantages
of operating in this spectral range. The rather poorer performance of detectors, relative to the photomultipliers employed in the
visible and ultra-violet region, represent a disadvantage common
to all infra-red methods. The development and incorporation of
heterodyne detection techniques will provide a major improvement
in this connection. The simplest approach would be to use the
Earth as a reflector and to estimate the total content of constituents giving rise to the differential absorption. On the assumption that the Earth behaves as a 10% Lambertian surface it can

be shown that measurements of N_2O are feasible with a 0.1 J laser
providing radiations at two suitable wavelengths near 4.5 μm can
be generated. Improvements in laser performance and available
Spacelab power should enable measurements of O_3 and CO by this
technique in the near future.

For range-resolved measurements, based on observations of
radiation scattered from aerosols, the weaker scattering efficiency compared with that from the reflecting earth surface demands
a much larger laser energy. It seems that the most feasible experiment at the moment is the measurement of N_2O at heights up to
about 10 km but this will require a 1 J pulse, and a heterodyne
detection system.

4.2.5 Wind velocity measurements

Velocity measurements if not considered a realistic possibility for FSLP, should be considered for future flights, depending
on the expectation that certain developments will take place,
especially in regard to the laser sources. Work in the development
of multichannel optical analysers should also help in reducing
the minimum measurement time.

In its simplest form, a lidar capable of a velocity measurement should utilize a stable, narrow band laser in the transmitter,
and/or a spectral scanning analyser (for instance, a spherical
Fabry-Perot interferometer) in the receiver, to determine the
Doppler shift of the echoes. The laser beam should point downwards making an angle to the vertical, whose value can be chosen
depending on

(a) the tracer utilized

(b) the spectral characteristics of the source

(c) the vertical stability of the platform.

Using aerosol as tracers, it is conceivable that vertical
angles of the order of a few degrees would be sufficient. Using
Na as tracer, wider angles should, in any case, be used, because
of the spectral width ($\Delta\lambda \approx 3 \times 10^{-3}$nm) associated with the
mesospheric atmosphere.

In order to obtain the velocity vector the beam should be
successively pointed to at least three different directions. If
the required beam offset from the vertical is small, substantial
advantages in weight will be obtained, because there will be no
need to move the telescope structure and the primary mirror.

Possibly, the three directions can all be simultaneously within the available image, or focal plane, of the telescope. If the field of view will not permit it, they can be obtained successively by properly positioning the secondary mirror. Alternatively, use could be made of the primary focus of the telescope. These considerations should be borne in mind in the design of the telescope. Also, to avoid switching the transmitter beam, more lasers could be used, simultaneously or in sequence.

As regards the source, it will take some research and development to establish whether the performances of the solid state and the dye lasers of present design are suited for the velocity measurement, and, in any case, bring the source performance to the necessary specifications.

A suitable source should emit a reasonably narrow line, or a set of equally spaced narrow lines: the line-widths should be of the order of 10 MHz. This performance is not incompatible with the pulse lengths of the laser sources of the type presently considered but their mode structure is probably not sufficiently stable.

A rather idealized analysis of signal strength and the time necessary to carry out a measurement is as follows. Consider a narrow-line laser emitting a continuous train of pulses, with average power of 6 watts. The scanning analyser has the minimum resolution which permits all the photons received to be either within or outside its pass-band. At periodic time intervals, the centre frequency of the analyser, during which a sufficient number of photoelectrons can be accumulated, is moved by an amount equal to its pass-band.

We assume that the other characteristics of the lidar are as mentioned elsewhere (cf. 4.2.1) and that the addition of a scanning analyser will reduce the efficiency by 50%.

We estimate that, from the stratospheric aerosol layer, at times of high concentration, an integration over a 3 km vertical interval will yield approximately 50 photoelectrons for a 10 km horizontal path, when the spectrum of the echoes will be within the analyser's pass-bands. By shifting the centre frequency of the analyser every 1.5 sec., one will be able to obtain essentially one component of the velocity vector, for a scan consisting of 10 steps. This means improving any "a priori" knowledge of the velocity field by a factor of 10. In most situations accuracies of the order of 1 m sec^{-1} could be achieved.

For the Na layer, the integration time required will be somewhat shorter, for the same resolution and accuracy.

In the lower troposphere the signal levels received from aerosols, in general, will be higher than in the stratosphere. Thus better accuracy or shorter integration times may be obtained.

5. OCEANOGRAPHIC AND GEODETIC APPLICATIONS

The ground echoes from a short pulse lidar system may provide detailed information on the surface structure at various scales. The lidar is in fact, in itself, a rather precise altimeter.

Over the sea surface the lidar will be able to give statistical information on wave heights with a vertical accuracy better than 1 metre without need of vertical stabilisation. The measurement can be obtained at a high rate - several times per second - thus permitting an accurate map with a high horizontal resolution.

The shape of the geoid can be similarly obtained over the sea after filtering tidal and salinity perturbations provided the orbit of Spacelab and the vertical direction are accurately known. Present methods for the determination of the gravity field based on orbit perturbations have limited horizontal resolutions, at present corresponding only to spherical harmonics of order about 10. The maximum amplitude of the undulations at such large scale is presently about 100 metres, but the fine scale detail is poorly known.

The determination of the gravity field of the Earth is essential because the irregularities of the potential reflect the inhomogeneities of the internal structure of the Earth, over which information comes mainly from seismic methods. As already mentioned, as of today only long wavelength terms in the potential are known. The higher terms are important because they are related to the structure of the litosphere and of the upper layers of the atmosphere; they should provide information on the characteristics of those regions at the edges of the tectonic plates.

Even during the first Spacelab flight the lidar will be able to provide information with a vertical accuracy of about 1 metre or possibly better. The accuracy is limited by the vertical stabilization of the spacecraft, a knowledge of atmospheric refractive index (which depends on pressure, temperature, H_2O concentrations), and laser pulse shape. At the present time, the effects due to imperfect knowledge of the atmospheric structure, affecting the refractive index, probably amount to about 20 cm, while a finite beamwidth of 10^{-4}r at an angle off the vertical of $1.5°$ (maximum expected for Spacelab) will give a spread of 0.65 m.

With a laser pulse length of 30 nsec, which is typical, it is certainly possible to recover the leading edge of the return signal with an accuracy better than 1 metre. Thus an estimate of 1 metre for the instrumental accuracy is well within the present capability. By further work on these aspects the accuracy could be improved by an order of magnitude.

It is worth pointing out that by the use of multiple beams, or by scanning the beam over a moderate angle and measuring the range of ground echoes, the direction of the local vertical axis on Spacelab can be obtained, when flying over sea, with an accuracy of a few tenths of a degree, depending on the state of the sea.

Much improvement can be obtained by measuring the Doppler shift of the ground echo. The technique required will probably not be available on the first flight and the integration time required is still to be ascertained.

Considering in more detail the applications of the technique to geodesy and geodynamics, it is important that knowledge about Spacelab orbit be improved. In this respect Spacelab can be used as a platform for observing other satellites whose orbit can be studied with better accuracy, or by observing well defined ground points.

Tracking of other satellites can be accomplished by microwave radar or Doppler techniques (e.g. the NASA Grovsat project, the French Diabole proposal), or by laser.

The accuracy of the Doppler is estimated today at 1 mm/T (T being the integration time in seconds). If terms to order 40 are desired in the geopotential, an accuracy on the velocity of 0.1 mm s^{-1} is needed. It is expected that optical interferometry could provide improvements of a few orders of magnitude on the accuracy of the measurement of relative velocities.

For positioning various ground stations by observations of satellites, two techniques are presently used:

(a) dynamical when the stations are positioned with reference to the orbit of the satellite.

(b) geomtrical when the stations are triangulated by simultaneous measurements, without necessarily knowing the orbit.

Both methods are limited by difficulties in synchronizing the observations and by the necessity of including in the data analysis angular measurements obtained by photographs of satellites, to an accuracy largely inferior to the limit set by the laser telemetry itself.

These difficulties will be removed if the whole telemetry
measuring unit can be placed in orbit as on board Spacelab. The
low altitude orbit limits the possibility of using the dynamical
method, the geometrical method on the other hand does not demand
a good orbit; it is sufficient to establish simultaneous optical
links between the Spacelab and different sites on the ground.
These sites cannot be very distant (up to a few 10^3 km). There
are many regions of such size of great geophysical interest
(Red Sea, the Alps, the Calabrian-Peloritan Arc).

6. CONCLUSIONS

The study presented here was based on realistic values of
laser performances. The main potential of the lidar lies in the
following improvements of lasers and detectors which might be
foreseen in different directions in the near future:

- possibility of using UV lasers (eximer type) in development
  now - instead of having to depend upon doubling in frequency,
- improvement of the efficiency of the laser which is far from
  its physical limits right now,
- existence of powerful tunable lasers in the whole wavelength
  range (they are mainly missing in the IR),
- increase of the spectral density of the laser and stability of
  the emitted wavelength,
- use of heterodyning technique,
- improvement of detectors.

These improvements might also lead to the use of different
methods mainly for the study of winds and turbulence which are
not possible now with the present state of the art.

It is also expected that more power for lidar operation will
be available on later Spacelab missions.

The work performed up to now has been limited to early missions assuming no use of subsatellites and limited pointing possibilities. These additions will increase the number of possible
experiments, and enlarge the use of the lidar to other fields of
interest, notably geodesy.

REFERENCES

Ref. 1 - Atmospheric Research Using Spacelab-Borne Lasers,
          ESRO MS(74)5, March 1974.

Ref. 2 - Atmosphere, Magnetosphere, Plasmas-in-Space, Report
          on the Mission Definition Study, ESRO, MS(74)31,
          December 1974.

SUBSATELLITES FOR ATMOSPHERIC RESEARCH

A. Pedersen

Space Science Department, ESA
ESTEC, Noordwijk, The Netherlands

1. INTRODUCTION

In 1974 NASA initiated a study of a Spacelab payload which would
be dedicated to the investigation of Atmospheric, Magnetospheric
and Plasma Physics. This payload is referred to as AMPS.
It became apparent very soon in this study that subsatellites
launched from Spacelab and used in connection with experiments
on Spacelab would be required. A brief summary of investigations
on AMPS requiring subsatellites are:

Atmospheric Physics

1. Chemical releases from locations removed from AMPS for
   scientific and safety reasons.
2. Optical reflectors on subsatellite allowing for a number of
   absorption/excitation experiments.
3. In situ measurements in several positions simultaneously.
4. Subsatellite on a long tether in the E-region: the Sky-hook.
   (not part of ESA study).

Magnetospheric Physics

1. Releases, e.g. Ba shaped charges from orbits different from
   AMPS for scientific and safety reasons.
2. Multipoint measurements of wave-fields and particles prior to
   and during perturbations from releases or plasma/electron guns
   on AMPS.

Plasma Physics

1. Study of beam-plasma interactions in connection with releases
   or plasma/electron gun.
2. Use ionospheric plasma for active experiments e.g. wave-
   wave and wave-particle interactions requiring transmission
   and reception.

In October 1974 ESA Headquarters started an in-house study of
Spacelab subsatellites; the purpose was to arrive at a definition
of low cost categories of subsatellites which could fit into the
AMPS concept.
    On one side the ESA study was aimed at subsatellites less
complex than the Atmospheric Explorer type of satellite which
has been mentioned as a possible NASA Spacelab subsatellite for
AMPS. On the other side the subsatellites to be studied were more
complex than the kind of simple ejectable package any European
national laboratory could develop.
    With this rather loose definition the study was started.
The situation would have been similar if sounding rockets did
not exist and a concentrated effort had to be performed with
the aim of producing scientific justifications for such a development and to define types of sounding rocket payloads.
    The problems with the scientific justification is that it
is practically impossible to have sufficient foresight to see all,
and possibly the most valuable results which can be achieved
with a new facility.
    On the technical side there are various motivations for the
basic design in terms of science, cost, and operations together
with Spacelab.
    The requirements of magnetospheric and plasma physics
experiments have influenced to a large extent the designs which
have by now been studied in some detail. However it will be
demonstrated in the following discussions that these types of
subsatellites are also well suited for many atmospheric physics
studies.
    Many atmospheric physicists have indicated that what is
required in particular for atmospheric physics is a number of
freeflying satellites launched from the Shuttle and being
operated over a period longer than 7 days. There are certainly
good scientific arguments for such satellites considering the
long time-constant in many atmospheric processes. However the
role of the Shuttle/Spacelab would be in this case only that of
a launcher and operations would require a ground network rather
than Spacelab facilities. The ESA subsatellite study is dedicated
to defining Spacelab facilities and is therefore aimed at subsatellites flying for 7 days near to Spacelab and being intimately
connected with the scientific investigations on Spacelab.

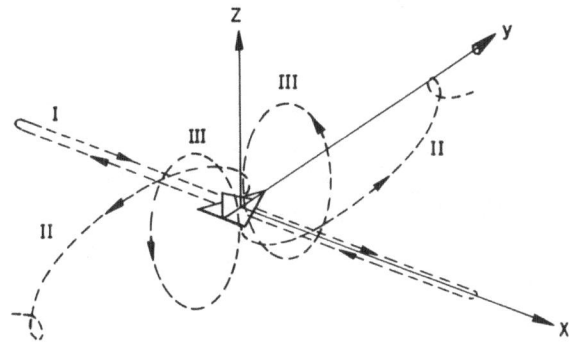

X : ⊥ ORBITAL PLANE   I : 1·18 M/S EJECT VELOCITY PER KM MAX SEPARATION

y :   VELOCITY VECTOR   II : 0.13/n

Z :   LOCAL VERTICAL   III: 0.59

Fig. 1 Orbits of subsatellites relative to the Shuttle for
       idealized conditions i.e. no differential drag.

It should however be emphasised that this subsatellite study
may form an important technical basis for any future study of
Shuttle-launched low-cost freeflyers.

2. CHARACTERISTIC FEATURES OF SUBSATELLITES

The purpose of this paper is not to give an exhaustive technical
description of the subsatellites which have been studied; only
those characteristics will be given which are necessary to
provide sufficient background for the discussion of scientific
programmes related to atmospheric physics.

Fig. 1 illustrates the orbits of subsatellites relative to
the Shuttle for idealized conditions i.e. there is no differential
drag. The ejection velocities required to achieve a maximum
separation of 1 km is given. The subsatellite orbit time is equal
to the Shuttle orbit time around the Earth. In practice variations of the orbits shown can be obtained by choosing different
ejection directions and in addition allowing for different
values of differential drag.

Fig. 2 shows the lifetimes of passive subsatellites in
circular orbits, and it is important to note that for a 7 day
lifetime the lowest altitude which can be reached is between
250 km and 300 km depending on area/mass ratio and orbit inclination. Because of differential drag, subsatellites tend to drift
away from the Shuttle/Spacelab and the data link cannot be kept
unless the altitude is about 320 km where differential drag is
sufficiently small. Obviously a satellite with an orbit- and
attitude-control system can reach lower altitudes at the expense
of gas for the orbit-control system and can also keep within
comfortable range for data transmission to Spacelab.

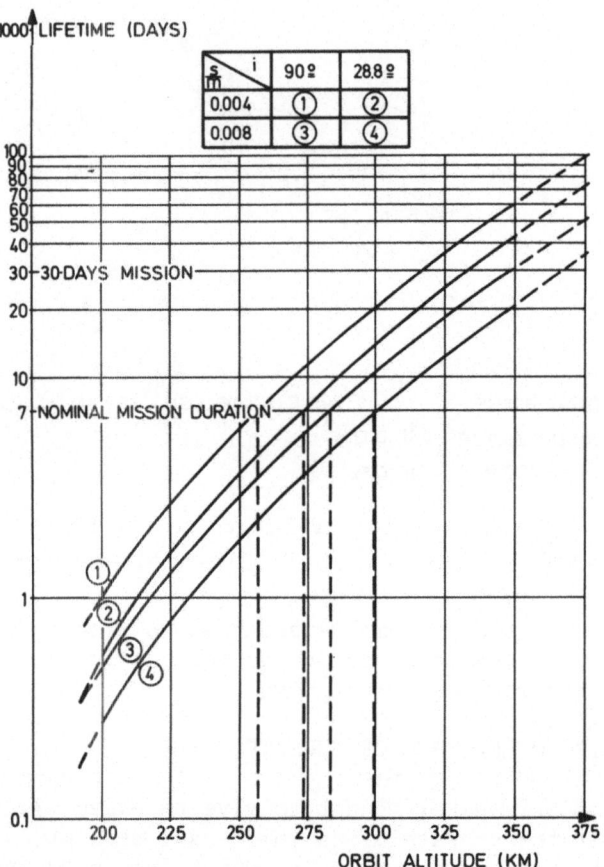

Fig. 2 Lifetime of typical subsatellites,

Launch of a subsatellite into a slightly elliptical orbit
can improve lifetime and visibility conditions. Such an orbit
is shown in Fig. 3 and it is interesting to note that the displacement cord altitude also varies and is passing through an
altitude range of importance for atmospheric physics.

Fig. 4 illustrates in a simple way the types of subsatellites considered in the ESA study.

The most complex subsatellite can be controlled in attitude
(spin axis position) and the orbit can be controlled by the use
of a total velocity increment, $\Delta V = 150$ m/s which allows a few
excursions down to approximately 150 km. The weight is largely
determined by the choice of power system, batteries, solar cells
or fuel cells.

By removing the orbit control system a simpler version with
only attitude control can be obtained. The same comments are valid
for the power system. The displacement distance will strongly
depend on differential drag.

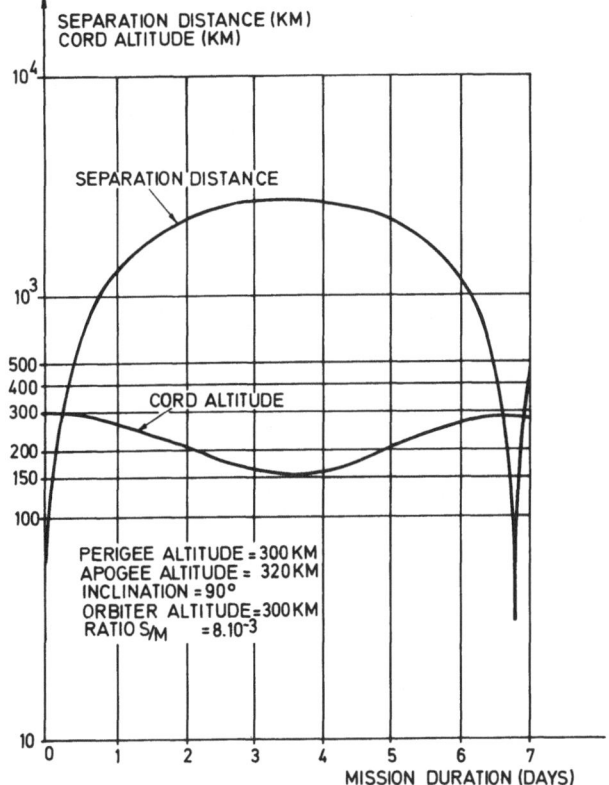

Fig. 3 Characteristics of a slightly elliptical orbit.

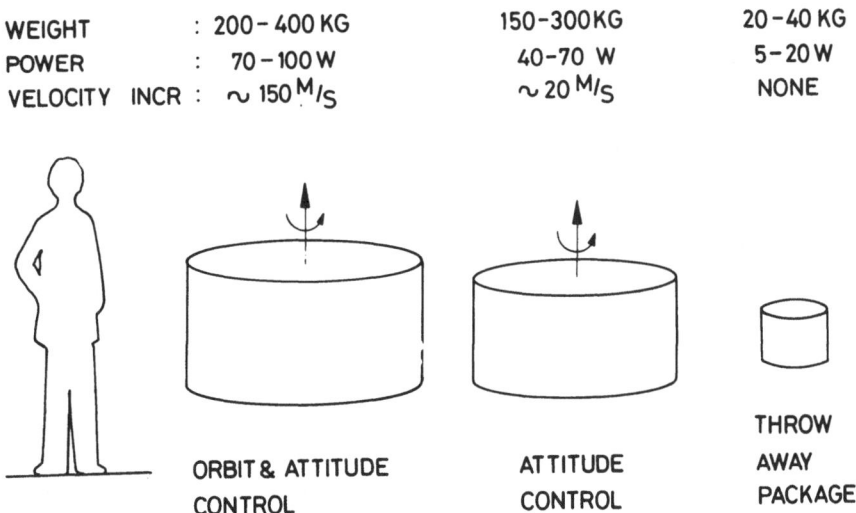

Fig. 4 Simplified technical summary of subsatellites studied
by ESA.

The simple Throw-Away-Package can most likely be made by principal investigators rather than by ESA, however it is meaningful to study standard telemetry and procedures for ejection and operation of such packages.

3. SUBSATELLITES AND ATMOSPHERIC PHYSICS

3.1 Ion and neutral chemistry in plasma or particle beams

The main purpose of plasma generators and electron/ion guns on Spacelab is to perturb the Ionosphere/Magnetosphere and to study beam-plasma interaction processes. Subsatellites with wave and particle detectors in and near the beam at different distances from Spacelab will be required to study these processes.

It has been recognized for some time that the atmospheric chemistry in the auroral regions is influenced by particle precipitation, e.g. excessive concentrations of NO has been measured with rockets in the auroral zone.

A plasma or particle beam generated on Spacelab will give opportunities in addition to the studies mentioned above of examining cross-sections of atmospheric processes. This can be done partly by remote sensing from Spacelab of excited atoms and molecules in and near the beam, but in situ measurements using mass spectrometers on subsatellites would probably yield more information.

One interesting possibility would be to aim the beam towards the earth and to use a subsatellite with orbit manoeuver capability to measure at altitudes lower than the Shuttle. This would allow the direct investigation of atmospheric excitation by auroral particles.

3.2 Active atmospheric experiments

The so-called passive remote sensing experiments on Spacelab use the Sun as a source; an example is a limbscanning experiment which measures absorption. Other sources are naturally-occurring phenomena (airglow) which can yield information about atmos- pheric processes. An artificial light source may allow for a con- trolled experiment where the intensity of the light source can be modulated. Such experiments for measurements of concentration of atomic and molecular species and cross sections of reactions have been performed on rockets by using a boom-mounted source and a "receiver" on the payload. On Spacelab this concept can be extended further with a "transmitter" on Spacelab and a "receiver" on a subsatellite or vice versa.

3.3 Optical reflector on subsatellite

It may be possible to pursue the type of atmospheric experiment mentioned under 3.2 by using an optical reflector on a simple subsatellite. In particular the use of a laser on Spacelab and a reflector on a subsatellite has been discussed in some detail. The subsatellite orbit with a cord altitude of approximately 150 km, shown in Fig. 3, may be of particular interest in this connection. It is interesting to note that the Soyuz/Apollo mission carried such a system the reflector being on Soyuz.

3.4 Subsatellites as carriers of chemical releases

Chemical releases have played an important part in the development of atmospheric physics, and Spacelab will offer new opportunities for continuing such experiments. It is virtually impossible to sum up briefly different schemes foreseen in connection with the AMPS study. However in practically all cases it will be necessary to use a subsatellite as a ferry for the chemical releases for safety reasons and in order to have the release at the right point relative to Spacelab.

3.5 Multipoint measurements

Atmospheric observations from satellites have so far allowed for in situ measurements at one point; the most advanced satellites so far are the Atmospheric Explorers with the capability to dip down to altitudes near 150 km.

Subsatellites can provide one or more measuring points in addition to Spacelab itself and thereby, to some extent, resolve temporal and spatial variations of phenomena with relatively short time and space scales e.g. gravity waves.

Initially subsatellites will be required for measurements near Spacelab to determine to what degree Spacelab "pollutes" the atmosphere. It may turn out that the more sensitive mass spectrometer measurements will require to be mounted on a subsatellite in order to avoid contamination.

A mass spectrometer on a subsatellite will require orientation of one axis parallel with the velocity vector and this immediately imposes certain requirements on the subsatellite so that it will fall within the "middle" or "most complex" category shown in Fig. 4.

3.6 The Sky-hook

As part of the AMPS study NASA has initiated a substudy on a long tether of the order 100 km long. With the Shuttle/Spacelab a little above 200 km it would then be possible to have a suspended platform for in situ measurements in the E-region. Such a tether

would be demanding in terms of weight and operations for
deployment. Another complication is heat generation from drag
forces if the subsatellite comes below 120 - 130 km.

There are however good reasons for continued studies of
such a system. The E-region cannot be surveyed on a global scale
neither by rockets nor satellites and the Sky-hook may be the
only way to obtain full information about this region where so
many important processes take place. It is sufficient to
illustrate this with one example; the global dynamo and current
system which is located in the E-region is far from being
understood today.

ESA has carried out studies on shorter tethered subsatellites with tether lengths up to 10 km. The idea was to investigate
if these tethered subsatellites could replace the ordinary
subsatellites in some cases. Certain parts of this study may
be useful for the further studies of the Sky-hook.

The conclusion of the ESA study on tethered subsatellites
is that there are a number of problems connected with deployment
and retraction both in terms of mechanics and operations
(time spent and restriction on Shuttle/Spacelab attitude) which
require further examination.

In other words the capability of ordinary subsatellites
is easy to demonstrate whereas tethers still need further
development before their feasibility is demonstrated.

ACKNOWLEDGEMENT

Material from ESA's study on subsatellites has been used in this
survey. Study manager has been Dr. G. Haskell and Mr. G. Duchossois
has been responsible for the engineering part of the study. The
following scientific consultant have been involved;
Dr. S. Cazes, Dr. A. Gonfalone, Dr. G. Haerendel, Dr. D. Krankowsky,
Dr. W. Riedler, Dr. T. Sandersson.

DISCUSSION

P.M. Banks: (Comment): With regard to chemical releases, the
ejection of small ablative pellets from AMPS can create
visible cloud trails in the E-region. Through ground
observations, the E-region wind field can be deduced to show
large-scale phenomena hidden in single-point measurements.

C. Muller: What is your estimate of the weight of a cable for a
tethered satellite?

A. Pedersen: The weight is of the order of 500 kg for drum,
cable and mechanisms.

THE ROLE OF PASSIVE SOUNDER PACKAGES ON SPACELAB

J. E. Harries

National Physical Laboratory
Teddington
Middlesex
UK

1. INTRODUCTION

This paper is based on the results of an ESA-sponsored study of passive sounders for Spacelab, the objectives of which were defined as:-

"To study the objectives, in atmospheric science, which can be reached using Spacelab-borne passive sounders and the definition, description, accommodation and operation of coordinated sets of instruments on Spacelab. The study should aim primarily at establishing one or several model payloads compatible with the constraints of the first Spacelab flight and consistent with its resources; particular emphasis should be put on (passive) experiments complementary to the LIDAR facility, in case this facility is included in the first Spacelab payload. A secondary aim should be a preliminary definition of integrated atmospheric payloads which could be flown in less-constrained later Spacelab missions."

In broad terms, the aims of the study were interpreted as follows:-

- To appraise the scientific objectives and priorities for atmospheric science studies using Spacelab.

- To define potential instrument packages to achieve these objectives.

(. To investigate the design, interfacing and cost of these packages).

J. J. Burger et al. (eds.), Atmospheric Physics from Spacelab, 265–276. All Rights Reserved.
Copyright © 1976 by D. Reidel Publishing Company, Dordrecht-Holland.

Only the first two items will be considered here.   The third item is largely an engineering accommodation study, the results of which may be found in the published study report.

The study has, of course, received contributions from several people in the UK, and I should mention their names.   They include:-

Dr Houghton and colleagues at Oxford University;
Dr Rees and colleagues at University College, London;
Dr Thomas, Dr Gibbins, Dr Street and Dr Courtier, at Appleton Laboratory;
Dr Farrow and colleagues at Hawker Siddeley Dynamics;
Myself and colleagues from National Physical Laboratory.

2.   METHODOLOGY OF AND APPROACH TO THE STUDY

Some general comments on our approach to the study are appropriate:-

(a)   The use of passive sounders on any spacecraft presents us with a bewilderingly large number of possibilities for combining experiments and their aims.   The examples we have considered are, therefore, only representative and should be understood not to be in any way a selection of the payload for Spacelab.

(b)   The arbitrary division of the atmosphere has been avoided wherever possible, and a continuous treatment, from surface to thermosphere has been attempted, though not always possible.

(c)   The study has been conducted by setting up scientific criteria: ie it has been "science oriented" rather than "instrument oriented" or, on SL, "facility oriented".

3.   CHARACTERISATION OF PASSIVE SOUNDERS

An extremely important development in atmospheric science in recent years has been the ability to observe the global atmosphere on a continuous basis, made possible largely through passive remote sounding from satellites.

The measurements that are required are of the three dimensional fields of atmospheric density, composition and motion.   So far passive sounding has concentrated on temperature measurement (to infer density); measurements of water vapour and ozone have also been made. Before 1980 a number of satellite missions are planned from which other minor constituents will be measured.

In satellite experiments, sounders are used in two basic geometrical configurations – nadir sounders which point vertically downwards, and limb sounders pointing towards the earth's limb.

Nearly all remote sounding measurements have so far been made with instruments viewing the nadir or near to the nadir. Measurements which have been made are:-

- Vertical temperature structure from the surface to 90 km. (IR emission from CO_2 or microwave emission from O_2).

- Distribution of H_2O in the lower atmosphere (IR and microwave emission).

- Distribution of O_3 in the stratosphere (backscattered UV and IR emission).

- Cloud cover and cloud height (Vis. and IR).

- Outgoing radiation (UV to IR).

- Precipitation (microwave).

- Surface properties such as pressure, temperature, albedo, ice cover, water content, (Vis. and IR).

There will be a continuing requirement for many of these observations to be made routinely, where possible, from free flying satellites. Spacelab is not suitable for the routine continuing programme because of its restricted coverage in space and time. Where Spacelab could be important is in the testing of new developments and new techniques, should these come along, and for experiments that make particular use of the Spacelab facilities.

For nadir sounders a contribution function can be defined which determines the height region of the atmosphere from which the radiation mainly originates. By careful choice of wavelength for the measured radiation, the contribution function can be made negligible outside a region roughly one scale height wide, centred on some chosen height. Horizontal resolution of nadir sounders is determined simply by the field of view of the instrument.

For limb sounders both vertical and horizontal resolution depend on geometrical factors as well as on the vertical distribution of the observed constituent. Limb sounders are particularly useful for measurements of temperature up to great height (~100 km) and for measurements of trace constituents where it is

necessary to maximise the optical depth of material in the line
of sight.   Limb sounders offer improved vertical resolution if
the line of sight is well defined, the horizontal resolution is
however worse than with nadir sounders.   Typical spatial resolution characteristics of the existing sounders of each type are
indicated below.

Sounder	Vertical Resolution	Horizontal Resolution
Nadir	10 km	25 km
Limb	2 km	200 km

Much thought is currently being given to the measurement of
minor constituents which may only be present in a few parts in 10^9
or 10^{10}.   The main difficulty with measurement of these, using
vertical sounding methods, is the presence of the variable radiation background provided by the Earth's surface and by clouds.
This problem is not present for limb-viewing instruments so far
as the stratosphere and above is concerned so that all the proposals for measuring composition of the stratosphere currently employ
limb sounding instruments.   In addition a much longer atmospheric
path length at a given level is available, so that weak emission
is enhanced and the altitude observed is determined by the
direction of view, and is hence largely independent of the spectral
selection of the instrument.   However, because of clouds and aerosols in the troposphere, and also because of refraction problems,
the limb view is not easy to apply to observations of tropospheric
constituents.

The limb-sounding technique can be used to sense (a) emission
from the atmosphere or (b) absorption by the atmosphere when a
suitable source (eg the Sun) is 'occulted' by the limb.

4.   SPACELAB IN RELATION TO PASSIVE SOUNDERS

The advantages and disadvantages of Spacelab have been discussed many times, including at this conference, so we will only
highlight a few points of particular importance and relevance to
passive sounders.

Advantages:-

.   Re-usable.

.   Large payload and power capability.

.   Considerable potential for carrying cryogenics.

Disadvantages:-

• Contamination of immediate environment.

• Short mission duration.

• Attitude stability poor.

Shortage of space precludes further discussion of these points here, but further consideration may be found in the study report. A summary of points particularly requiring further attention for passive sounders is given in Sect. 9.

5. INSTRUMENTS AND FACTORS INFLUENCING THEIR USE

In considering passive sounding instrument types, a number of factors were taken into account including

• Wavelength

• Source of radiation

• Field of view

• Sensitivity.

Thus, a variety of radiative sources are possible, eg thermal emission, fluorescence, excited state decay, and solar or lunar transmission. Wavelengths between the UV and the microwaves were considered. And considerations of field of view and sensitivity have obvious implications in determining the usefulness and accuracy of a particular device or measurement.

It was, of course, essential to consider specific instrument types since technical instrumental limitations often are the greatest ones to a precise measurement. In order to avoid unfairness or parochialism we attempted to choose a representative set of instruments from which we should eventually select example packages. This set is given below.

Instrument Types

Filter Photometer	Gas Correlation Radiometer (PMR, SCR)
Filter Radiometer	Heterodyne Radiometer
Grating Spectrometer	Microwave Radiometer
Grille Spectrometer	Polarimeter
Fabry-Perot Interferometer	Albedometer
Michelson Interferometer	TV System.

6.   SCIENTIFIC STUDIES

     The basic aims of a passive sounder package on Spacelab should
be the determination of composition, temperature structure and
dynamics of the atmosphere over all heights, and on a continuous
global basis.   For each of these three areas, the study has
carried out a survey of previous work, a definition of requirements as at 1980, and a study of how specific passive sounder
packages might meet these objectives.   This part of the study was
fairly extensive, but will be considerably abbreviated here and
given in summary form only.

6.1 Composition

     Measurements of composition by passive sounders are carried
out, of course, by selecting in some way a spectral band specific
to a given constituent and quantitatively measuring the atmospheric
emission or absorption in that band.

     Recent research programmes, spurred by well-known problems
associated with stratospheric O_3, have produced a vast amount of
new data.   This production can be expected to continue up to 1980.
Some issues which may not be settled by then, and which may require
further work are

(a) Aerosol/Cloud in troposphere.

(b) Minor composition in stratosphere.

(c) Aerosol in stratosphere (Junge) and mesosphere (NLC).

(d) Neutral and ionic composition in mesosphere - thermosphere.

(e) Radiation from constituents and energy balance (CO_2, OI).

     The sensitivities required for such measurements were assessed
from current knowledge and varied from mixing ratios of about 10^{-7}
(eg for water vapour, ozone) to 10^{-12} (for certain minor constituents such as CF_2Cl_2, ClO etc).

     The study showed that a number of the passive sounder types
continue to demonstrate great sensitivity for composition measurements, and a number were therefore included in the packages
(see Section 7).

     Two special cases, thought to be of particular importance at
present, were selected for special attention - these were atomic
oxygen and aerosols.   Potential methods for measurements of these
were defined (see Sect. 7).

     The attitude stability of Spacelab was found to be insuf-
ficiently accurate to meet the requirements of passive

determinations of composition, particularly for limb-sounding.
Attitude needed to be known to within 0.03° in roll and 1° in
pitch and yaw for the period of one measurement, for limb-sounding
directed at 90° to the velocity vector.   Attitude sensing methods
(including infrared horizon sensors, star trackers etc) for
achieving these precisions were discussed.   Certain passive
instruments are capable of producing such information internally.

6.2 Temperature

    Infrared and microwave methods of atmospheric temperature
sounding which depend on absolute measurements of atmospheric
radiance in the 15 μm CO_2 band or the 2 mm O_2 band are well known.
In the visible, temperatures in the mesosphere and thermosphere
may be determined from measurements of Doppler line width.

    Even with the considerable amount of prior work which has been
done, several areas of scientific interest remain for temperature
measurements from Spacelab, including the study of planetary waves
and stratospheric warmings, tidal effects, and also to try to
improve the absolute accuracy of temperature determinations
to better than $\pm$ 1 K.

    Instruments capable of making significant contributions by
1980 were assessed; these will be mentioned in the next section.

    The attitude stability requirements for temperature deter-
minations are similar to those for composition measurements.

6.3 Winds

    The importance of dynamics in atmospheric physics has been
made clear at this conference, in tropospheric meteorology and
forecasting.   Also jet streams are of particular interest.   In
the stratosphere, the transport of minor constituents is of great
importance.   In the mesosphere/thermosphere motions (gravity
waves, tides) act as important energy inputs.

    Wind measurements may be carried out by passive sounders
capable of distinguishing the Doppler shift of a spectral line due
to the combined motion of the air and of the spacecraft: knowledge
of the latter then allows a measurement of wind.   Only instruments
capable of resolving a Doppler width can be used, which restricts
the potential devices to Fabry-Perot interferometers, Gas correlation radiometers and Microwave heterodyne radiometers.

    A survey of current knowledge of atmospheric winds led to
desired accuracies of $\pm$ 3 m s^{-1} in the troposphere and $\pm$ 3 to

$\pm$ 10 m s^{-1} in the stratosphere and above, though for the latter
accuracies of $\pm$ 30 m s^{-1} would still be useful.   Passive instruments seem just capable of achieving these levels of accuracy.

Because the spacecraft motions have to be subtracted from
the measured Doppler shift, the spacecraft attitude must, in some
cases, be known with greater precision than for temperature or
composition measurements.   Thus, for measurements perpendicular
to the velocity vector attitude need to be known to

$\pm$ 0.025° in yaw and pitch;

$\pm$ 0.003° in roll;

for measurements parallel to the velocity vector

$\pm$ 1.5° in yaw and pitch;

$\pm$ 0.15° in roll;

(all estimated for an uncertainty of $\pm$ 3 m s^{-1})

Because of the particular methods used to obtain high spectral
resolution, the gas correlation radiometer can only be used in the
first configuration.   The Fabry-Perot and Microwave radiometers
can operate in either configuration.

7.   EXPERIMENT PACKAGES

To arrive at different packages a number of different scientific and technical considerations have been taken into account.
From the scientific point of view three regions of the atmosphere
have been selected, namely (a) the 70-140 km region, (b) the
15-70 km region and (c) the 0-15 km region, in all of which there
are important scientific problems which need to be tackled.   Two
major technical considerations are (a) the large size of a microwave radiometer, and (b) the requirements of very good attitude
stability and measurement if the motion field is to be accurately
measured.

Four main objectives (A,B,C and D) have been defined.   Three
(A,B and D) are aimed respectively at the three regions of scientific interest, and the fourth (C) is aimed exclusively at the
measurement of the motion field.   The microwave radiometer is
mainly of value for measurements in the higher region (70-140 km)
and in order to make room for other instruments it is excluded
from package B.

All measurements except where otherwise stated are made on
the atmospheric limb, and for all limb viewing cases accurate

knowledge of the position of the limb being viewed will be
required. For microwave and infrared radiometer measurements
this can be achieved over a considerable altitude range by
including appropriate channels within the instruments themselves.

Details of the packages and selection procedures are given
in the study report. A summary of the packages follows here.

Package A

Temperature: kinetic - 70-120 km (with microwave O_2)
                           70-140 km (with Fabry-Perot Interferometer)

                rotational O_2 - (microwave)
                           CO_2 - (15 µm)

                vibrational O_2 - (microwave)
                           CO_2 - (4.3 µm and 15 µm)

Composition: Atomic O - airglow (130.4, 557.7, 630.0 nm)
                               far IR (63 µm and 147 µm)

                    O_3 - microwave
                    CO_2 - IR fluorescence
                    CO - microwave
                         IR fluorescence
                    H_2O - microwave
                         IR fluorescence emission

Important airglow
emissions: OH - near IR
                    Atomic O (as above)
                    O_2 $^1\Delta$ (1-27 µm)
                         $^1\Sigma$ (700 µm)

Instruments Required for Package A

• Microwave radiometer O_2 - 118 GHz
                       O_3 - 184 GHz
                       H_2O - 183 GHz
                       CO - 115 or 230 GHz

• High resolution gas CO_2 - 15 µm and 4.3 µm
  correlation radio-   CO - 4.7 µm
  meter H_2O - 2.7 µm + other bands

• UV-NIR spectrometer O
                       O_2
                       OH emissions

Package B

Temperature: CO_2 – IR (15 μm)

Composition: aerosol (occultation or emission)
                     O_3
                     H_2O, CO, CH_4
                     NO, NO_2, HNO_3, N_2O
                     HCl, HF, ClO, CH_2, ClF etc.
                     SO_2 etc.

Instruments Required for Package B

- Gas correlation radiometer for temperature and some gases.

- Grille spectrometer looking at sun (also emission?) for near IR spectra of some gases.

- Occultation photometer for aerosol determination.

- IR filter radiometer for observing aerosol emission.

- Grating spectrometer (Backscatter UV downward looking for O_3 determination).

- Michelson interferometer for near or far IR emission spectra of some gases.

Package C

Temperature: with microwave
                     IR
                     UV and visible?

Motion field: with microwave
                     IR
                     UV and visible.

Instruments Required for Package C

- Microwave radiometer with several channels (wind 80–120 km, temperature 50–120 km).

- Pressure Modulated Radiometer with azimuth scan and several channels (wind 50–130 km, temperature 50–120 km).

- Fabry Perot Interferometer (wind 50–140 km).

Package D

Polarisation and spectral distribution of backscattered radiation in the range approximately 0.4 to 2.2 μm for study of clouds and aerosols.

Instruments Required for Package D

- Spectral albedometer for cloud measurement (nadir sounding).
- Multi-channel polarimeter for aerosols (nadir sounding).

8. COMPLEMENTARITY WITH LIDAR

Most of the experiments in packages A to C are limb-sounders so that direct comparison of results with a nadir-viewing lidar experiment are not directly applicable (exceptions are the nadir sounders of package D). Some areas of complementarity may be defined, however, bearing this point in mind.

Package A:-   • Comparison of temperatures with those measured from Na emission by lidar.

Package B:-   • Comparison of measurements of aerosols and related constituents (eg SO_2) with aerosol measurements by lidar.

Package D:-   • Measurements of tropospheric aerosol to compare with lidar determinations.

9. POINTS REQUIRING FURTHER ATTENTION FOR PASSIVE SOUNDERS

We may define several areas which require further attention in order to completely assess the potential of passive sounders on Spacelab. These, as this study has shown, include

- Details of cryogenic-carrying capability.
- Ambient contamination levels and the effects on the more sensitive (and particularly the cooled) passive sounders.
- The possible provision of a simple one-axis stabilised platform for passive sounders to improve attitude stability. Such a platform would be much simpler than the Instrument Pointing System planned for astronomical experiments on later Spacelab flights.

10. CONCLUSIONS

10.1 In the areas of atmospheric composition, temperature and dynamics, this study has shown that the impressive past record of passive sounders can be continued into the Spacelab era, and that even on Spacelab 1 a powerful range of accurate measurements can be made by a passive sounder package probably in all three areas. This would allow a very large advance in our knowledge of the properties of the whole atmosphere.

10.2   We have shown that all spectral regions from UV − μWave
(100 nm − 1 cm) offer different advantages, and also a variety of
instrumental methods should be used.

10.3   Example packages of instruments have been put together to
study the various areas of atmospheric science.   The instruments
in these packages would complement and support each other.

10.4   If a lidar experiment were to fly, its main aims (aerosol
and Na measurements) could be complemented and supported, to a
greater or lesser extent, by each passive sounder package A to D.

DISCUSSION

A. *Vidal-Madjar:*   Is there any physical or historical reason that
   explains why you seem to neglect in your study the passive
   sounding techniques in the EUV wavelength range (below 2000 Å
   to 3000 Å)?

J.E. *Harries:*   There is no such reason.   There are two points:
   the first is that it was clear that much useful atmospheric
   science could be studied at longer wavelengths.   The second
   is that the study did not, in fact, completely ignore the
   EUV; for example, in some sections, NO emission at 215 nm
   and OI emissions at 130 nm were considered, for wind and
   composition measurements.

A REVIEW OF THE EVOLUTION OF
ATMOSPHERIC SCIENCE EXPERIMENTS FOR AMPS

Andrew F. Nagy

Atmospheric and Oceanic Science Department,
University of Michigan, Ann Arbor, USA

A brief review was given of the conceptual development of atmospheric science experiments, covering the various studies carried
out during this decade (e.g. NASA-JSC/Martin Marietta, 1972;
NAS-SSB, 1973; Martin Marietta, 1973; AMPS Science Definition
Working Group, 1975). The presentation was concluded with a
summary of the present science objectives and instrument con-
cepts for the early AMPS flights.

Further information can be obtained by contacting Prof. Nagy.

J. J. Burger et al. (eds.), Atmospheric Physics from Spacelab, 277. All Rights Reserved.
Copyright © 1976 by D. Reidel Publishing Company, Dordrecht-Holland.

SESSION 5

EXPERIMENTS FROM SPACELAB

FIELD ALIGNED CURRENTS MEASURED BY INCOHERENT SCATTER ON BOARD
THE SPACELAB

M. BLANC and P. BAUER
CNET/CRPE, 3 avenue de la République
92131 Issy-les-Moulineaux, France.

and

G. LEJEUNE
CEPHAG, Campus Universitaire de Saint Martin d'Hères
38040 Grenoble, France

1. INTRODUCTION

The incoherent scatter technique has in recent years played
an important role in the study of the ionosphere and the thermosphere. Used as a remote sensing experiment, this technique has
made possible over the whole range of ionospheric heights, and at
several very different latitudes from the equator to the auroral
zone, the study of daily and seasonal variations of the bulk parameters of the neutral and ionized atmospheres. Used on board the
Spacelab, with the corresponding energy and size limitations, an
incoherent scatter radar cannot cover similar ranges in space and
times. Our purpose is to show that, used in a different way, aiming
at a quasi-in-situ measurement (only one or two kilometers away
from the spacecraft), it could permit with a very satisfactory
accuracy the mapping of a parameter which has not been studied in
ground-based experiments : the bulk velocity of thermal electrons.
Associated with a simultaneous measurement of bulk ion velocities,
this gives access to the study of field-aligned charge transports,
which is of strong geophysical interest at all latitudes.

J. J. Burger et al. (eds.), Atmospheric Physics from Spacelab, 281–296. All Rights Reserved.
Copyright © 1976 by D. Reidel Publishing Company, Dordrecht-Holland.

2. MEASUREMENTS PRINCIPLES

For wavelengths that are large with respect to the Debye
length of the plasma, the incoherent scatter spectrum of electromagnetic waves is characterized by three lines : the first, which
is known as the ion line, is centered on the transmitted frequency
except for a small Doppler shift due to the mean motion of the
ions with respect to the observer ; the two other lines, known as
the plasma lines, are approximately located at $\pm f_p$ (plasma frequency) except also for a global Doppler shift characteristic of
the Doppler shift induced by a motion of the electrons relative to
the observer (BAUER et al, 1976). Therefore, in first approximation,
the measurements of these two Doppler shifts provide the mean motions of the thermal ions and of the thermal electrons, which are
a fundamental and in many instances dominant contribution to the
currents.

3. FEASIBILITY ESTIMATES

In the following sections we study, for each component of the
backscatter spectrum, the instantaneous signal-to-noise ratio and
the optimal velocity measurement errors that can be obtained as
function of the main parameters of the transmitting system. The
pulse schemes and spectral analysis techniques capable of actually
providing these optimal errors are discussed, and a first proposition is made for the general structure of the measurement system.

3.1. Main parameters of the experiment

In an incoherent scatter experiment, the received signal is a
superposition of two signals of the same statistical nature. Backscattered signal due to the ionosphere and sky noise are both random functions of time, with a gaussian distribution centered around
zero. They differ only by their power spectral densities. It is
these spectral power densities, or their Fourier transforms, the
autocorrelation functions, which must be measured. One can derive
the errors on the measured parameters influencing the shape of the
observed spectra provided that the signal-to-noise ratio is known,
as well as the family of theoretical shapes, by use of formulas
derived from the theory of incoherent detection (PETIT, 1968).

Basic parameters considered in that study and their interrela-
tionships, are shown in figure 1. They are :

(i) the parameters of the transmitting system :

- f_e, operating frequency, and λ, wavelength

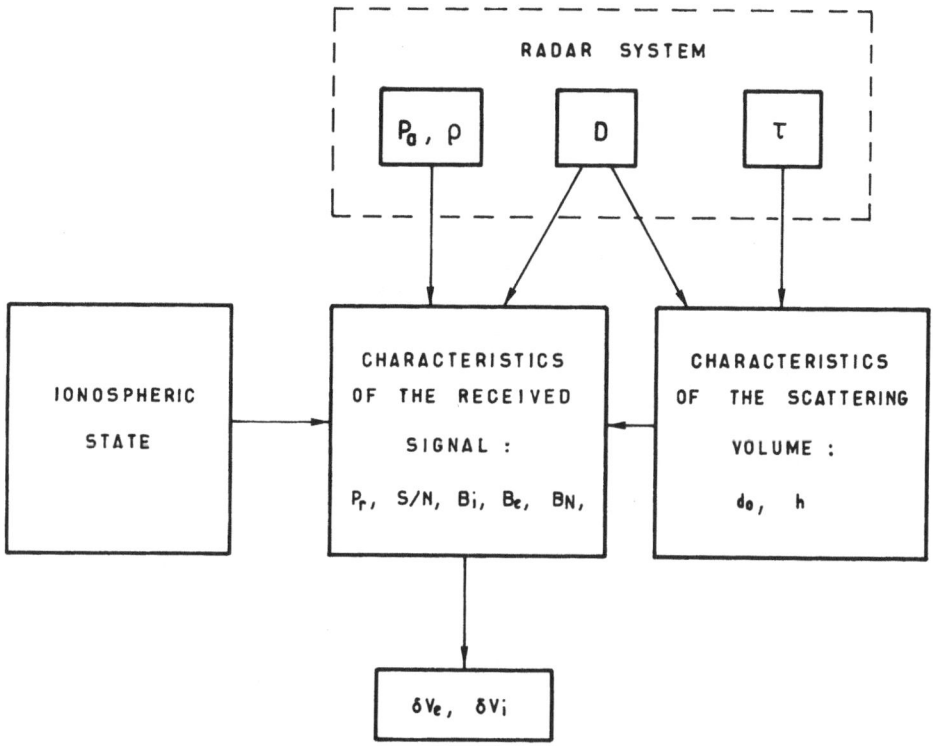

<u>Figure 1</u>. Interrelation ships of the various parameters of a
"quasi-in-situ" incoherent scatter experiment from
Spacelab.

- P_a, average radiated power

- P_p, peak radiated power, and ρ, ratio of P_p to P_a

- τ , elementary pulse length

- D , diameter of the antenna

(ii) the characteristics of the scattering volume :

- d_{MIN}, distance from the antenna to the closest point of
the scattering volume. That distance will be taken equal
to the smallest possible one, i.e. that beyond which the
approximation of a purely active and radial propagation of
the transmitted signal becomes valid. One has :

(1) $d_{MIN} = \dfrac{D^2}{\lambda}$

- h, the length of the volume, is classically defined as

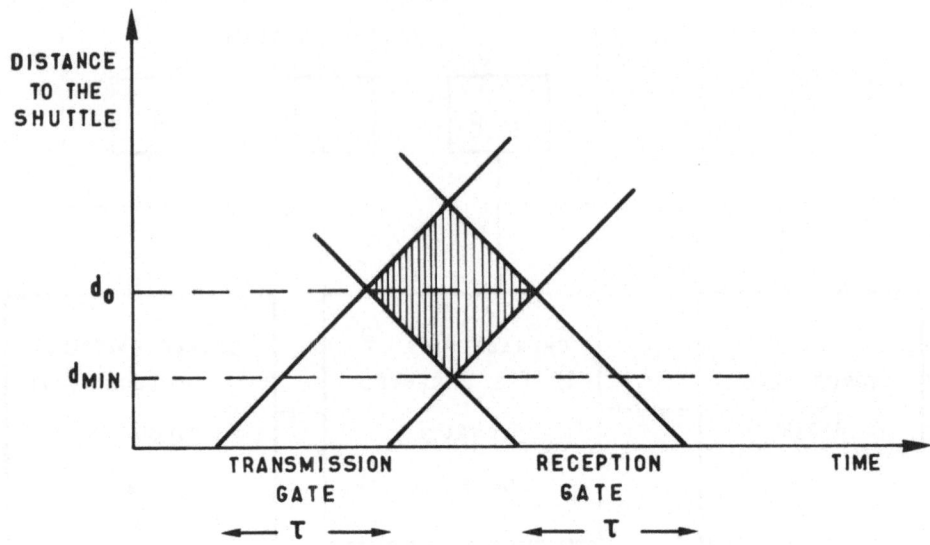

<u>Figure 2</u>. Classical time-distance representation of an elementary
pulse transmission-reception sequence. d_{MIN} is the minimum distance where a purely active and radial propagation from the antenna can be assumed.

(2) $\quad h = \dfrac{c\tau}{2}$

- d_o, distance from the antenna to the mean point of the
volume. Figure 2 shows the classical time-distance representation of an elementary pulse transmission-reception
sequence. It appears that :

(3) $\quad d_o = d_{MIN} + h$

An important parameter of the problem is the ratio d_{MIN}/d_o,
which will be extensively used in the following calculations.

(iii) the characteristics of the received signal :

- B_i and B_e, bandwidths of the ion and electron lines (after
integration over the scattering volume). For the ion line,
centered around the transmitting frequency, the integration
of the elementary power spectrum over the scattering volume
changes only negligibly the bandwidth, which is directly
proportional to the transmitting frequency. For typical
F-region conditions :

(4) $\quad B_i = \dfrac{f_e}{3.10^4}$

For the electron lines, on the contrary, the width of an elementary spectrum is negligible with respect to the variation of its central frequency position within the volume. The apparent width is therefore equal to the variation of the plasma frequency in that volume. For typical electron density gradients, one finds :

(5) B_e(kHz) = 20 h

where h is expressed in kilometers.

- B_N, noise bandwidth, is nothing but the bandwidth of the receiver. Considering that it must be large enough both to receive the ionospheric echo and to define a time gate equal to the transmitted pulse duration, one finds for each receiving channel :

(6) $B_{N_{e,i}}$ = MAX $\left(B_{e,i} , \dfrac{1}{\tau} \right)$

- $P_{e,i}$, instantaneous received powers for each of the lines, can be expressed as :

(7) $P_{e,i}$ = $\dfrac{1}{2} \rho P_a$ $\sigma_{i,e}$ $\dfrac{\Pi D^2}{4}$ $\dfrac{h}{d_o^2}$

if $\dfrac{\Pi D^2}{4}$ is taken as the efficient surface of the antenna, and D is correlatively defined as an "efficient diameter". (7) is an underestimate of the actually received power, because it neglects variations of the distance to the receiver within the scattering volume which are not negligible for the proposed experiment. The incoherent scatter cross-sections per unit volume are given by :

(8) σ_i (m^{-1}) = 3.98 10^{-30} N_e

(8') σ_e = 1.21 10^{-26} $k^2 T_a$

with N_e, electron density expressed in m^{-3}, $k = \dfrac{4\Pi}{\lambda}$, the incoherent backscatter wave number, and T_a, the apparent temperature of the electron gas in the vicinity of the resonant electron velocity ω_p/k (ω_p is the plasma pulsation).

3.2. Signal-to-noise ratios

As shown by equation (8'), and provided that T_a does not very drastically with λ, it is of interest to increase the transmitted frequency as long as the requirement of λ being large compared to λ_D is satisfied. f = 900 MHz, which is close to the operating frequency at Saint-Santin, is from that point of view a good basis

TECHNICAL INPUT PARAMETERS

```
AVERAGE RADIATED POWER :        P_a = 1 kW
ANTENNA DIAMETER :              5 to 20 m
EXTERNAL NOISE TEMPERATURE :    T_E = 20 K
AMPLIFIER NOISE TEMPERATURE :   T_A = 50 K
TRANSMITTING FREQUENCY :        f_o = 900 MHz
PULSE SCHEME : SINGLE PULSE MODULATION
```

TABLE 1

for the numerical estimates which follow, and offers the opportunity to compare theoretical signal-to-noise ratios to those measured in the French experiment.

Main technical input parameters used in that study are summarized in Table 1. Considering that half of the power supplied to the transmitter is actually radiated, and taking 2 kW from the Spacelab energy sources, lead to an average radiated power of 1kW. If the antenna is not oriented towards the ground, a 20 K sky noise can be expected, to which 50 K must be added for the amplifier noise. For the following calculations, ionospheric conditions found in the vicinity of the F2 peak are considered : $N_e = 5.10^{11}$ m^{-3}, $T_a = 20$ T_e, where T_e is the temperature of cold electrons.

To completely determine the calculations, one must indicate the type of pulse scheme to use (see FARLEY, 1969, for details on the various correlation measurement techniques). (6) shows two possibilities. First, if

$$(9) \quad \frac{1}{\tau} < B_{e,i}$$

one comes to the single pulse technique : the various autocorrelation points are calculated within each elementary received pulse. Using equations (1) to (7) and introducing the indicated values of the various imposed parameters gives the signal-to-noise ratios for each of the lines. In order to emphasize the influence of d_o and D, they are considered as the only independent parameters of the problem and are used to express h, τ, and d_{MIN}, which leads, with help of $r = d_{MIN}/d_o$, to :

$$(10) \quad \left(\frac{S}{N}\right)_i = 0.27 \ 10^2 \ \frac{\rho D^2}{d_o} \ (1 - r)$$

$$(10') \left(\frac{S}{N} \right)_e \;=\; 1.7 \; 10^4 \; \frac{\rho D^2}{d_o^{\,2}}$$

Conversely, if :

$$(11) \quad \frac{1}{\tau} \;>\; B_{e,i}$$

one comes to the multiple pulse technique. The various points of the autocorrelation function are calculated by multiplication of one received pulse with the others. Signal-to-noise ratios to be considered are :

$$(12) \quad \left(\frac{S}{N} \right)_i \;=\; 0.54 \quad 10^{-2} \quad \rho D^2 \quad (1 - r)^2$$

$$(12') \left(\frac{S}{N} \right)_e \;=\; 2.3 \quad 10^{-3} \quad \rho D^2 \quad (1 - r)^2$$

3.3. Accuracy of velocity measurements

The error on velocity measurements (PETIT, 1968 ; EVANS, 1969) is finally :

$$\delta V_{e,i} \;=\; \frac{\lambda}{\sqrt{2}} \; \frac{1}{(S/N)_{e,i}} \; \left(\frac{\Delta F_{e,i}}{T} \right)^{1/2}$$

Taking T = 10 s, in order to allow for a satisfactory space resolution (80 km) on board the satellite, then introducing $\Delta F_{e,i}$, the portion of the bandwidth over which the spectrum decreases from its maximum value to zero, by :

$$(13) \quad \Delta F_i \;\simeq\; 5 \text{ kHz}$$

for the ion line, and finally, assuming a simple triangle shaped electron line, corresponding approximately to the pulsed radar technique :

$$(13') \quad \Delta F_e (\text{kHz}) \;\simeq\; \frac{B_e}{2} \;=\; 10 \text{ h}$$

with h in kilometers, for the electron line, one finds :

$$(14) \quad \delta V_i \;=\; 0.174 \; \frac{d_o}{\rho^{1/2} \, D^2 \, (1 - r)}$$

$$(14') \quad \delta V_e \;=\; 0.124 \; 10^{-4} \; \frac{d_o^{\,5/2}}{\rho^{1/2} \, D^2} \; (1 - r)^{1/2}$$

for the single pulse technique, and :

$$(15) \quad \delta V_i = 8.8 \quad 10^2 \quad \frac{1}{\rho^{1/2} \, D^2 \, (1 - r)^2}$$

$$(15') \quad \delta V_e = 0.9 \quad 10^2 \quad \frac{d_o^{1/2}}{\rho^{1/2} \, D^2 \, (1 - r)^{3/2}}$$

for the multiple pulse technique.

Figure 3 shows the variations of B_e, B_i, $\frac{1}{\tau}$ as function of τ (or $h = \frac{c\tau}{2}$). For pulses longer than 33 μs, corresponding to distances longer than 5 km, one can use the single pulse technique for both ion and electron lines. For pulses shorter than 18 μs, corresponding to distances smaller than 2.7 km + d_{MIN}, one can use the multiple pulse technique for all lines, provided that supplementary conditions to be discussed later are fulfilled.

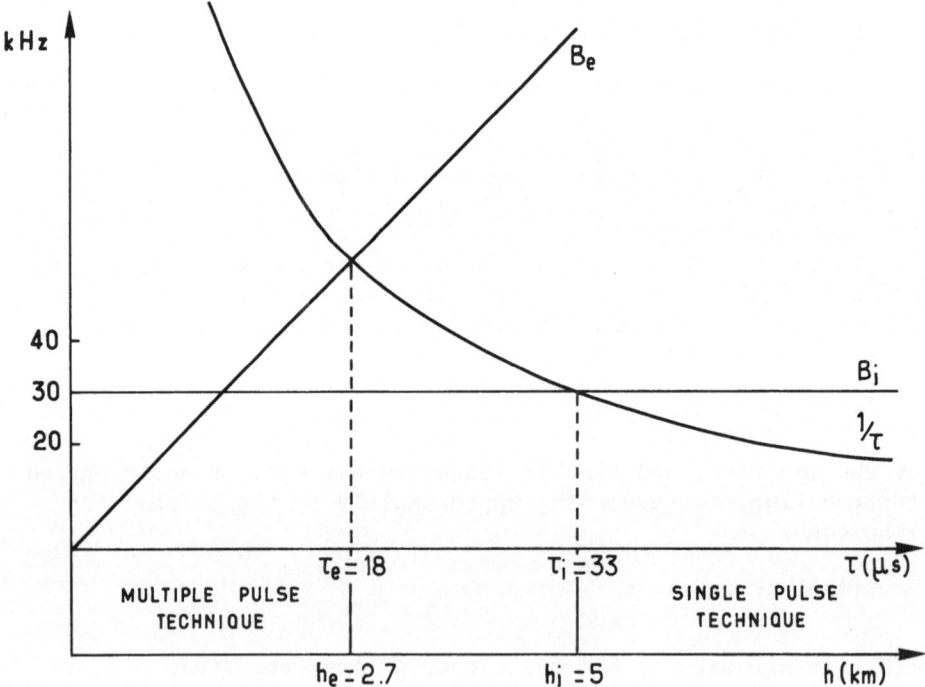

Figure 3. Variation of electron and ion spectra bandwidths B_e, B_i, compared to the bandwidth $1/\tau$ of an elementary pulse, as functions of the length h of the scattering volume, or the pulse length τ and corresponding validity domains of multiple pulse and single pulse techniques.

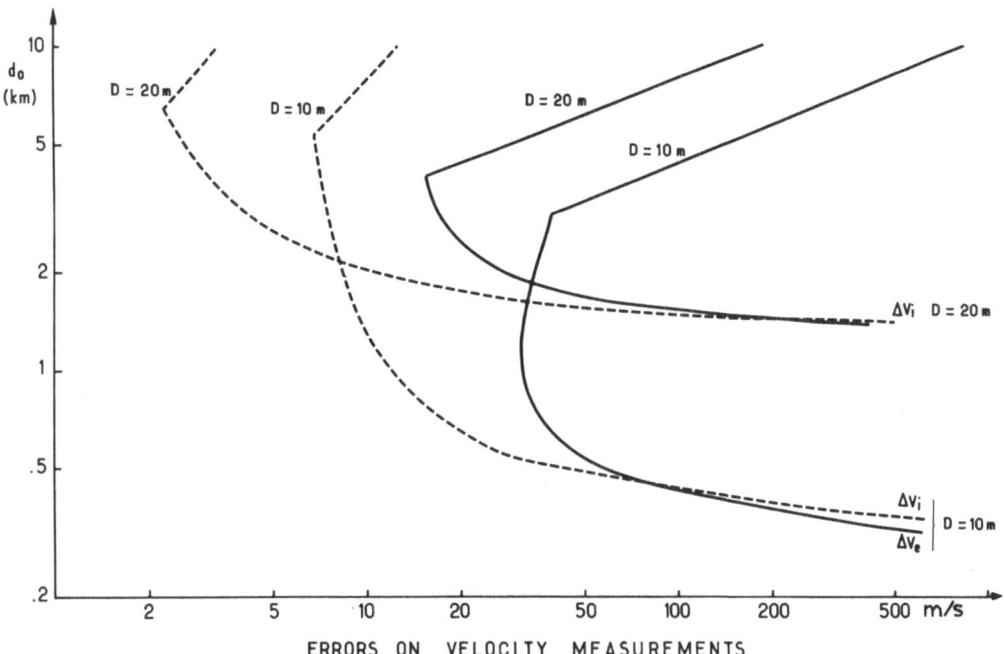

<u>Figure 4.</u> Variations of ΔV_e (full curves) and ΔV_i (dashed curves)
as functions of distance of the mean point of the scat-
tering volume to Spacelab, for 10 m and 20 m diameter
antennas.

Figure 4 gives the resulting variation of δV_e and δV_i as function
of d_o, for 10 m and 20 m diameters antennas. The minimum value of
ρ compatible with the assumed average radiated power, equal to :

$$\rho_{MIN} = \frac{2d_o - d_{MIN}}{d_o - d_{MIN}} = \frac{2 - r}{1 - r}$$

has been introduced for that calculation, so that (Iδ) and (Iδ')
can be rewritten as :

$$(16)\ \delta V_i = 6.2\ 10^2\ \frac{1}{D^2\ (1 - r/2)^{1/2}\ (1 - r)^{3/2}}$$

$$(16')\ \delta V_e = 0.64\ 10^2 \frac{d_o^{1/2}}{D^2\ (1 - r/2)^{1/2}\ (1 - r)}$$

to represent the accuracy obtained when no amplification of the
peak radiated power over its minimum level, defined as $\rho_{MIN}\ P_a$,
is performed.

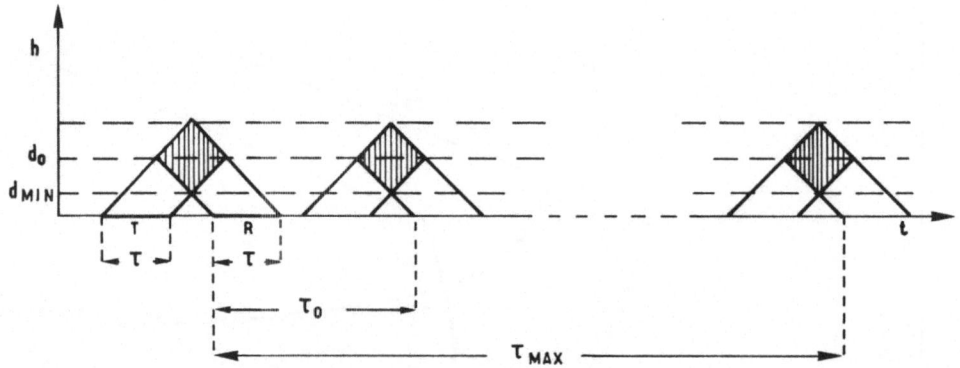

<u>Figure 5</u>. Fundamental pulse scheme used for a multiple pulse corre-
lation function measurement. T and R are the transmitting
and receiving gates, τ_o and τ_{MAX} the elementary and maximum time lags.

3.4. <u>Some elements on the definition of a multiple pulse system</u>

Figure 5 shows the multiple pulse scheme, and its fundamental
parameters, with the transmission gate (T) and the reception gate
(R) of an elementary pulse. The elementary time delay τ_o (first
non-zero point in the correlation function) is related to the observed bandwidth B_{OBS} by :

$$(17) \quad \tau_o = \frac{1}{B_{OBS}}$$

whereas τ_{MAX}, the maximum time delay for which the autocorrelation
function is computed, is related to the spectral resolution δf of
the analysis by :

$$\delta f = \frac{1}{\tau_{MAX}}$$

In order to make the measurement possible, the following une-
qualities must hold :

$$(18) \quad \delta f \ll B_{i,e}$$

$$(19) \quad B_{i,e} < B_{OBS}$$

$$(20) \quad 2\left(\tau + \frac{2d_{MIN}}{C}\right) < \tau_o$$

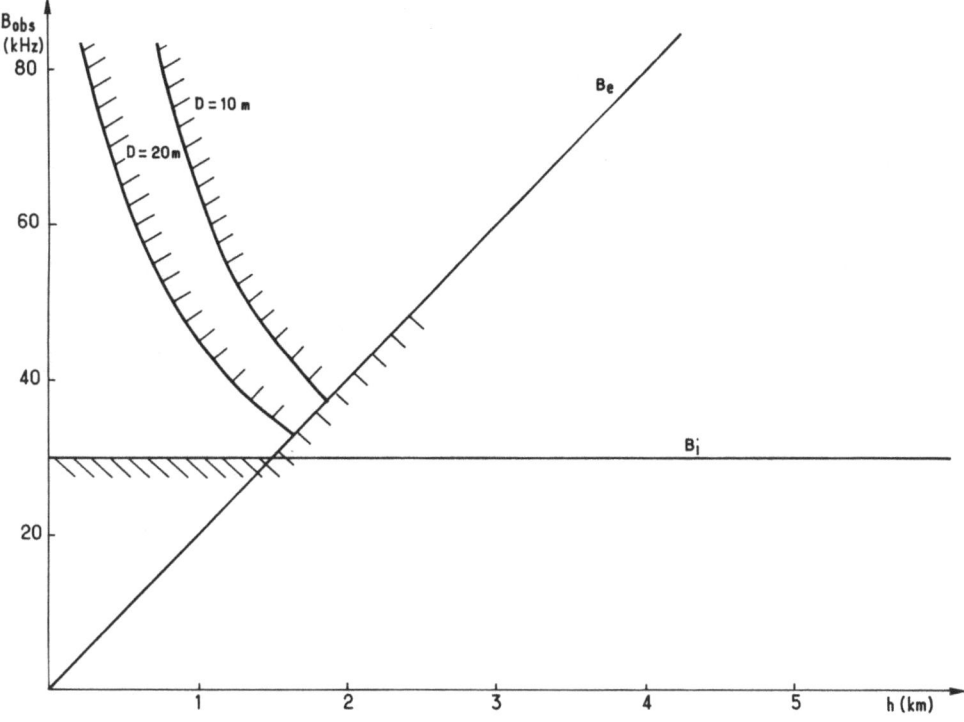

Figure 6. "Feasability domain" of multiple-pulse correlation function measurements in the (h, B_{OBS}) plane.

For the plasma line, (20) gives, by use of (5) and (17) :

$$(21) \quad B_{OBS} < \frac{C}{2(h + d_o)}$$

It is that equation together with (19) which finally determines
the pulse scheme (τ, τ_o, τ_{MAX}) or equivalently (h, B_{OBS}, τ_{MAX}).
Figure 6 shows the domain determined by (19) and (21) in the (h,
B_{OBS}) plane. (21) depends on the antenna diameter D in such a way
that, in order to increase D (which is favorable to the accuracy)
one has to reduce B_{OBS} and h. Reasonable sets of parameters in the
domain of figure 6 are, for example :

$$\begin{cases} D = 10 \text{ m} \quad, \quad d_{MIN} = 300 \text{ m} \quad, \quad h = 800 \text{ m} \\ \tau = 5 \text{ μs} \quad, \quad B_e = 15 \text{ kHz}, \quad B_{OBS} = 75 \text{ kHz} \\ \quad \delta V_e = 31 \text{ m/s} \quad, \quad \delta V_i = 11 \text{ m/s} \end{cases}$$

or

$$
\left\{
\begin{array}{l}
D = 20 \text{ m} \quad , \ d_{MIN} = 1200 \text{ m} \quad , \ h \quad = 1200 \text{ m} \\
\tau = 7 \ \mu s \ , \ B_e = 21 \text{ kHz} \ , \ B_{OBS} = 40 \text{ kHz} \\
\quad \delta V_e = 21 \text{ m/s} \qquad \qquad \delta V_i = 6 \text{ m/s}
\end{array}
\right.
$$

As seen here, one cannot expect to observe more than 80 kHz for the plasma line measurement. It is therefore necessary that this narrow frequency window can be displaced to cover the actual location of the plasma line, which varies a priori by several MHz according to space and time variations of the ionospheric medium. The adequate displacements of the window will be obtained by means of a plasma line tracking system, comprising a real time plasma frequency measurement controlling the local oscillators of the two plasma line acquisition channels.

Figure 7 shows a schematic block-diagram of the resulting radar system : the received signal, after HF amplification, is treated in three different channels for the acquisition, by transposition to zero frequency, low-pass filtering, and digital correlation, of the complex autocorrelation functions of :

- the ion spectrum (local oscillator F_o + filter 1)
- the two plasma lines (local oscillator $(F_o - F'_p)$ and $(F_o + F'_p)$ + filters 2 and 3).

F'_p is synthesized by a local oscillator adjustable by steps, and chosen to be the frequency of the synthesizer spectrum which is closest to F_p, the plasma frequency measured at the considered time. This plasma frequency measurement can be achieved in real time by means, for instance, of a relaxation sounder (HIGEL, 1975 ; HIGEL and DE FERAUDY, 1976).

4. GEOPHYSICAL SITUATIONS OF INTEREST

One of the key elements needed for the understanding of the electrodynamic state of the Earth's ionosphere - magnetosphere system is the current distribution. While the purely ionospheric components (horizontal S_q current system, equatorial and auroral electrojets) and the purely magnetospheric components (magnetopause and tail currents, ring current) are relatively well-known, their interchanges through field-aligned currents are poorly described by presently available data. Their critical importance in many geophysical situations has been reviewed by BAUER et al (1976) : among others, Birkeland currents, connecting the auroral electrojets to the geomagnetic tail, of strengths up to 5.10^{-5} A/m^2, currents in the cusp up to 10^{-4} A/m^2, interhemispherical currents associated

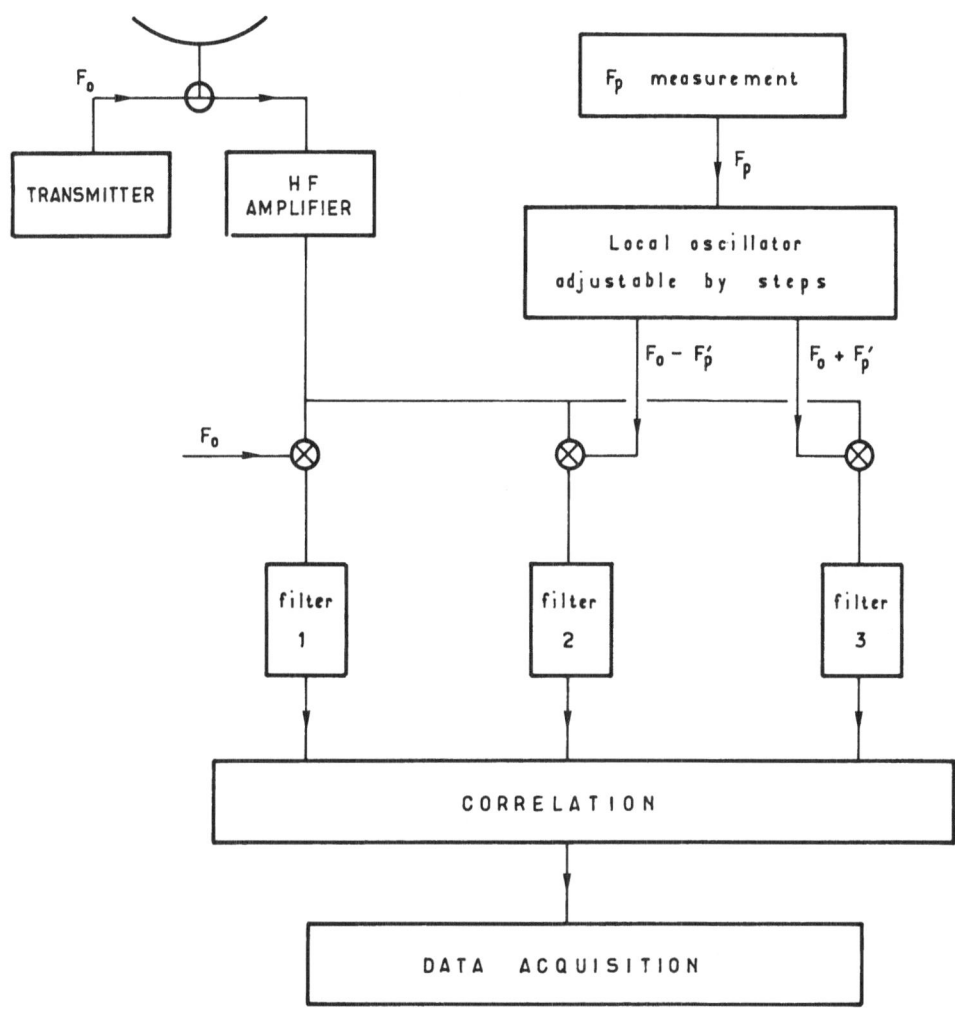

Figure 7. Schematic block-diagram of the radar system, showing the main operations to be performed in the data processing.

with asymetric dynamo action with strengths between 5.10^{-6} and 5.10^{-5} A/m^2. In addition to the estimation of net currents, the knowledge of contributions from particles of various signs and various energies (in particular distinction between thermal and non-thermal) is very important for the analysis of the involved mechanisms, and will be provided by the proposed radar if only :

- measurements can be performed with various carrier frequencies, giving access to various points in the spectrum of energetic electrons,

- they are associated with suitably chosen energetic particle detectors.

Within such a framework, the incoherent scatter technique could lead to the exploration of the structure of current and particle interchanges between the ionosphere and the magnetosphere in space, species and energies.

5. CONCLUSION

The interest of using the incoherent scatter technique for measurements on board Spacelab has been examined. In contrast to other parameters, which have been extensively studied for years by ground-based experiments, measurements of thermal electron velocities appear as a new field of investigation, and could be fruitfully performed during a Spacelab mission. Associated with other parameters deduced from the same incoherent scatter experiment, such as ion drift velocities and photoelectron energy distributions, one could study the morphology of field-aligned currents, of field-aligned transport of the thermal plasma, and come into many exciting questions related to the dynamical coupling between thermal and non-thermal components of the ionospheric-magnetospheric plasma.

Despite radiated powers much smaller than those used from ground level, the possibility of quasi in-situ measurements leads to a convenient accuracy on the determination of ion and electron drift velocities, for a use of the multiple-pulse technique. With a 10 meter diameter antenna, it is proposed to measure one kilometer away from the shuttle in a one kilometer long scattering volume. A first diagram of the associated radioelectric chain, for a simultaneous acquisition of the ion spectrum and the two plasma lines, is proposed.

The scattering volume, very close to the shuttle though far enough to avoid any contamination by it, offers an easy possibility of active experiments in which incoherent scatter could be used as a means of plasma diagnosis, therefore extending the interest of the proposed experiment to questions of plasma physics, and in particular to those relevant to the description of high-latitude phenomena.

REFERENCES

BAUER, P., K.D. COLE and G. LEJEUNE, Field-aligned electric currents and their measurement by the incoherent backscatter technique, to appear in Planet. Space Sci., 1976.

EVANS, J.V., Theory and practice of ionospheric study by Thomson scatter radar, Proc. IEEE, 57, 496-530, 1969.

FARLEY, D.T., Incoherent scatter correlation function measurements, Radio Sci., 4, 935-953, 1969.

HIGEL, B., Non-stationnarités des signaux de résonance des plasmas ionosphérique et magnétosphérique, Ann. Télécom., 30, 239-246, 1975.

HIGEL, B., and H. DE FERAUDY, Space Plasma diagnosis through electrostatic waves : a review of experimental and theoretical results, invited paper at Symposium on active experiments in space plasmas, Boulder, June 1976.

PETIT, M., Mesures de températures, de densité électronique et de composition ionique dans l'ionosphère par diffusion de Thomson. Etude du déséquilibre thermodynamique dans l'ionosphère diurne, Ann. Géophys., 24, 1-38, 1968.

DISCUSSION

P.M. Banks: Measurement of electron velocities parallel to $\vec{B}$ will be extremely important to determine interhemispheric currents. Expected current densities of 10^{-7} A/m^2 require knowledge of the electron velocity to better than a few m/s (within the F_2-region). Can incoherent-scatter in-situ measurements achieve such accuracy?

Also, to measure thermal plasma fluxes of $\simeq 10^8$ cm^{-2} s^{-1} requires v_i" to be known to 5m/s or better (in the F-region). Can this be achieved? (Note that topside fluxes begin only some distance above the F_2-peak and that production of O^+ between 300 and 500 km contributes significantly to the net outward flux.

M. Blanc: Indicated accuracies correspond to a 10 s integration time, and to a system which does not consider any amplification of the peak power above the level of the mean power

by more than a factor 2. It is therefore the accuracies
related to the unamplified levels which have been indicated
here. Now, for the type of phenomena that you mention, which
require improvement of the accuracies by typically a factor 10,
their spatial extent and spatial variation is such that it
should allow for integration times much larger than 10 s. So
short an integration time is in fact only required for
phenomena of short spatial extent, namely auroral phenomena.
But then both involved electric currents and hot particle
fluxes enhancing the plasma line are much stronger, and
should be detectable provided that the plasma line tracking
system is very efficient.

S.A. Bowhill: Will the asymmetry of the plasma lines not make it
difficult to match the upper and lower lines with sufficient
accuracy to measure the plasma drift Doppler velocity?

M. Blanc: This is a very important point, which must be examined
very carefully. However, we don't need an <u>absolute</u>
symmetry of both lines, but a symmetry <u>in shape</u>, that is
except for a normalisation constant, which will be determined
by the regression analysis, provided that regression can
assume the same shape for both lines. Departures from that
relative symmetry will be due only to asymmetries in the
relative variations of the photoelectron distribution
function within the domain of plasma-wave phase velocities
covered by each line separately. For typical F-region daytime conditions, and a 1 km long scattering volume, the
relative variation of v_ϕ within each line is only 3×10^{-3},
and I think the resulting asymmetry should be very small.
Anyway, this point will be investigated experimentally at
Saint Santin, where we intend in the very near future to
look simultaneously at both lines. We think that a definite
answer to your question will be available very soon.

MONITORING OF THE TROPOSPHERICAL AEROSOL USING A SPACELAB BORNE LIDAR SYSTEM

Gerhard H. Ruppersberg and Wolfgang Renger

DFVLR - Institut für Physik der Atmosphäre
D-8031 Oberpfaffenhofen, Post Wessling, FRG.

1. INTRODUCTION

Aerosol particles are reaction partners in physical chemical processes in the atmosphere, they are tracers in atmospheric transport phenomena, indicators for air pollution, condensation nuclei
with the hydrometeors, scatterer and absorber of any radiation
transfer in the atmosphere. With respect to GARP, its influence
on the atmospheric heating rates is of special interest.

Fig. 1 is taken from a survey by BULLRICH and shows the cooling and heating rates of different atmospheric absorbers according to an investigation by DOPPLICK, recalculated and extended
by GELEYN, KORB and PANHANS. On the left the long wave radiation
flux divergence is shown, on the right there is the short wave
(solar) divergence. It is remarkable that in the lower regions
of the atmosphere, below 3 km, the aerosol causes a slight cooling
in the long wave region which however is to be neglected with respect to the effect of water vapour. This is different in the solar spectral range, here the heating rate according to the aerosol is equivalent to the rate due to water vapour and thus is of
considerable influence.

The exponential decrease with height of the aerosol heating
rate corresponds to the aerosol concentration profiles given by
ELTERMAN. Frequently in the boundary layer another vertical distribution of aerosol concentration is found, as e.g. it has been
measured above Germany by DUNTLEY et al. and by PETERSON above
Northwest India etc. (Fig. 2).

The stratospheric aerosol, too, shows significant heating
rates, see Fig. 3 given by KERSCHGENS, RASCHKE and REUTER.

Nevertheless, it would be wrong to value the influence

J. J. Burger et al. (eds.), Atmospheric Physics from Spacelab, 297–313. All Rights Reserved.
Copyright © 1976 by D. Reidel Publishing Company, Dordrecht-Holland.

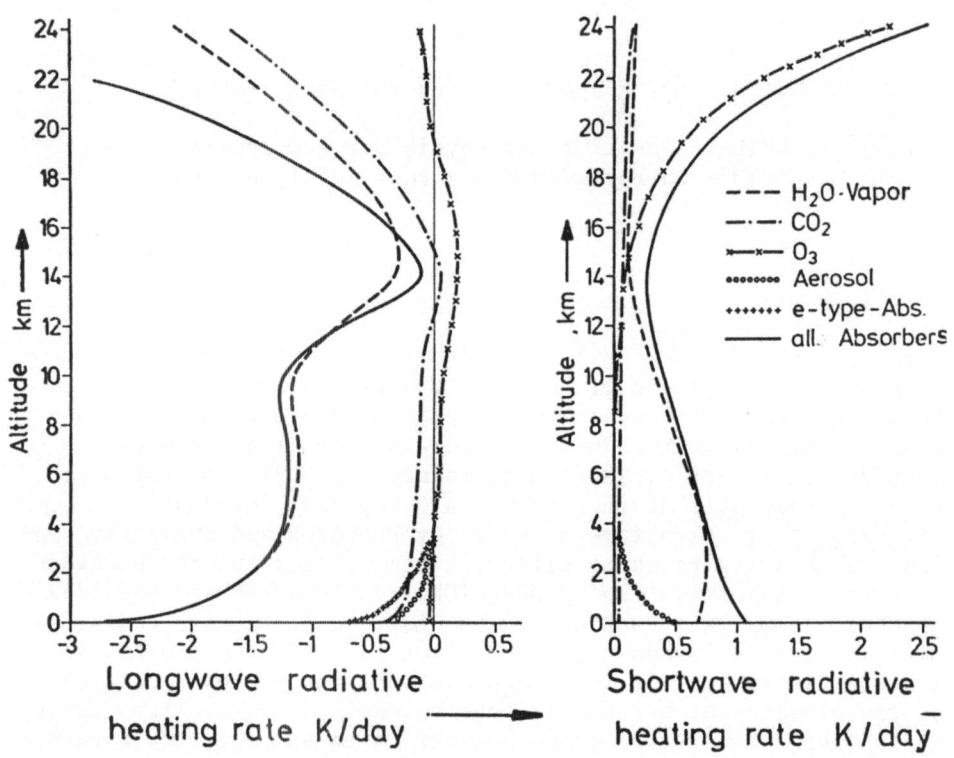

Fig. 1

Radiative heating rates in a cloudless standard atmosphere due to
the absorption by H_2O, CO_2, O_3 and below 3 km altitude e-type
and aerosol absorption; see DOPPLICK, new calculated and extended
by GELEYN, KORB and PANHANS. H_2O-vapor concentration between
1.8 cm pw (ground) and 0.5 pw (3000 m altitude). In the short-
wave range a mean albedo of the earth of A = 0.1 and a mean sun
elevation of 15° was assumed (BULLRICH).

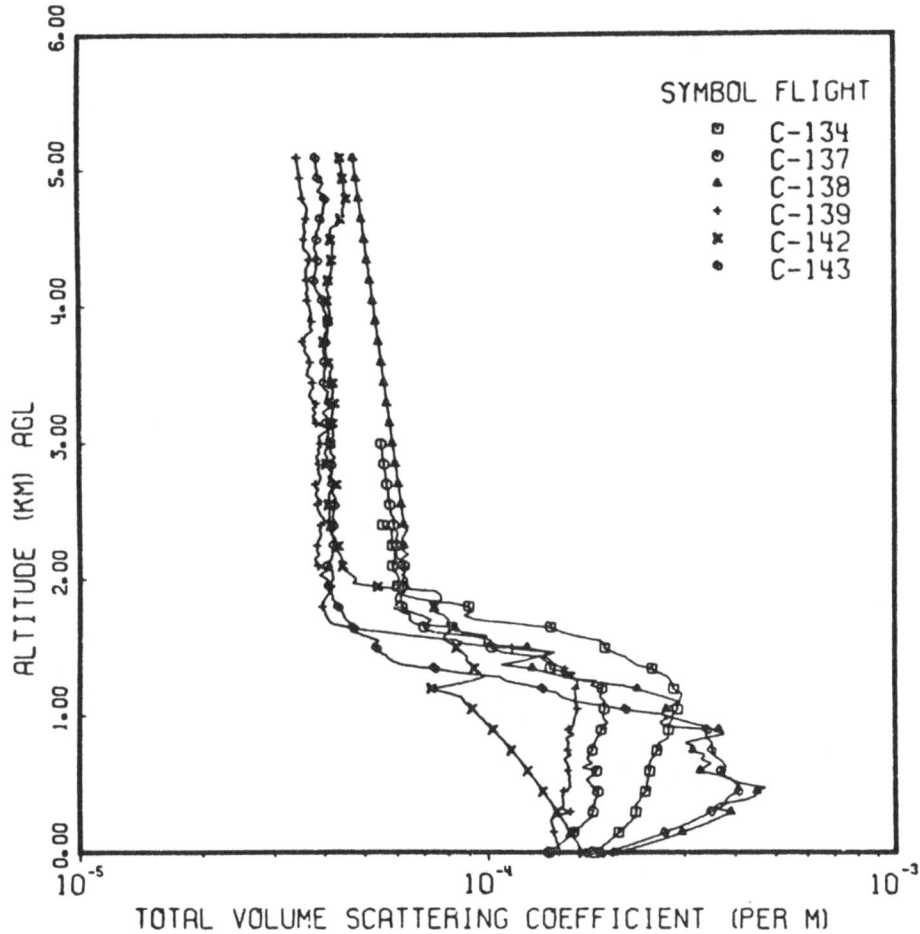

Fig. 2

Vertical distribution of aerosol concentration measured by the
total volume scattering coefficient. Measurements near Memmingen,
Germany, March 1970 (DUNTLEY).

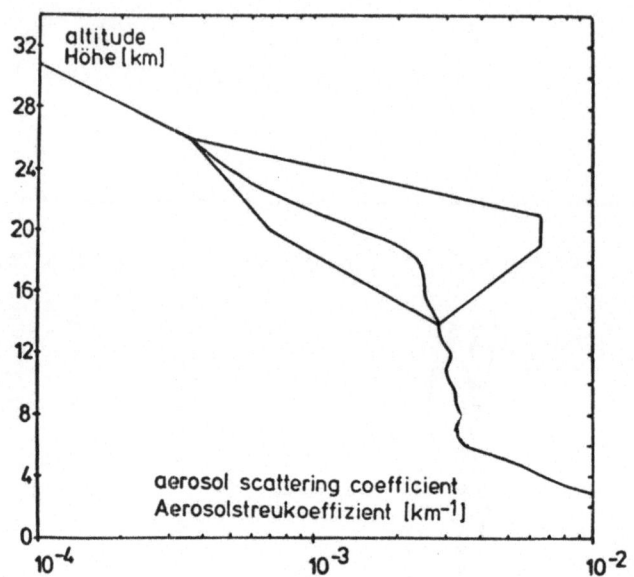

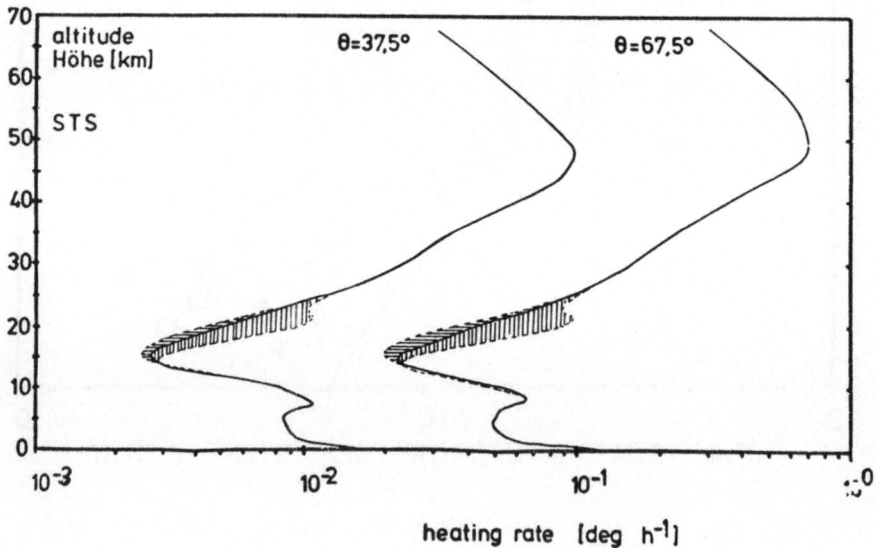

<u>Fig. 3</u>

The effect of lower stratospheric dust on heating (KERSCHGENS, RASCHKE and REUTER).

of aerosols, only looking at these heating rates in haze. Aerosols are the condensation nuclei for cloud droplets; they affect the fraction of absorption and consequently the heating rate and the albedo of clouds (GRASSL); cloud formation and cloud frequency also depend on the aerosol concentration. The heating rate in haze again is depending on the albedo of the surface below. Finally the global albedo (R in Fig. 4) is affected by the aerosol in such a way that bright regions on the surface of the earth (A = 0.8 in Fig. 4) appear to be darker with increasing turbidity T and dark regions on the surface of the earth (A = 0 in Fig. 4) appear to be brighter with increasing turbidity. Taking the assumptions on which calculations in Fig. 4 are based, a ground albedo A = 0.25 appears unchanged by the overlying aerosol. Under these conditions the aerosol is undetectable by visual or photographic methods.

On the whole, there is no doubt that the tropospheric aerosol has an substantial influence on climate and in a larger sense on the biosphere. In addition to global measurements from spacelab, and owing to complicated interrelations, beside further theoretical work, series of accompanying measurements on ground and on board of aircrafts are necessary. As much as it is known today, one can assume that such additional measurements are sufficient for a better interpretation of the lidar dates.

2. OBSERVATIONAL POSSIBILITIES OF THE LIDAR SYSTEM

2.1 General Remarks

Tab. 1 - GARP Publ. Ser. No. 16 - gives a summary of tentative observational requirements for a better understanding of aerosol-processes in the atmosphere (see also BOLLE et al.). Details of some values indicated in the left hand column possibly will become measurable on the ground in the near future within the postulated accuracy; it seems to be questionable from a ballon and impossible from a satellite. In paragraph 5.9.4.2 (right column) these difficulties have been considered and with regard to the Global Observation Program, it has been stated:" It is too early to specify in detail an observational program for validation of global models. A program along the following lines is suggested. It seems essential to obtain the aerosol size distribution and its vertical profile for all regions of the globe. For the other parameters listed in Table (1), such as the refractive index, etc., it is perhaps possible to derive satisfactory average values which can be applied for the different geographical regimes. Chemical analyses for the important constituents will certainly be required ...".

A monostatic lidar - a system in which transmitter and receiver are close together or even coaxial - measures the product

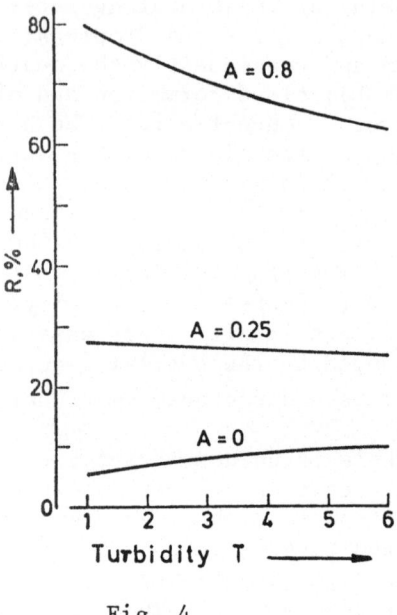

<u>Fig. 4</u>

The global albedo R as a function of LINKE's turbidity T and the albedo of the earth A. R is the percentage of the incident sun radiation reflected back into space. Calculations are valid for 0.55 μm wavelength, m = 1.5 – 0.02 i, latitude 53° (BULLRICH).

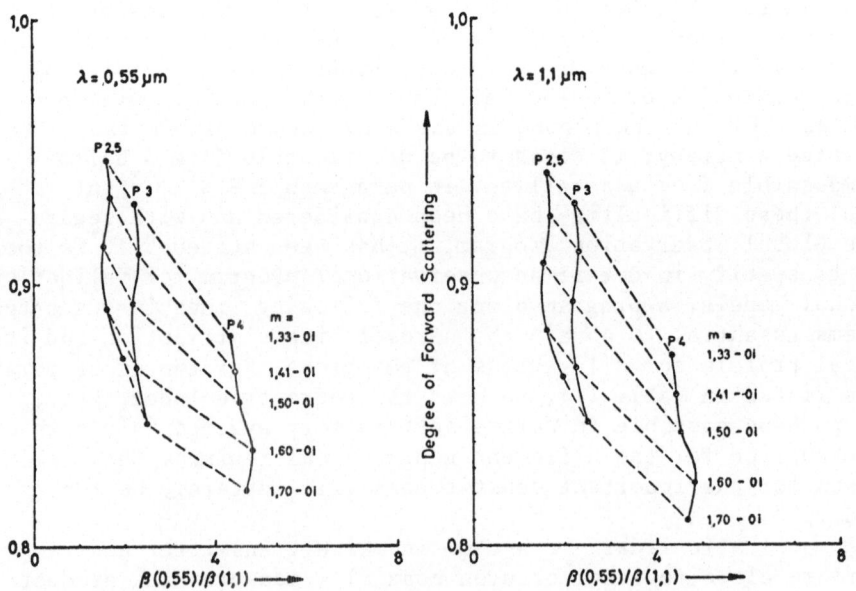

Tab. 1 - (GARP Publ. Series)

Aerosol Processes — Summary of Tentative Observational
Requirements (see text for detail)

I. *Study of Processes*

(a) Radiative effects of aerosols.

Required aerosol parameter for troposphere and stratosphere	Observational requirement and accuracy
Size distribution	
$\frac{dn}{dr}$ in cm^{-4} STP	5%
Vertical profile of size distribution	5% Required vertical resolution generally 0.5 to 1.0 kilometer
Real refractive index of bulk material n	1% over the range $1.0 \leqslant n < 2$
Imaginary part of the refractive index k	10% over the range $0.001 < k < 0.1$
Bulk density δ of aerosol particles, in g cm^{-3}	5% over the range $1.0 \leqslant \delta < 3.0$
Solubility of aerosol particles and/or growth characteristic with relative humidity	Use of 3 to 4 typical growth curves

For necessary data to calculate energy balance of the atmosphere see 5.2.5 (albedo atlas)

Required aerosol parameter for troposphere and stratosphere	Observational requirement and accuracy
(b) Aerosol cloud interaction	Cannot be specified at this time
(c) Volcanic events — stratosphere	See detailed specification in 5.9.4.1

II. *Global Observation Programme*

Global aerosol climatology required — see 5.9.4.2

III. *Monitoring*

Variables to be monitored (see also 5.9.4.3)	Space resolution	Time resolution	Accuracy
(1) Total number concentration (2) Concentration of optically important particles (3) Total mass concentration (4) Concentration of gaseous precursors	about 20 baseline stations distributed over the globe	daily	5%

For stratospheric aerosols see 5.9.4.3

◀ Fig. 5

Degree of forward scattering as a function of the quotient of the measured backscattering coefficients for different power law distributions and different refractive indices of the aerosol particles.

The degree of forward scattering is defined as

$$\int_{0}^{\pi/2} \beta(\Theta) \sin\Theta \, d\Theta \Big/ \int_{0}^{\pi} \beta(\Theta) \sin\Theta \, d\Theta$$

$\beta\tau^2$. β is the backscattering coefficient of the actually regarded
scattering volume of the atmosphere depending on divergency and
gate interval. τ is the transmittance between instrument and
volume. Scattering objects are molecules, aerosol particles, cloud
elements or the surface of the earth. Measuring downward into the
atmosphere, τ^2 in general does not much differ from 1. Only in the
case of high aerosol concentration normally existing in the boundary layer, τ^2 may be significantly smaller than 1. It is assumed
that the lidar is working out of resonant absorption lines of atmospheric gases.

The suggested spacelab borne lidar system primarily measures
the backscattering coefficients at the wavelengths 1.06 μm and
0.53 μm. It is not very promising to proceed from these backscattering coefficients via the particle size distributions to get
the radiation divergences. It would be more advisable to examine
the correlation of the backscattering coefficients with those
factors which become part of the calculation of the radiation divergences. These factors are: The normalized scattering function,
the volume scattering coefficient and the volume absorption coefficient.

2.2 Normalized Scattering Function

The normalized scattering function with the scattering angle Θ
as variable indicates the probability that a photon after a collision with the particle, is being scattered into the direction
defined by Θ. Similar to the scattering coefficient, it can be
assumed, that the wavelength dependence of the backscattering
coefficient stands in close relation to the type of scattering
function: Steep particle size distributions (those with many small
particles) lead to strong wavelength dependences of the scattering
and of the backscattering coefficient and to smooth scattering
functions, less steep particle size distributions (in which large
particles predominate) lead to minor wavelength dependences and
to a remarkable higher forward scattering.

Fig. 5 shows the relations for quite some different particle
size distributions and refractive indices. It is based on the
study by QUENZEL, RUPPERSBERG and SCHELLHASE as well as on some
additional scattering functions by QUENZEL. The constant refractive indices are proceeding as expected. But the different refractive indices entail displacements of the curves. In the example
of Fig. 5 a relationship of the measured backscattering coefficients of e.g. 4 leads to the statement that the concerning aerosol produces between appr. 82% and 88% forward scattering (measured from the total scattering). If the relationship of the backscattering coefficient is 1, a 85% to 96% forward scattering can
be estimated. The spread of results is remarkably limited if informations exist about the refractive index. They should be provided
by random test measurements (see paragraph 2.3). The relation of

the backscattering coefficients at different wavelengths measured
with lidar, in this case informs, whether one is still in the range
where the results of single in-situ measurements are valid.

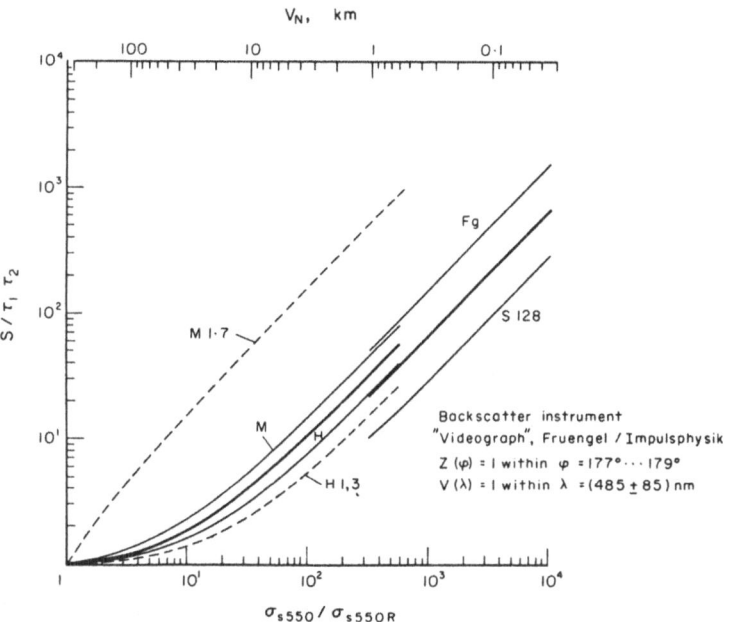

Fig. 6

(See QUENZEL, RUPPERSBERG and SCHELLHASE). Reduced input signal of a
backscatter instrument "Videograph" versus the reduced scattering
coefficient at 550 nm wavelength (lower abscissa) and the standard
visibility near the ground (upper abscissa). For standard visibilities greater than 0.6 km the curves stand for 21 different
haze particle size distributions. For standard visibilities less
than 1 km the curves are calculated for 252 different cloud and
fog particle size distributions. The additional (dashed) lines are
maximum and minimum curves for all size distributions for extreme
refractive indices m = 1,33 - 0 i and m = 1,70 - 0 i.

2.3 Volume Scattering and Absorption Coefficient

By the question of correlation between the backscattering coefficient β and the scattering coefficient σ_s, one is touched by an
about 20 years lasting conflict between those people recommending

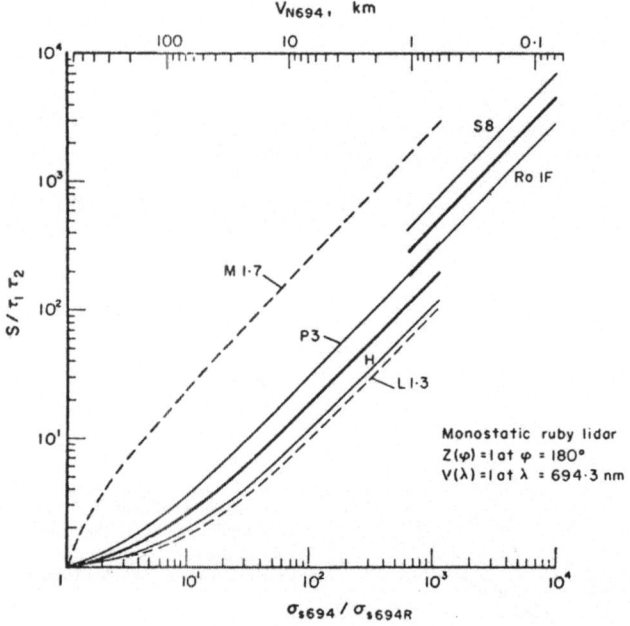

Fig. 7

As for Fig. 6 but for Ruby Lidar.

the backscatter principle for visibility measurements and those
opposing it. It is reasonable to start with the results which have
been achieved in course of the discussion.

On the one hand the relation β/σ_s was calculated by means of
the MIE-theory based on experimentally or empirically found particle size distributions and complex refraction indices. Fig. 6
shows the result for the (assumed) modification of a well-known
backscattering instrument (see QUENZEL, RUPPERSBERG and SCHELL-
HASE). It is obvious that results are slightly different for haze
(left curve) and for clouds and fog (right curve); as already demonstrated in the past, there is a typical error given by a factor
of 2 towards smaller and larger values. Much larger error limits
are resulting from varying the real component of the refractive
index between the observed extreme values 1.33 and 1.7 (HAENEL).
Fig. 7 shows a similar trend for the Ruby-Lidar which is valid for
the Neodymium-Lidar as well.

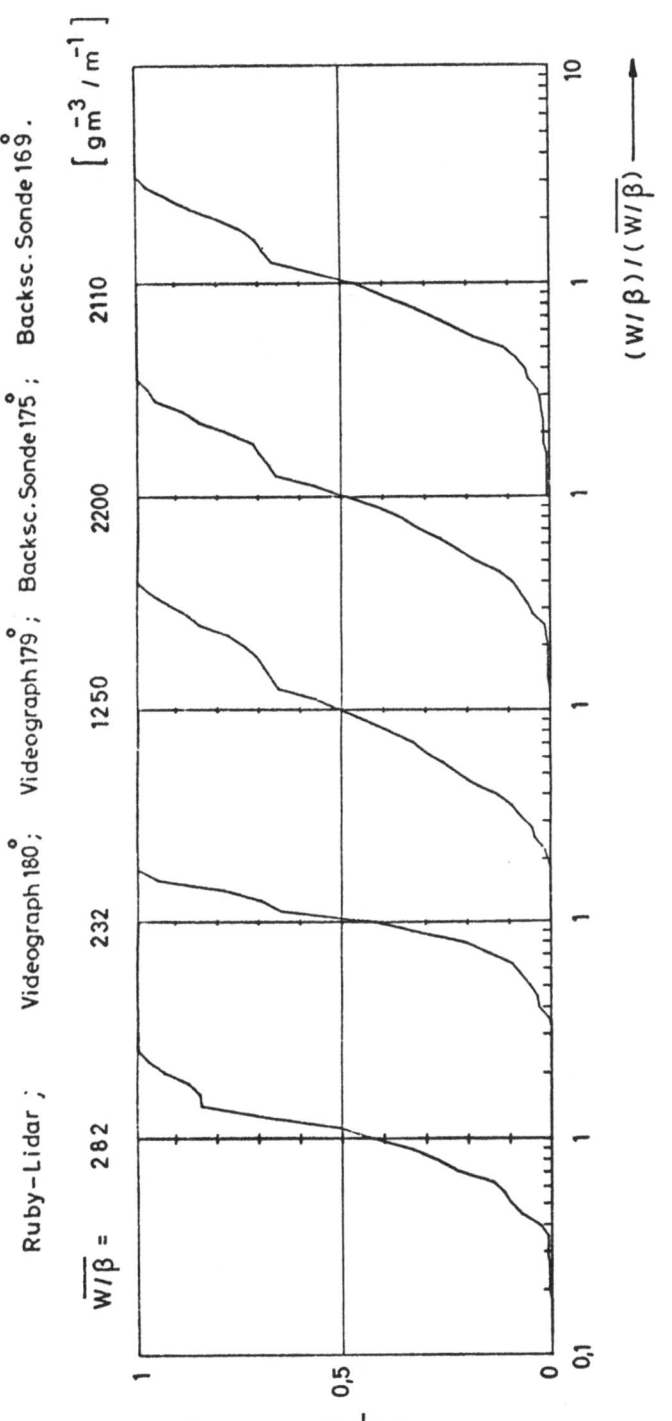

Fig. 8

Integrated frequency of water content W [g m^{-3}] over backscattering coefficient β [sτ^{-1} m^{-1}] for ruby lidar and some different backscatter instruments. Calculations are based on 232 particle size distributions for clouds and fog.

It is very surprising, on the other hand, that by direct comparisons of backscattering instruments with instruments measuring
the scattering coefficient σ_s or the extinction coefficient σ_e,
indeed different relations between $\beta/\sigma_{s,e}$ and σ_s or σ_e have been
found, however, that there have never occured those large devia-
tions which due to variations of particle size distributions and
refractive indices would have been expected (CURCIO and KNESTRICK.
EINGRIEBER; VOGT; HOCHREITER). Maximum deviations in the order of
30% and about 20% average have been reported.

These direct comparisons are extremely difficult and therefore not very frequent. Nevertheless, there is every reason to believe in these apparently contradictory results - as also shown
in other investigations (see paragraph 3) - that there exist in
the various regions characteristical aerosol regimes which do not
have the whole range of variation as one should expect.
As already mentioned in the introduction, the accompanying ground
based and airborne measurements therefore are necessary for yielding reasonable values of the generally not very good correlation
between β, σ_s and σ_e.

It is interesting that for clouds and fog the backscattering
coefficient β is proportional to the water content $W/[g/m^3]$. For
a great number of particle size distributions (SCHICKEL) due to
clouds and fog, deviations by a factor of 2 from the average must
be taken into account (Fig. 8).

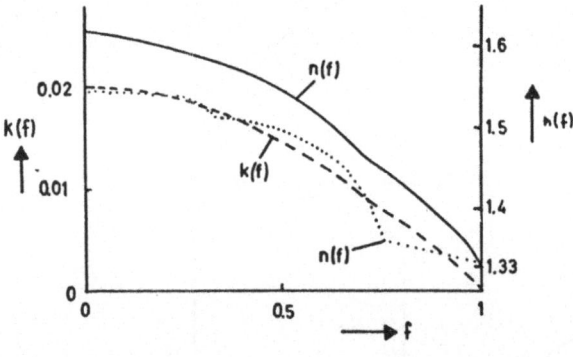

Fig. 9

Real part n (solid line) and imaginary part k (dashed line) of

the mean refractive index of atmospheric aerosol particles as func-

tions of the relative humidity f (mean values for aerosol particles

collected at Mainz, Germany, during summer 1966). Real part n (dotted)

line) of the refractive index of sea-water aerosol as function of

the relative humidity f. All values for increasing f from f = 0

(HAENEL).

The volume absorption coefficient cannot be measured using lidar instruments; from our present knowledge it seems to be utopic to believe that perhaps this could be done by means of N_2 Raman scattering. So far, the absorption coefficient either could be determined indirectly from measurements of sun- and sky-radiation using complicated numerical models (LEUPOLD, QUENZEL), or it could be measured from aerosol samples (FISCHER; see EIDEN and ESCHELBACH too). The received results make believe that it is possible to derive satisfactory average values which can be applied for the different geographical regimes.

3. REDUCTIONS

The variety of observation results can be reduced and more simply interpreted by considering the known relation with the given boundary conditions and the meteorological parameters. Well known are the correlations between the aerosol properties and the humidity U. In 1971 HAENEL found as mean values for aerosol samples collected at Mainz, Germany, during summer 1966 for U = 0 ... 100% a monotonic decrease of the real component of the refractive index slightly above 1.6 to 1.33. At the same time the imaginary part decreases from 0.02 to 0. The real component of the refractive index of sea-water-aerosol decreases from values of somewhat above 1.5 to 1.33 (Fig. 9). For the haze extinction coefficient as a function of the humidity U between 40 and 99.9%, HAENEL (1972) finds - in extending relations found by KASTEN (1968 and 1969) - with significant approach the term

$$(1) \quad \frac{\sigma_{eD}(U_1)}{\sigma_{eD}(U_2)} = (\frac{1 - U_2}{1 - U_1})^{2\varepsilon^{\ast}} \qquad (\lambda = 0{,}55 \ \mu m)$$

For 6 different aerosol types the exponent has values in the range $2\ \varepsilon^{\ast}$ = 0.46 ... 0.69. Only in the range of the last percents before U = 100%, the curve becomes so steep that reductions in praxi are problematic because of the measuring accuracy of the relative humidity.

WINKLER gave an explanation for the fact that even though the nature of the atmospheric aerosol is very complicated, its behaviour can be described quite good by such a simple relation. He was able to show with continental aerosol tests that the aerosol particles at least at radii r > 0.1 μm have the same composition of different substances.

Reductions of this kind are required e.g. for estimations of the aerosol particle mass too. This becomes evident from the fact that continental aerosol particles at U = 70% consist of water up to an amount of 50% (mass).

Among the natural sources for aerosol particles, the sea- and earth surface on one side and on the other side the photochemical process in the atmosphere with participation of H_2S, NO_x, NH_3 and organic substances are of the same order of magnitude (JUNGE; ROBINSON and ROBBINS). A local aerosol concentration due to the first mentioned sources is depending on the production rate on the surface of the earth, on the horizontal and vertical transport and on the wash-out and deposition. It should be possible to simulate these processes with numerical models, so that one can discern in a local aerosol concentration between the influence of changes e.g. of the structure of the surface and the influence of the synoptic situation. The adequate treatment of the photochemical production is to some extent more complicated but none the less important. The same is valid for the anthropogenic production, presently being 1/8 of the total aerosol particle production (appr. 2.5×10^9 t/a).

Aerosol monitoring and other problems connected with the presence of aerosols definitely cannot be solved without these kinds of numerical models. The aerosol-lidar in spacelab in consequence would have to be used for the verification of these models. For this problem surely several missions during different seasons are desirable. (Short missions, as the first spacelab flight, should be recognized as a first but promising step).

4. CONCLUSIONS

It is possible with the Spacelab Borne Lidar to draw a conclusion from the measured two spectrally different backscattering coefficients on the type of scattering function and on the scattering coefficient and especially its vertical and horizontal variation. Technical requirements to the lidar instrument itself as well as to the organization of data processing, nevertheless, are very demanding; they claim a rather considerable effort by industries and scientists involved in application of instruments and analysis of data.

For the analysis accompanying ground based and airborne measurements and perhaps passive aerosol sounding from Spacelab are necessary and adequate in order to seize the special properties of the prevalent aerosol type, required for the reduction and interpretation of the lidar data.

The aerosol monitoring and problems coherent with the aerosol most likely only can be solved to everyone's satisfaction in connection with accordingly adapted theoretical-meteorological simulation models. In consequence, a main advantage of a spacelab borne lidar giving global coverage, seems to be the possibility to verify these models, rejecting the disadvantage of temporarily limited missions.

5. LITERATURE

BOLLE, H.J. REDEMANN, E. KUNKEL, B. VOLZ, F.	(1975)	Air quality measurements from space platforms. ESA CR-577.
BULLRICH, K.	(1975)	Die Rolle der Spurenstoffe im Strahlungshaushalt der Atmosphäre. Promet 3, p. 19-22.
CURCIO, J.A. KNESTRICK, G.L.	(1958)	Correlation between atmospheric transmission and back-scattering. Journal of the Optical Society of America 48, p. 686-689.
DOPPLICK T.G.	(1971)	Radiative heating, climatic effects. Report of the study of critical environmental problem. SCEP-Rept., Cambridge/Mass., MIT-Press, p. 86.
DUNTLEY S.Q.	(1975)	Measured visible spectrum properties of real atmospheres. AGARD conference print on optical propagation in the atmosphere (to be published).
EIDEN, R. ESCHELBACH, G.	(1973)	Das atmosphärische Aerosol und seine Bedeutung für den Energiehaushalt der Atmosphäre. Zeitschrift f. Geoph., 69, p. 189-228.
EINGRIEBER, G.	(1966)	Über Erfahrungen mit Sichtmeßgeräten in Meppen. Sonderbücherei der Ortung und Navigation der Deutschen Gesellschaft für Ortung und Navigation, Düsseldorf, B.-Nr. 1055.
ELTERMANN, L.	(1966)	An atlas of aerosol attenuation and extinction profiles for the troposphere and stratosphere. AFCRL-66-828, Env. Res. Pap. No. 241, 128 p.
FISCHER, K.	(1970)	Bestimmung der Absorption von sichtbarer Strahlung durch Aerosolpartikel. Beiträge zur Physik der Atmosphäre 43, p. 244-254.
GARP	(1975)	The physical basis of climate and climate modelling. GARP Publication Series No. 16, 265 p.
GELEYN, J. KORB, G. PANHANS, W.	(1975)	to be published; see Bullrich.
GRASSL, H.	(1975)	Albedo reduction and radiative heating of clouds by absorbing aerosol particles. Contributions to Atmospheric Physics (Beitr. Phys. Atmosph.) 48, p. 199-210.
HAENEL, G.	(1971)	New results concerning the dependence of visibility on relative humidity and their significance in a model for visibility forecast. Contributions to Atmospheric Physics (Beitr. Phys. Atmosph.) 44, p. 137-167.
HAENEL, G.	(1972)	Computation of the extinction of visible radiation by atmospheric aerosol particles as a function of the relative humidity, based upon measured properties. Aerosol Science, 3, p. 377-386.
HOCHREITER, F.C.	(1969)	Analysis of visibility observation methods. Weather Bureau, Sterling, Va., Test and Evaluation Lab. WBTM T/EL g, 47 p.
KASTEN, F.	(1968)	Der Einfluß der Aerosol Größenverteilung und ihrer Ände-rung mit der relativen Feuchte auf die Sichtweite. Beiträge zur Physik der Atmosphäre 41, p. 33-51.
KASTEN, F.	(1969)	Visibility forecast in the phase of pre-condensation. Tellus 21, p. 631-635
JUNGE, Chr.	(1971)	Der Stoffkreislauf der Atmosphäre. Jahrbuch der MPG zur Förderung der Wissenschaften c.V. Tab. 2, p. 171.

KERSCHGENS, M.J. (1975)   Theoretical studies of the transfer of solar radiation
RASCHKE, E.   in the atmosphere.
REUTER, U.   AGARD CPP-183, p. 15.1-15.10.

LEUPOLT, A.   (1966)   Bestimmung der Kontinuum-Absorption im Spektralbereich
                            von 0,5 bis 2,5 um.
                            Optik 23, p. 538-558 and p. 567-588.

PETERSON, J.W.   (1968)   Measurement of atmospheric aerosols and infrared radi-
                            ation over Northwest India and their relationship.
                            Technical Report No. 38, University of Wisconsin,
                            Dept. of Meteorology, Madison, Wisconsin 53706, 165 p.

QUENZEL, H.   (1967)   Optische Bestimmung der Kontinuum-Absorption maritimer
                            Luftmassen im Spektralbereich der Sonnenstrahlung.
                            "Meteor"-Forschungsergebnisse Reihe B Nr. 1, p. 36-40.

QUENZEL, H.   (1976)   Normalized scattering functions for different size dis-
                            tributions and complex refractive indices.
                            Private communication, Met. Inst. of the University
                            Munich, Theresienstraße.

QUENZEL, H.   (1974)   Calculations about the systematic error of visibility-
RUPPERSBERG, G.H.   meters measuring scattered light.
SCHELLHASE, R.   Atmospheric Environment 9, p. 587-601.

ROBINSON, E.   (1971)   Emissions, concentrations, and fate of particulate atmo-
ROBBINS, R.C.   spheric pollutants.
                            Final Report, SRJ Proj. 8507 for American Petroleum
                            Institute.

SCHICKEL, K.P.   (1975)   Sampling of droplet distributions from water clouds.
                            DLR-Mitt. 75-25, 177 p.

VOGT, H.   (1968)   Visibility measurement using backscattered light.
                            Journal of Atmospheric Sciences 25, p. 912-918.

WINKLER, P.   (1973)   The growth of atmospheric aerosol particles as a func-
                            tion of the relative humidity-II. An improved concept
                            of mixed nuclei.
                            Aerosol Science, 4, p. 373-387.

WERNER, Chr.   (1976)   Airborne Lidar measurements near Munich.
                            Private Communication, DFVLR, D-8031 Oberpfaffenhofen.

DISCUSSION

L. Thomas: I should like to draw attention to the application of
   laser experiments on Spacelab to studies of clouds. Depending on the particle concentration within the cloud, and
   therefore the type of cloud, information could be obtained on
   the geometrical and optical thickness of clouds, as well as
   their topside heights.

W. Renger: For the treatment of signals coming out of thick
   clouds (if ever), multiple scattering has to be taken into
   account which makes interpretation difficult. As shown in
   my talk, it should be possible to determine the water content
   in g/m^3 at the topside of clouds.

G. Hunt: You showed examples of the effect upon atmospheric
   heating rates of tropospheric and stratospheric aerosols.
   What effect do these aerosols have upon the surface temperature?

W. Renger: Due to scattering and absorption along the path,
   surface temperature during the day is lower, due to backscattering radiation. Cooling of the surface at night is
   not so effective as in the absence of aerosols. I can't
   give numbers for these effects.

INVESTIGATION OF PLASMA INSTABILITIES IN EQUATORIAL
'SPREAD F' BY LANGMUIR PROBE USING SPACELAB

S.P. Gupta

Institut für Physikalische Weltraumforschung,
Freiburg, Germany

It is planned to study equatorial F-region plasma instabilities
using rockets and Spacelab, including near-simultaneous measurements of the horizontal and vertical variations of the parameters
relevant to the generation of electron-density irregularities in
the F-region. The horizontal parameters can be measured using
Spacelab and the vertical parameters using rockets at Thumba.
Some of the parameters to be measured by Langmuir Probe using
Spacelab were discussed during this presentation.

J. J. Burger et al. (eds.), Atmospheric Physics from Spacelab, 315. All Rights Reserved.
Copyright © 1976 by D. Reidel Publishing Company, Dordrecht-Holland.

ATMOSPHERIC EXPERIMENTS WITH THE SPACELAB LIDAR:
A DISCUSSION OF THE MAJOR FEATURES OF THE METHOD

J. Blamont, G. Mégie and M.-L. Chanin

Centre National de la Recherche Scientifique
Centre National d'Etudes Spatiales
Service d'Aéronomie du CNRS
Verrières-le-Buisson
France.

The lidar should not be considered for routine observations, which can be better made with passive remote sensing devices operated on board smaller spacecraft, but should be devoted to specific tasks.

DESCRIPTION

1. System

Three different methods have to be employed concurrently;

- vertical sounding with time (that is, space) resolution which provides the structure
- ground echo reflection which provides the integrated vertical total density
- reflection over a subsatellite which provides the integrated horizontal density : this method provides also the vertical structure.

2. Laser

For atmospheric experiments dye lasers appear as the only solution; they provide today a total coverage of the spectral domain from 300 nm to 900 nm with the following nominal performances:

- energy   : 1 Joule or 3×10^{18} photons per pulse in the visible

                     0.1 Joule or 10^{17} photons per pulse in the UV

J. J. Burger et al. (eds.), Atmospheric Physics from Spacelab, 317–327. All Rights Reserved.
Copyright © 1976 by D. Reidel Publishing Company, Dordrecht-Holland.

- repetition rate : 1 to 10 Hz
- line width   : 1 pm
- duration of the pulse : 1 μ sec.

For the FSLP[1] the available output power may be 10 times smaller.

We have presented in our proposals to ESA two new ideas in order
to solve two of the difficult points.

a) Modulation

The wave vector $\vec{k}$ of the laser wave is modified by diffraction in
the Bragg angle Θ created by an electroacoustic crystal placed
inside the cavity and excited with the acoustic wave $\vec{K}$

$$\sin \Theta_0 = \frac{|\vec{K}|}{2|\vec{k}|}$$

The wave vector $\vec{k}'$ of the laser light is now

$$\vec{k}' = \vec{k} \pm \vec{K}$$

therefore

$$\lambda' = \lambda_0 \pm \Delta\lambda$$

if λ_0 is the wavelength without modulation, λ' the wavelength
of the modulated laser and $\Delta\lambda$ is a shift in wavelength.

In the cavity the two lateral wavelengths are generated and can
even be separated spatialy. We have obtained this effect by
using a commercial cavity dumper at 470 MHz ($\Delta\lambda = 0.544$ pm).

It is therefore possible to build transmitters with a bandwidth
of 1 pm (0.4 pm is easily achieved in the laboratory) displace-
able by $\pm$ 0.5 pm. This displacement is known with six digits
accuracy.

b) Power available on board

The efficiency of a laser lies usually in the 10^{-3} range; 2 kWatts
available cannot therefore provide more than about 2 Watts or
2 Joules/s of laser light (excluding CO_2 laser). During the FSLP,
the power is limited to 1 kWatt or 1 J/s and if low efficiency
systems as laser-pumped lasers are used, the available laser light
energy falls to 0.1 Joules/s.

(1) First Spacelab Payload

We have therefore proposed to use a solar-pumped laser. Solar-pumped CW Yag lasers have been made to work in the US and also in our laboratory. The efficiency achieved up to now is 5 Watts multimode (CW) at 1.06 μ with a mirror of diamter 30 cm. With a solar energy of 1.4 kWatts m^{-2} this corresponds to an efficiency of 5%. Such a system with a 30 cm diameter mirror is a possible candidate for the FSLP and would be used to pump a dye laser, providing a CW 0.5 Watt power of light in the visible. The continuous beam can be pulsed at repetition frequences 1 - 10 kHz and also modulated at 500 MHz if needed.

It is realistic to envisage a 1 m^2 mirror for AMPS and with an increase of efficiency which can be obtained by doping the rod in order to maximize the effect of the solar spectrum a possible number of photons of 2.10^{19} per second can be thought of, without drain on the power available on board.

3. Major advantages of the system

In our view the value of any remote sensing method using a part of the electromagnetic spectrum depends upon the spectral density of the signal that is the number of photons available per frequency unit : this parameter characterizes the possibility of determining temperatures and wind velocity.

Therefore the principal characteristic of the FSLP lidar is that it can provide, during 10^{-6} sec, 3×10^{18} photons/pm.

Advantages : High spectral density
             No dispersive system needed
             Large entrance pupils
             Time resolution

DISCUSSION OF THE ACCURACY OF THE EXPERIMENT

1. The flux Φ received by the lidar can be written as a function of the transmitted flux Φ_0 (both expressed in a number of photons) by

$$\Phi = \Phi_0 \, G \, \eta \, R$$

where η is the optical efficiency of the receiver, G and R are factors describing the type of experiment to be performed.

- G is a geometrical factor taking care of the relative position
  of the spacecraft and of the scatter
- R is the efficiency of the scattering process.

Supposing that the signal follows Poisson statistics, the precision p can be written:

$$p = 1/\sqrt{\Phi}$$

Therefore

$$\boxed{1/p^2 = \Phi_o\, G\, R\, \eta}$$

A single estimate of the order of magnitude of the relevant parameters follows:

$$\eta \sim 10^{-1}$$

$$\Phi = 3 \times 10^{18}$$

$$p^2\, R\, G \sim 3 \times 10^{-18} \qquad \text{in the visible}$$

$$\sim 10^{-16} \qquad \text{in the UV}$$

From the Spacelab at 250 km of altitude with a one-meter telescope:

$$G \simeq 10^{-1}$$

$$p^2\, R \sim 3 \times 10^{-7}$$

and therefore the values for p obtained with one shot are:

Resonance $\qquad\qquad\qquad\qquad$ $p \simeq 0.2$

Differential absorption $\quad$ $p \simeq 0.6$

Reflection $\qquad\qquad\qquad\qquad$ $p \simeq 0.6 \times 10^{-3}$

In the important case of an experiment of absorption by an element of optical thickness

$$\tau = n\, l\, \sigma$$

n = number of absorber per cc
l = length over which the absorption if measured
σ = absorption cross-section

If Φ_0 is the incident flux and Φ the flux after absorption

$$\Phi = \Phi_0 \, e^{-\tau} = \Phi_0 \, (1 - \tau)$$

$$\frac{\Phi - \Phi_0}{\Phi_0} = \frac{\Delta\Phi}{\Phi_0} = \tau$$

Suppose that we can measure $\frac{\Delta\Phi}{\Phi_0}$ then this term is equal to the accuracy of the measurement, and therefore the detectable optical thickness τ is equal to the accuracy p. We choose l in our measurement and therefore the detectable n. Thus n_d is

$$\frac{1}{n_d^2} = (1\sigma)^2 \, \Phi_0 \, G \, R \, \eta$$

These considerations apply to <u>one echo</u>. It is obvious that if N echoes are received, the accuracy will be multiplied by $\sqrt{N}$.

	Metals	Molecules	Rayleigh
σ cm^{-2}	10^{-13}	10^{-19}	10^{-28}
n cm^{-3}	10^3	10^9	10^{17}
$n\sigma$ cm^{-1}	10^{-10}	10^{-10}	10^{-11}
l cm	10^5	10^6	10^5
τ	10^{-5}	10^{-6}	10^{-6}

We shall now apply these considerations to different possible experiments.

EXAMPLES

1. Vertical sounding of atomic sodium

Na atoms are distributed from 80 to 100 km with a peak density of 5×10^3 cm^{-3} at 87 km. We will take as typical, $n \simeq 10^3$ cm^{-3}. In this case, we have

$$G = \frac{S}{h^2}$$

where

S = surface of the receiving telescope $\simeq$ 1 m^2
h = distance between Spacelab and Na-layers $\simeq$ 150 km.

$$G \approx 5 \times 10^{-11}$$

$$R = \tau_{Na} = n\, l\, \sigma = 10^3 \times 10^5 \times 10^{-13} \approx 10^{-5}$$

with a length resolution of 1 km.

Therefore the product GR characterizing the experiment is

$$GR \approx 5 \times 10^{-16}$$

with typical value of $\eta \approx 10^{-1}$ and $\Phi_0 = 3 \times 10^{18}$ photons

$$\frac{1}{p^2} = 3 \times 10^{18} \times 10^{-1} \times 5 \times 10^{-16} \approx 150 \text{ photoelectrons}$$

$$p \approx \frac{1}{12} \approx 8\%$$

The accuracy of 10^{-2} is obtained with less than 100 echoes. At the peak of the Na density with a width of 0.4 pm, $\sigma = 3 \times 10^{-13}$, $n = 5 \times 10^3$ and an accuracy of 2% is obtained with only one echo. It is obvious that the experiment is accurate. Over the daypart of this orbit, the sky background has to be taken into account and the accuracy is decreased by approximately a factor of 5 depending on the detector field of view and the repetition rate of the laser.

The high accuracy of the method provides the possibility of performing temperature measurements by an observation of the lineshape at different wavelength, using the modulation technique.

2. Vertical sounding of molecules by differential backscattering

The laser transmits 2 wavelenghts separated by a few nm; this is obtained by a suitable rotation of the system of wavelength selection inside the cavity (grating - F.P.). The couple of fluxes backscattered at one altitude has to be compared to another couple backscattered at another altitude in order to obtain the vertical distribution. Each point therefore needs 4 measurements.

A comparison of the signal intensities at λ_1 and λ_2 leads to

$$n_{O_3} = \frac{\log \Phi}{2(\sigma_1 - \sigma_2)\Delta z}$$

- σ_1, σ_2 are the absorption cross-section of the constituent at λ_1 and λ_2
- Δz is the altitude resolution.

$$\varphi = \frac{\Phi_{\lambda_1}(z, z-\Delta z)}{\Phi_{\lambda_1}(z+\Delta z, z)} \cdot \frac{\Phi_{\lambda_2}(z+\Delta z, z)}{\Phi_{\lambda_2}(z, z-\Delta z)}$$

where $\Phi_{\lambda_1}(z+\Delta z, z)$ is the signal obtained by Rayleigh scattering in the altitude range z, $z+\Delta z$ at the wavelength λ_1

$$\Phi_{\lambda_1}(z+\Delta z, z) = \Phi_0 \frac{S}{h^2} \tau_{atm} \eta e^{-2\tau_{O_3}}$$

with

$$\tau_{atm} = n_{atm}(z)\sigma_k \Delta z \qquad \sigma_k \simeq 10^{-19} cm^2$$

The error is therefore

$$\Delta n_{O_3} = \frac{\Delta\varphi}{\varphi} \frac{1}{2\Delta\sigma\Delta z} \simeq 4 \frac{\Delta\Phi}{\Phi} \frac{1}{2\Delta\sigma\Delta z}$$

since 4 measurements of Φ is required to obtain φ.

In order to compute the accuracy over Φ we remark that without absorption, we would receive:

$$\Phi \simeq \Phi_0 \tau_{atm} \frac{S}{h^2} \simeq 3 \times 10^{18} \times 10^{-6} \times 10^{-11}$$

for the altitude of 50 km $\sim$ 30 photoelectrons (50 times less than for sodium).

In order to achieve the experiment the optical thickness of the absorber has to be less than 0.1 in the height range of the measurement. If we consider the decrease of concentration of O_3 with altitude, this allows us to increase the value of the absorption cross-section for the higher altitudes. The mean accuracy which can then be obtained is 5% with 5.10^4 shots or 100% in 100 laser shots.

3. Total integrated content of a minor constituent

It is obtained by observing the differential absorption on the ground echo.

The factor G is

$$G = \frac{S}{(\alpha z)^2}$$

z = altitude of S/c over the Earth's surface
α = aperture angle of the telescope

$$G \simeq 1.5 \times 10^{-7}$$

For R, we have to take the albedo for a regular reflection or the no factor divided by 2π for scattering. We can for a preliminary estimation use $R \simeq 1$ and

$$\Phi = \Phi_0 \, G \, R \, \eta = 3 \times 10^8 \times 1.5 \times 10^{-7} \times 1 \times 10^{-1}$$

$$= 5 \times 10^{10} \text{ photoelectrons}$$

This gives an accuracy of 10^{-5} and therefore a theoretical detectable optical thickness of 10^{-5}.

In the UV, the signal is 20 times smaller and the corresponding accuracy falls to 2×10^{-4}.

Since $\tau_{det} = N\sigma l$, for a length of 10 km and an accuracy of 10^{-5} the detectable σN is of the order of 10^{-11}. For instance, the number density of C10 detectable at 305 nm is $3 \times 10^7 \text{ cc}^{-3}$.

The measurable optical thickness is thus limited by the calibration of the relative value of the laser intensities emitted at the two wavelengths whose accuracy can be increased by taking the mean value of these intensities for a number of shots.

	Species	τ integrated detectable	λ nm
Troposphere	NO_2	10^{-1} to 10^{-3}	450
l = 10 km	NO	1 to 10^{-2}	200
	SO_2	1 to 10^{-3}	300
	H_2O	10^{-4} to 10^{-5}	750
Stratosphere	NO_2	10^{-3}	450
l = 20 km	NO	10^{-2}	200
	C10	10^{-3}	300

4. Vertical sounding of minor constituents by horizontal probing using a subsatellite

The subsatellite is a passive system covered with reflectors launched from Spacelab.

The observed flux is

$$\Phi = \Phi_0 \, e^{-\tau} = \Phi_0 (1 - \tau)$$

$$\frac{\Delta\Phi}{\Phi_0} = \tau = n \, l \, \sigma \equiv \text{accuracy}$$

If we receive 10^5 photons or 10^4 photoelectrons, this corresponds to an accuracy of 10^{-2}. $n l \sigma$ detectable $\simeq 10^{-2}$, or for $l = 10^3$ km

$$n\sigma \simeq 10^{-10}$$

Molecules with a density of 10^8 cm-3 can be detected for $\sigma \simeq 10^{-18}$.

Suppose that the half aperture of the laser beam is α and that 1 cm^2 of mirror is seen from the spacecraft under an angle Θ. A mirror of 1 cm^2 will provide $(\alpha/\Theta)^4$ photons at the receiver: the efficiency is 10^{-17} cm^{-2} at a spacecraft-subsatellite distance of 3500 km. Therefore with a 10 cm^2 reflector and a one-meter telescope 3 x 10^{18} photons will provide 3 x 10^6 photons.

Therefore the number detectable becomes 2 x 10^7 for $\sigma \simeq 10^{18}$. This is already better than a solar occultation experiment. Furthermore it is possible to integrate over one orbit, obtaining for instance with one pulse per sec for 100 minutes, a gain of 100: the method is better than a solar occultation by 2 orders of magnitude in the UV, a factor of $\sqrt{20}$ is best (efficiency and less photons for the same energy).

Furthermore a vertical structure can be attained by varying the distance between the satellite and the reflector.

Example: ClO at 300 nm, $\tau = 3$ x 10^{-18}, $n_{detectable} = 2$ x 10^7 per pulse.

DISCUSSION

G. *Fiocco:* Regarding the use of a scanning laser, doing without a scanning receiver, it should be pointed out that they will work with scatterers having a definite line shape response. In experiments where, as mentioned, the ground echo is also required to correct for Doppler shifts, the use of a scanning receiver of some sort cannot be avoided.

J.E. *Blamont:* In the case of non-selective scattering, one can correct for the Doppler shift using the ground echo, only if the measurement can be performed within one pulse, which does not seem to be very realistic. In the case of a selective process, on the contrary, it is possible to use a modulation to control the emitted frequency with the resonant frequency

in the Earth frame of reference; this is done by correcting
at each pulse the Doppler shift due to the ground with
respect to an absolute reference. A modulation technique
is therefore twice as useful in the case of a resonant process.

The accuracy of the measurement and consequently the integration time are therefore improved if one uses a scanning
transmitter rather than a scanning receiver. As far as the
ground echo is concerned, it is assumed that the level of
the received signal is sufficient to allow one measurement
per pulse.

L. Thomas: In your measurement of temperature from the Doppler
broadening of the signals resonantly scattered from sodium
atoms you plan to use simultaneously three pulses of spectral
width 10 mA and wavelengths separated by 5 mA. I am surprised
that you expect an accuracy of 2 K in temperature measurement
in view of the fact that the Doppler width of the D_1 or D_2
lines is only 30 mA.

J.E. Blamont: The accuracy is given by the ratio of the intensities
at two wavelengths within the absorption line. The existence
of a fine structure within the line allows one to choose the
appropriate wavelength spacing and to achieve the stated
accuracy (variations in width of the laser emission line have
no critical influence upon the convolution of the two spectral
shapes at the level of the pm).

J.T. Houghton: In the measurement of constituents by differential
absorption, what accuracy do you think is possible taking all
factors into account? You mentioned I think one part in 10^5
which sounds very low indeed to me.

J.E. Blamont: The real accuracy is that of the relative calibration of the intensities of the emitted waves, which can
be less than 10^{-3} if one averages out over a large number of
pulses.

E. Schanda: What kind of pump can be used on the Spacelab to
operate a CW dye laser emitting 1W? I assume that for a
differential absorption experiment a dye laser emitting
simultaneously at two wavelengths can be as good as the
modulation technique proposed by the author?

J.E. Blamont: (1) A solar pump.

(2) The optical-acoustic technique actually envisaged does not
allow one to achieve a wavelength separation of several nm,

which is necessary for differential absorption. A variation
of the emitted laser wavelength can be obtained by varying
the interference order of the selecting system.

W. Renger: In the case of temperature measurement using resonant
backscattering of sodium you proposed an acoustically
modulated laser. This must necessarily be a CW laser. What
is the average output? Which type of laser are you thinking
of?

J.E. Blamont: A CW modulated laser is in fact used to monitor
a pulsed laser whose emission spectral characteristics are
then those of the CW laser without any reduction in emitted
energy (1 Joule/pulse).

S.A. Bowhill: On what theoretical basis is it possible to
deduce eddy diffusion coefficient from two successive sodium
profiles taken at the same place?

J.E. Blamont: The values of the eddy diffusion coefficient
obtained in this way are lower limits because the theoretical
model does not account for the existence of a sodium source
at higher altitudes. This source will influence the time
variation of the scale height. The effect due to turbulent
processes is to slow down the evolution of the scale height
towards that of the atmosphere.

FAR INFRARED FOURIER SPECTROSCOPIC OBSERVATION OF THE EARTH'S ATMOSPHERE FROM SPACELAB

A. Bonetti, B. Carli and F. Mencaraglia
Institute of Physics
University of Florence - Italy

J. E. Harries and M. J. Bangham
National Physical Laboratory
Teddington - U.K.

D. H. Martin
Department of Physics
Queen Mary College-London-U.K.

ABSTRACT: Recently developed techniques allow measurements of the atmospheric emission in the far infrared, with very high spectral resolution over a large frequency range and with short acquisition time. Monitoring of minor atmospheric constituents from Spacelab using far infrared measurements appears extremely promising.

1. INTRODUCTION

In the long wavelength part of the infrared spectrum, thermal sources, which are in Rayleigh-Jeans tail of the Planckian distribution, emit relatively little energy. For some time this has been a limitation to applications of the far infrared; measurements were either obtained with poor signal-to-noise ratio or required long integration times. However far infrared spectroscopy has rapidly

changed in the last few years and the quality of the
spectra measured in this region and their acquisition
time are now becoming comparable with those at shorter
wavelengths. These improvements are mainly due to the
development of helium cooled germanium bolometers with
N.E.P. less than 10^{-13} Watts $Hz^{-\frac{1}{2}}$, and to the use of the
interferometers with high spectral efficiency.

In particular the polarising interferometer (1) is
proving to be particulary effective. The configuration
of this instrument is similar to that of a classical Michelson interferometer, but it has an optics based on
polarisers made with grids of free standing wires. The polarising interferometer has the classical luminosity and
multiplex advantages; furthermore, a wire polariser, as
a beam splitter, has a very high spectral efficiency
over a very wide spectral range (5-150 cm^{-1}). A second
important property is the existence of two independent
and easily accessible input ports. Observing through one
port the atmosphere and through the second port viewing
alternately two calibration black bodies, it is possible
to obtain an absolute measurement of the atmospheric
spectrum, without requiring post-hoc calibration.

In order to specify the potential of an interferometer operating in the far infrared for atmospheric sounding, it is necessary to consider the spectral resolving
power, the instrumental signal-to-noise ratio, the intrinsic strength of atmospheric lines, and instrumental
limitations to spatial resolution such as field of view
and diffraction. These features will be discussed in the
following sections.

2. RESOLVING POWER

It can be demonstrated that, in the case of white
noise in the interferogram, the noise level (not the signal-to-noise ratio) in each spectral channel depends
only on the total measurement time and not on the spectral
resolution. In the case of a single feature of half-width
$\delta\sigma$, smaller or equal to the spectral resolution $\Delta\sigma$, the si-

gnal contained in the spectral channel is ind**e**pendent of $\Delta\sigma$; as a consequence also the signal-to-noise ratio is independent of the spectral resolution. The reduction of the spectral resolution to a value as small as the half width of the lines which are being studied, without reducing the signal-to-noise ratio allows a better assignment of the lines and avoids blending effects. Therefore the measurement of small atmospheric features, which have tipically a half-width at aircraft altitude of the order of 10^{-2} cm^{-1}, requires a spectral resolution of the same order.

Far infrared atmospheric measurements have been limited in the past by the available signal-to-noise ratio to measurement of the stronger features at an intermediate resolution. In recent measurements made from the CV-990 aircraft during the ASSESS I Mission (2), the polarising interferometer was used and the enhancement of the signal level, due to the efficiency of the instrument, made possible, for the first time in the far infrared, a resolution of 0.01 cm^{-1}.

The recorded spectra show numerous reproducible features. The features already identified with lower resolving power can be better measured with less blending and the spectra are now being analysed for the spectral assignment of the new features. On the basis of the experience of the ASSESS measurements, and in view of the following signal-to-noise considerations, a resolution of 0.01 cm^{-1} is considered indispensible for the monitoring of minor atmospheric constituents.

3. SIGNAL-TO-NOISE RATIO

The intensity of the atmospheric emission per unit solid angle and per unit area in the spectral interval $0 - \sigma_{max}$ is given by:

$$I = \int_0^{\sigma_{max}} \varepsilon(\sigma)\, B(\sigma,T)\, d\sigma \simeq 8.3 \cdot 10^{-6}\, \bar{\varepsilon}\, T\, \sigma_{max}^3/3 \quad ergs/cm^2/sec/ster$$

where $B(\sigma,T)$ is the Planckian distribution, σ is the

wavenumber and T is the temperature of the atmosphere in
°K (of the order of 200 °K), $\varepsilon(\sigma)$ is the atmospheric
emissivity and $\bar{\varepsilon}$ is the weighted mean of $\varepsilon(\sigma)$.

A helium cooled bolometer with cold filters will
receive a total power of

$$P = F \, A \, \Omega \, I \, ,$$

were $A\Omega$ iis the throughput of the bolometer, optically
matched to the interferometer and masked according to
the limb scanning requirements, and F is the transparency
of the cold-filters weighted and averaged over the spectrum in the same way as $\varepsilon(\sigma)$ in the pass band region.
For example,[+] $A\Omega$ is assumed equal to 0.016 cm^2 sterad
and F to be of the order of 0.1, the total power reaching
the detector in the spectral interval, 0-150 cm^{-1}, is:

$$P = \bar{\varepsilon} \; 3.0 \; 10^{-7} \; \text{watts} \, .$$

This radiation input causes a photon noise which can be
calculated using the following formula (3):

$$\left(N.E.P. \right)^2_{photon} = \frac{4}{3} \left(k\,T \right)^2 c \, \sigma^3 \left(A\Omega \, F\bar{\varepsilon} \right)$$

For the numerical examples given above this is:

$$\left(N.E.P \right)_p = \sqrt{\bar{\varepsilon}} \; 7.0 \; 10^{-14} \; \text{watts} \, .$$

Except when $\bar{\varepsilon} \ll 1$, this noise is comparable with the intrinsic noise of bolometric detectors at the present sta-
ge of development and it's likely to be the dominant noise
in experiments performed in the next few years.

The interferometric method achieves the multiplex
recording of many different spectral channels by means
of a cosine modulation and the following demodulation
process during the Fourier transformation reduces the
signal in the spectrum relative to the noise level by
$1/\sqrt{8}$ (4). In addition, a factor 0.5 must be introduced
since only one plane of polarization is used by the interferometer. Hence the average signal-to-noise ratio in

the spectral distribution is given by:

$$\overline{\left(\frac{S}{N}\right)_s} = \frac{P_{0.5}}{\sqrt{8}} \cdot \frac{\sqrt{t}}{(N.E.P.)_P} \cdot \frac{1}{n}$$

where n is the number of independent spectral channels and t is the measurement time. Thus, for the frequency interval 0-150 cm^{-1}, for AΩ = 0.016 cm^2 sterad, F = 0.1 and t = 40s,

$$\overline{\left(\frac{S}{N}\right)_s} = 320 \sqrt{\overline{E}}$$

This high signal-to-noise ratio is in fact in accord with the results obtained in the measurement made during the ASSESS Mission (2).

In order to define the minimum number of molecules which can be detected, it is necessary to specify the minimum emissivity which can be measured, $\overline{\varepsilon}_{min}$. This is given by

$$\overline{\varepsilon}_{min} = \overline{\varepsilon} \Big/ \overline{\left(\frac{S}{N}\right)_s} \; ;$$

for a frequency interval 0-150 cm^{-1}, AΩ = 0.016 cm^2 sterad and F = 0.1,

$$\overline{\varepsilon}_{min} = \sqrt{\overline{\varepsilon}} \Big/ 320 .$$

4. HEIGHT RESOLUTION

Limb scanning of the atmosphere, from an orbit at an altitude of 250 km, requires the use of a beam with a very small vertical angular aperture and, at the long wavelength end of the infrared spectrum, this imposes the use of optics of sufficiently large diameter because of diffraction. This requirement could be satisfied by means of a large-diameter input optical system focussed at the entrance of a smaller diameter interferometer; but there is the second requirement, to reach high spectral resolving power with a large throughput, and this requires optics of relatively large diameter throughout the interferometer.

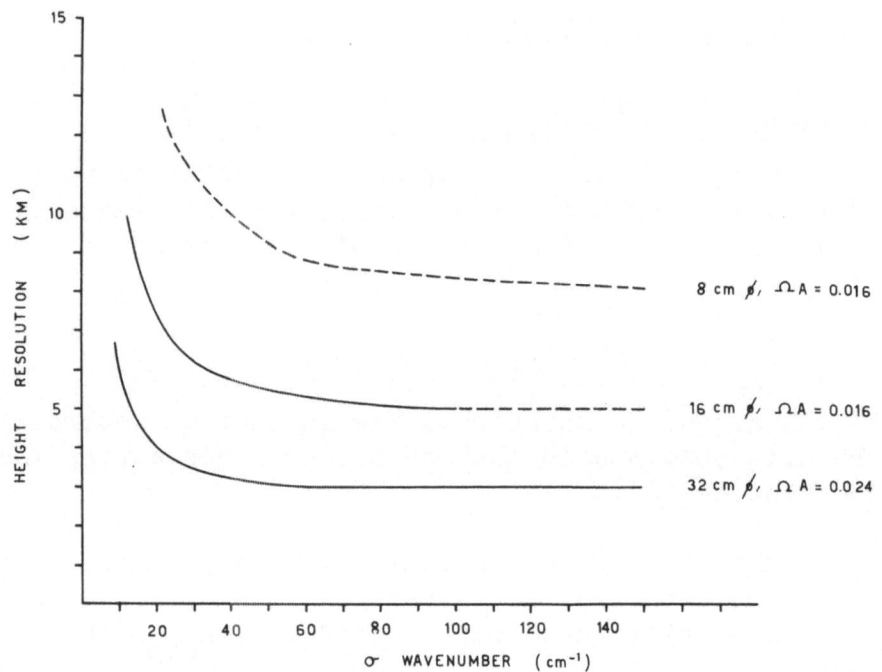

Fig. 1 Height resolution as a function of wavenumber for interferometers of different cross section.

Fig. 1 shows the height resolution as a function of frequency obtainable from a 250 km orbit, with interferometers of different sizes and optimized throughputs. The solid part of the curves indicate the spectral intervals over which a resolution $\Delta\sigma = 0.01$ cm^{-1} would be obtained. As expected, the height resolution is limited by diffraction at low frequencies; on the other hand, at high frequencies, the maximum resolution requires a small solid angle, which implies, either a large diameter optics, or a small throughput. A 16 diameter optics appears at the moment to be a useful compromise between the needs to achieve high performance and small volume. The value $A\Omega = 0.016$ cm^2 sterad is that used earlier in the numerical examples.

A linear array of detectors may be used at the output of the interferometer. Each detector measures the emission from a different layer of the atmosphere, and the system thereby provides multiplex measurement instead of sequential limb scanning.

T A B L E 1

Species	σ cm^{-1}	s cm^{-1}mol^{-1}	n_{min}	Detectable Mixing Ratio 50 Km	30 Km	20 Km	Expected Mixing Ratio 50 Km	30 Km	20 Km
H_2O	88.08	$1.1 \cdot 10^{-18}$	$5.4 \cdot 10^{14}$	$3.5 \cdot 10^{-10}$	$3.5 \cdot 10^{-11}$	$8 \cdot 10^{-12}$	$3 \cdot 10^{-6}$	$3 \cdot 10^{-6}$	$3 \cdot 10^{-6}$
O_3	34.33	$6.0 \cdot 10^{-20}$	$1.7 \cdot 10^{16}$	$1.1 \cdot 10^{-8}$	$1.1 \cdot 10^{-9}$	$2.8 \cdot 10^{-10}$	$2 \cdot 10^{-6}$	$3 \cdot 10^{-6}$	$1 \cdot 10^{-6}$
O_2	25.81	$3.2 \cdot 10^{-25}$	$3.1 \cdot 10^{21}$	$2.0 \cdot 10^{-3}$	$2.0 \cdot 10^{-4}$	$5 \cdot 10^{-5}$	0.21	0.21	0.21
HNO_3	21.96	$3.5 \cdot 10^{-19}$	$2.8 \cdot 10^{15}$	$1.9 \cdot 10^{-9}$	$1.9 \cdot 10^{-10}$	$4.7 \cdot 10^{-11}$	?	2.10^{-7}	5.10^{-9}
NH_3	80.10	$3.7 \cdot 10^{-18}$	$2.7 \cdot 10^{14}$	$1.7 \cdot 10^{-10}$	$< 1.7 \cdot 10^{-10}$	$4.2 \cdot 10^{-12}$	?	?	?
SO_2	39.77	$1. \cdot 10^{-18}$	$1. \cdot 10^{15}$	$6. \cdot 10^{-10}$	6.10^{-11}	$1.5 \cdot 10^{-11}$	?	$< 10^{-9}$	$< 10^{-9}$
HCl	41.76	$5.4 \cdot 10^{-19}$	$1.9 \cdot 10^{15}$	$1.2 \cdot 10^{-9}$	$1.2 \cdot 10^{-10}$	$3. \cdot 10^{-11}$	?	6.10^{-10}	3.10^{-10}
H_2S	35.84	$4.1 \cdot 10^{-19}$	$2.5 \cdot 10^{15}$	$1.6 \cdot 10^{-9}$	$1.6 \cdot 10^{-10}$	$4. \cdot 10^{-11}$	?	?	?
CO	42.18	$5.0 \cdot 10^{-21}$	$2.0 \cdot 10^{17}$	$1.3 \cdot 10^{-7}$	$1.3 \cdot 10^{-8}$	$3.2 \cdot 10^{-9}$	$10^{-7}?$	10^{-7}	3.10^{-8}
NO_2	52.77	$3.8 \cdot 10^{-19}$	$2.7 \cdot 10^{15}$	$1.8 \cdot 10^{-9}$	$1.8 \cdot 10^{-10}$	$4.5 \cdot 10^{-11}$	$10^{-8}?$	2.10^{-8}	10^{-9}
N_2O	21.77	$9.0 \cdot 10^{-22}$	$1.1 \cdot 10^{18}$	$7.6 \cdot 10^{-7}$	$7.6 \cdot 10^{-8}$	$1.9 \cdot 10^{-8}$	$5.10^{-9}?$	10^{-7}	2.10^{-7}
NO	31.77	$1.5 \cdot 10^{-21}$	$6.7 \cdot 10^{17}$	$4.4 \cdot 10^{-7}$	$4.4 \cdot 10^{-8}$	$1.1 \cdot 10^{-8}$	$\sim 10^{-9}$	5.10^{-9}	$\sim 2.10^{-10}$
HF	82.35	$\sim 5. \cdot 10^{-18}$	$1.8 \cdot 10^{-14}$	$1.2 \cdot 10^{-10}$	$1.2 \cdot 10^{-11}$	3.10^{-12}	?	5.10^{-11}	$< 10^{-10}?$

5. MOLECULES WHICH CAN BE MONITORED

Table 1 shows a preliminary list of the molecules which can be monitored in the 10–150 cm^{-1} region. Most of the information available for the compilation of this list was relative to the long wavelength part of the spectral interval. This list must be considered only indicative since much information is still missing and blending effects have not been taken into account. Nevertheless, considering that the minimum number of molecules which can be detected in the line of sight, n_{min}, has been calculated assuming a conservative minimum detectable emissivity of 10^{-2} (see sec. 2), it can be seen that concentration measurements and monitoring of many interesting minor constituents are possible with good precision.

6. CONCLUSION

The principal characteristics and parameters of a far infrared interferometer for monitoring of the atmospheric minor constituents from Spacelab are summarized in Table 2. These parameters are derived from experience of aircraft and balloon flights, and a dedicated system for Spacelab will almost certainly be lighter and smaller than indicated.

The considerations of this paper have shown that the use of a high resolution polarising Michelson interferometer in the far infrared enables the achievement of very high signal-to-noise ratio (several hundreds) in the 10–
–150 cm^{-1} region, using currently available liquid helium cooled bolometric detectors. Consideration of the intrinsic strength of atmospheric lines in the far infrared shows that accurate limb scans of many minor constituents are possible simultaneously in quite a short time ($\sim$40 sec), including H_2O, O_3, O_2, NO_2, N_2O, NH_3, HF, HCl, CO and others.

T A B L E   2

Total volume	: $0.9\ m^3$
Total weight	: 180 kg
Power	: 50 Watts
Detector	: array of 10 bolometric detectors, liquid helium cooled
Spectral interval	: $10 - 150\ cm^{-1}$
Scanning period	: 40 sec to record in parallel 10 interferograms
Recording bit rate	: 64 kbit/sec
Calibration	: obtained from two successive spectra, time 80 sec
Spectral resolution	: $0.01\ cm^{-1}$ in the spectral interval $10-100\ cm^{-1}$
Height resolution of the earth's atmosphere	: 5 km from 10 km up to 60 km, in the spectral interval $40-150\ cm^{-1}$
Horizontal resolution of the earth's atmosphere	: 300 km.

REFERENCES

1 D H Martin and E F Puplett, Infrared Physics, 10, 105, 1970
2 B Carli, D H Martin, E F Puplett and J E Harries, Nature, 257, 649, 1975
3 R A Smith, F E Jones and R P Chasmor, The Detection and Measurement of Infrared Radiation, 2nd edition, Oxford University Press, 1968
4 B Carli, Infrared Physics, 12, 251, 1972.

DISCUSSION

L.H. Meredith: (a) What would you consider to be the major
scientific question that the flight of your instrument on
Spacelab could help answer?

(b) How many individual limb measurements would need to be
made for this experiment?

B. Carli: (a) The aims of our experiment are the determination
of the horizontal, vertical and temporal variations of the
concentration of a large number of stratospheric constituents
in order to study the sources, life cycles and sinks of these
constituents.

The major atmospheric problems we can study are:

    (a) ozone photochemistry
    (b) H_2O cycle
    (c) nitrogen photochemistry.

(b) The number of limb scannings for each molecule will be
a compromise between the horizontal resolution and the
precision of concentration measurement, the maximum horizontal
resolution being 300 km.

Martin: All the molecular species are seen together - one does
not have to decide which to look at and assign time to each
separately.

INFRARED OBSERVATION OF NON THERMAL EMISSIONS FROM SPACELAB

G. Moreels* and C. Muller**

* Service d'Aéronomie du C.N.R.S. (France)

** Institut d'Aéronomie Spatiale de Belgique.

Observations of the emissions of atmospheric ozone and nitric acid in the 10 µm window have already been used for the monitoring of these constituents in the stratosphere. We propose to generalise these methods to the emissions of mesospheric constituents using non thermal emissions. Thermal emission $E(\nu)$ can be expressed by the formula :

$$E(\nu) = A(\nu).B(\nu)$$

where $A(\nu)$ is the absorption of the emitting gas and $B(\nu)$ is the blackbody emission at frequency ν.

We shall not consider the cases where self-absorption occurs and shall limit ourselves to the so-called optically thin approximation, taking into account the fact that we have to consider long optical paths, the previous expression could be rewritten

$$E(\nu) = \int_0^\infty A(\nu,s) \, B(\nu,s) \, ds$$

for an optical path s divided into differential segments ds. Infinity occurs when no more emission features can be detected ; for thermal emission, this limit is rather rapidly attained. In fact, the absorption spectrum $A(\nu,s)$, for zenithal paths, can be considered as nearly flat above 50 km with the exception of the 15 µm CO_2 bands used for temperature soundings. If limb scanning is used in the stratosphere, the advantage of altitude resolution, still given by absorption

J. J. Burger et al. (eds.), Atmospheric Physics from Spacelab, 339–342. *All Rights Reserved.*
Copyright © 1976 by D. Reidel Publishing Company, Dordrecht-Holland.

spectrometry, is lost because a large optical field is needed
in order to have a detectable signal with light instrumentation.

Mesospheric non thermal emissions may, on the contrary,
be observed with rather high precision as has been proved by
the auroral observations of Stair et al. (1975). The process
is always the same, a molecule is brought in an excited state
and desexcites itself in one of its vibration modes. These
emissions have no relation with the blackbody temperature
emission of the unexcited gas. We shall not consider the purely
rotational transitions which could be also observed from much
higher altitudes (Simpson, 1976).

We propose to limit ourselves to the near infrared
(1-15 μm) because of the availability of simple existing
instrumentation (Ackerman et al., 1976). We shall discuss
especially the emissions of excited ozone. The chemistry of
excited ozone is governed by the equations compiled by
Moreels (1975) :

$$O_3^+ + h\nu \rightarrow O + O_2 \qquad\qquad J_2 = 8 \times 10^{-13} \text{ sec}^{-1}$$

$$O + O_2 + M \rightarrow O_3^+ + M \qquad k_3 = 10^{-34}\, e^{450/T} \text{ sec}^{-1} \text{ cm}^{-3}$$

$$O + O_3^+ \rightarrow O_2 + O_2 \qquad\quad k_4 = 1.3 \times 10^{-11}\, e^{-2250/T} \text{ sec}^{-1} \text{ cm}^{-3}$$

$$O_3^+ \rightarrow O_3 + h\nu \qquad\qquad A_6 = 11 \text{ sec}^{-1} \text{ cm}^{-3}$$

$$O_3^+ + M \rightarrow O_3 + M \qquad\quad k_7 = 2 \times 10^{-14} \text{ sec}^{-1} \text{ cm}^{-3}$$

$$O_3^+ + H \rightarrow OH^+ + O_2 \qquad k_8 = 2.2 \times 10^{-11} \text{ sec}^{-1} \text{ cm}^{-3}$$

This system leads to a photochemical equilibrium value of

$$n(O_3^+) = \frac{k_3\, n(O)\, n(O_2)\, n(M)}{A_6 + J_2 + k_4\, n(O) + k_7\, n(M) + k_8\, n(H)}$$

This amount of vibrationnally excited ozone leads to an
emission process distributed among the different vibration
rotation bands of the molecule.

These modes are usually the most intense in infrared
absorption spectra. However, laboratory and theoretical studies
are still needed in order to determine the relative emissions
of the bands.

In the case of excited ozone, Von Rosenberg and Trainor (1974) show that the ozone structure in the 9.6 μm band does not exactly correspond with the absorption pattern of the same band at room temperature : emissions tend to be stronger than expected at the longer wavelengths. This correspond to emissions from the hot bands $\Delta v_3 = 1$. It means that the ozone producing process corresponds to a much higher temperature than the atmospheric ambiant air. According to Von Rosenberg and Trainor (1974), if every O_3 formed by $O + O_2 + M$ is in the $\Delta v_3 = 1$ state, it means an equivalent temperature of 2160 K but if only one of 2 formed O_3 turns to be vibrationally excited, the distribution is characterized by $T_v = 1080$ K. These temperatures and the now well known ozone molecular parameters would permit to build a model if the emission in the considered bands.

The observed intensity will be, in any case, related to the concentration of excited ozone which, in its turn, is directly related to the atomic oxygen concentration by its formation process. As a result of chemical reaction with atomic oxygen above 80 km and collisional quenching below 60 km, the oxygen related ozone emission should peak between 65 and 75 km.

The main advantage of this kind of experiment is to be performed from space using light instrumentation and parts from other optical experiments when these are not in use. As an instrument, we propose a circular variable filter radiometer scanning the 10 μm atmospheric window such as the one we have built for stratospheric monitoring.

In order to assess the variation of the v_3 ozone emission related to sunset or sunrise, we intend to perform this year a balloon flight using the stratospheric monitoring radiometer above the ozone layer.

REFERENCES

ACKERMAN, M., LIPPENS, C. and MULLER, C., Emission features in the 10 μm atmospheric window related to nitrogen and halocarbon compounds, to be presented at the E.G.S. meeting, Amsterdam (1976).

MOREELS, G., Oxygène atomique mésosphérique, Proposition au service d'aéronomie du C.N.R.S. (France) (1975).

SIMPSON, J., Infrared emission from the atmosphere above 200 km., Nasa Technical Note TN D-8138, Ames Research Center (1976).

STAIR, A.T., ULWICK, J.C., BAKER, K.D. and BAKER, D.J., Rocketborne observations of atmospheric infrared emissions

in the auroral region, Atmospheres of Earth and the
planets, McCormac ed., Reidel, Dordrecht (1975).

von ROSENBERG, C.W. and TRAINOR, D.W., Vibrational excitation
of ozone formed by recombination, J. Chem. Phys., $\underline{61}$,
2442 (1974).

IMAGE INTENSIFYING TV OBSERVATIONS IN

THE 1975 SPACELAB ASSESS MISSION

P. Rothwell, J. Crawford and N. Benjamin
Physics Department, University of Southampton.

ABSTRACT

An image intensifying isocon TV camera and associated video recording system was flown on the NASA Convair 990 aircraft in the first Spacelab Assess Mission in June 1975. Two different types of experiment were performed during the mission.

1)   Observation of hydroxyl airglow structures in the near infrared

2)   Observation of meteors.

For the airglow measurements, a Wratten filter, which cut out optical emissions with wavelengths less than ~ 670 nm was placed in front of the camera lens. The long wavelength cut off of the isocon tube was about 900 nm. Good pictures of hydroxyl airglow structures were obtained through the aircraft 15° window and the altitude of the emissions was estimated at ~ 85 km. The horizontal dimension of a typical airglow "cloud" was about 20 km. It is likely that the structures arise from gravity waves generated in the lower atmosphere.

The meteor observations were made without filters. The faintest meteors observed with the isocon TV system were fainter than 9th magnitude

J. J. Burger et al. (eds.), Atmospheric Physics from Spacelab, 343–355. All Rights Reserved.
Copyright © 1976 by D. Reidel Publishing Company, Dordrecht-Holland.

stars. 84 meteors were recorded in about 14 hours of observing time in
the local time interval 21.00 - 03.00 hours. During much of this time the
moon was in the sky.

We hope to fly a similar but improved image intensifying TV system on
Spacelab. As well as making more extensive observations on airglow and
meteors, it could monitor the dynamics of the Shuttle's own atmosphere in
the light of the optical emissions which could be excited by e.g. electron
guns or lasers carried on Spacelab.

1. INTRODUCTION

"Spacelab Assess" missions simulate some of the conditions under
which experiments will be performed on Spacelab, using the NASA Convair 990
Flying Laboratory as an experiment platform. In the 1975 NASA/ESA Assess
mission, particular interest was centred on the "experiment operators", four
of whom had to operate seven different experiments installed in the aircraft.
To simulate operating conditions on Spacelab, the operators flew a 4-6 hour
mission each evening for a week, and slept in trailers alongside the aircraft, isolated from other scientists except through telecommunications
links. One of the operators was a NASA trained scientist-astronaut, two
were postdoctoral research fellows (one each from NASA and ESA) and one was
a European research student, (from the Southampton group). Principal
Investigators were given time to operate their own equipment on the aircraft for a period following the confined mission, enabling them to compare and contrast data they obtained themselves with data obtained by the
operators.

An image intensifying television camera flown on either an aircraft
or on Spacelab can take advantage of clear skies above the tropopause to
make observations of faint or transient emissions of light from the upper

atmosphere, such as airglow or meteor trails. From such observations information can be obtained both on the chemistry and on the dynamics of the upper atmosphere.

Our equipment consisted of an image isocon TV camera and associated video recorder and TV monitor. The sensitivity of the isocon tube could be further enhanced by integration of the picture within the TV tube for variable periods. Time lapse video recording was used when the TV camera was operating in the integrating mode to reduce video tape consumption and play back time. This equipment needed the intervention of an experiment operator to decide e.g. when to change integration times, or when to remove or replace filters to look at airglow or meteors, as well as to perform such routine operations as changing video tapes.

2.   THE EXPERIMENTS

The most intense atmospheric airglow emissions lie in the near infrared, and arise from transitions between the vibrational energy levels of the hydroxyl molecule (OH). Peterson and Kieffaber (1973) reported that they had photographed 'clouds' of hydroxyl airglow from a mountain in New Mexico with an ordinary camera and Kodak infrared sensitive film. It seemed likely, therefore, that an image intensifying TV system should be able to record hydroxyl airglow structures. We were interested to make a survey of OH airglow "clouds", from a platform where there are few meteorological clouds, to see whether they are sufficiently well defined, and widespread in time and space, to be used as "tracers" of upper atmosphere motions. For the hydroxyl airglow observations, a Wratten filter with a short wavelength cut off at 670 nm was placed in front of the camera lens. The TV tube had an S25 extended red photocathode with cut off at 900 nm, so the equipment responded primarily to the airglow emissions from the $(7,2)$, $(8,3)$, $(4,0)$, $(9,4)$, $(5,1)$ and $(6,2)$ bands of OH. The camera had a $30°$ field of view and could be positioned at aircraft windows looking out at 60° and at 15° to the horizon.

To give additional image intensification the TV camera was adapted so that the picture could be integrated inside the TV tube for variable periods up to several seconds. When the camera was integrating, and TV pictures were produced at a slower rate than the normal 25 per second, we made synchronised time lapse video recordings, so in this way, one hour's video tape could last for much of the flight, and data from the flight could be examined quite rapidly afterwards by playing back the tapes at normal speed. Petersen and Kieffaber (1975) flew their cameras with image intensifiers and photometers on the same flight, so we were able to check that we were observing the same phenomena with the different instruments.

The CV990 made 16 flights in the period May 28th – June 21st 1975. On many of these flights, conditions were unsuitable for airglow observations because it was twilight, or the moon was in the sky. A total of 18 hours good airglow observing time was obtained during the Assess mission. On several occasions when it was not possible to make airglow observations, we removed the filter, operated the TV camera and video recorder in real time without integration, and made observations of meteors.

An image intensifying TV camera and associated video recorder is an excellent instrument for recording meteor trails. It is much easier to spot a meteor moving on the TV screen, than on a still photograph. The video tape can be played many times over at variable speeds to ensure that few meteors are missed. It is difficult to spot meteors fainter than 4th magnitude with photographic techniques, but with the TV system meteors fainter than 9th magnitude stars can be observed. An estimate of the amount of cosmic dust entering the upper atmosphere in any particular period can be obtained from meteor observations. Cosmic dust near the earth's orbit varies considerably in both quantity and particle size distribution, particularly when the earth's orbit intersects the orbits of comets and meteor showers are observed. (Hughes, 1971).

3.   RESULTS

(i)   Hydroxyl Airglow.

Hydroxyl airglow structure were particularly well marked when viewed at low elevation angles. Figure 1 shows a good example of these structures seen on June 18th 1975, looking East from an altitude of 39,000 ft. from the 15° window of the aircraft. The aircraft position was 45°N, 122°W, and the local time was 01.22 (09.30 U.T.). The picture was obtained by integrating over seven TV frames (i.e. for 0.28 seconds). The atmosphere was not sunlit below about 400 km when this recording was made, which excludes the possibility that we were viewing noctilucent clouds. Moreover, the photometers flown by the New Mexico group confirmed that bright hydroxyl emissions were present at this time.

Although the airglow patterns moved relative to the star background, due to the aircraft motion the patterns themselves changed very little as they drifted across the TV screen. Moreover, the apparent velocity of the airglow structures did not change when the aircraft changed direction. We can therefore assume that the aircraft speed (250 m/sec) was much greater then the velocity of the airglow structure at this time (which is reasonable at mid latitudes), and hence we can estimate the altitude (80-85 km) and horizontal dimensions ($\sim$ 20 km) of individual bright airglow patches from their apparent velocity.

Hydroxyl airglow intensity varied considerably from day to day. When viewed through the 60° aircraft window, airglow structures were diffuse

FIGURE 1

Airglow clouds seen looking East from an altitude of 39,000 feet on June 18th 1975, at 09.30 U.T., latitude 45°, longitude 122°W, local time 01.22. The picture was obtained by integrating over seven TV fields (i.e. for 0.28 seconds).

and not so bright as at the 15° window, and small scale structure was not easy to identify. However, even at 60°, there were some variations in intensity over the field of view of the TV camera, particularly when well defined structures were visible at lower elevations through the image intensifiers of the cameras of the New Mexico group. (A cloud 20 km across at an altitude of 80 km would, of course, fill much of the 30° field of view of the TV camera when viewed at a 60° angle of elevation.) The lower airglow intensities observed at the high elevation window indicate that the thickness of the airglow structures is small compared with their horizontal dimensions.

Observations of fluctuations in the intensity and temperature of upper atmosphere emissions made from mountain observatories have led to suggestions (Krassovsky, 1975) that the fluctuations are due to the passage of gravity waves, with periods, typically, of some tens of minutes, and velocities of the order of 100 m/sec, originating from low pressure centres in the lower atmosphere. The wavelength of these waves should be some tens of kms. The dimensions of the bright airglow patches which we have observed are of this order of magnitude, but the well marked airglow structures observed at the low elevation window on the night of June 17/18 suggest a standing wave pattern rather than waves moving with velocities of $\sim$ 100 m/sec. It is interesting that on this night, when the most well marked structures were observed, a volcano was erupting in Costa Rica (10°N, 85°W). On the other hand, the intensities were highest at the most northerly point reached by the aeroplane that night (47°N) and it was moderately disturbed magnetically ($\measuredangle K_p = 20^+$).

(ii) Meteors.

In a total of 14 hours observing time, much of it with the moon in the sky, we recorded 84 meteors in the 30° field of view of the camera in

TABLE I

METEORS RECORDED IN 14 HOURS OBSERVING TIME ON ASSESS I

METEOR MAGNITUDE (M)	NO. OF METEORS OF MAGNITUDE $\leqslant$ M
$>$ 9	84
9	82
8	80
7	68
6	54
5	32
4	18
3	15
2	9
1	2
0	1
-1	0

the local time interval 21.00 - 03.00. To determine the magnitude of a
meteor after spotting one on the TV monitor, we stopped the video recorder
at the frame showing the maximum meteor brightness, and then gradually
turned down the brightness control until it disappeared. We then looked
up the magnitude of a nearby star which disappeared at the same time on a
star atlas. The faintest meteors spotted had a magnitude greater than 9,
but when the moon was bright, no meteors fainter than magnitude 7 were

observed. An important limitation in the observation of meteors from the aeroplane was the brilliant star background. Table I lists the number distribution of meteors with magnitude, for the 84 sporadic meteors observed during the Assess mission in the magnitude range -2 to +9. The mass distribution index S, estimated from these numbers is ~1.5, which is lower than the value 2.0 obtained from ground based observations for meteors of magnitude fainter than 3 (Hughes, 1971). However, we have not attempted to introduce any correction factors for e.g. variations in sensitivity across the TV screen, or differences in background light level. It is likely that many of the fainter meteors may have escaped detection, and that the mass distribution index should really have a higher value.

4. DISCUSSION

The first Spacelab Assess mission has shown us that an image intensifying TV camera, flown on a moving high altitude platform is indeed a useful tool for study of light emissions from the upper atmosphere, both for observations of airglow structure and dynamics, and as a monitor of meteor influx. The experiment operators were able to deal quite well with our equipment even when looking after several other experiments at the same time, although, not surprisingly, more date could be collected when one or two principal investigators could dedicate all their time to their own experiment. In the second Assess mission we plan a number of changes to the equipment to simplify operation and automate it where possible.

On Spacelab itself, a TV camera could study the structure and dynamics of various airglow emissions, looking down on some (e.g. hydroxyl, 557.7 nm Oxygen), and up at others (e.g. 630 Oxygen SAR arcs). It could monitor cosmic dust influx into the upper atmosphere, which would provide useful backup information for experiments measuring atmospheric sodium and other metallic atoms.

A TV camera flown on Spacelab should also prove very useful in the study of the atmosphere immediately surrounding the Shuttle/Spacelab complex. A Martin Marietta report (Rantanen and Ress, 1975) on the major sources of contamination of the Shuttle suggests that the Shuttle fuel cells will be generating between 7 - 14 kg water per hour which will be flash evaporated through nozzles and ejected with a velocity of 10^4m/sec. The Vernier control engines will also be squirting out a mixture of gases in bursts, (mainly CO_2, CO, N_2 and water) at a rate of $\sim 2\frac{1}{2}$ kg per hour. The Shuttle itself will thus be providing an interesting active experiment in the upper atmosphere, and it will be interesting to study both the chemical reactions of the ejected gases and the general dynamical flow of gas around the Shuttle from optical emissions excited in the gas. The TV system would be most useful for the latter type of experiment. If a magnetic field could be introduced into the Spacelab environment, the dynamics of the gas surrounding it would become even more interesting, particularly to the magnetosphere and plasma physics members of the AMPS community. Introduction of a magnetic field of a few tens of gauss by e.g. running current through coils wound around Spacelab, would be within the power requirements of the system, particularly if the current was only turned on from time to time. A magnetic field would trap soft electrons around Spacelab and they would excite more optical emission from the gas surround the vehicle, thus making the gas gynamics easier to study. It might even be possible to study reconnection between the Spacelab magnetic field and the earth's field. Finally, an image intensifying TV system could detect the optical emissions which might be excited locally by electron guns or lasers carried on Spacelab.

5.   ACKNOWLEDGEMENTS

We thank Dr. D.W. Hughes of Sheffield University for helpful discussions on meteors, Mr. N. Wells, Dr. J. Beckman and Dr. K. Dick, Assess experiment operators, gave great assistance in operating our equipment. NASA Ames Airborne Science Office, and in particular Mr. L. Haughney provided invaluable help throughout the Assess mission. The financial support of E.S.A. and the skilful management of Mr. J. De Waard are gratefully acknowledged. The U.K. Science Research Council provided studentships for two of us (J.C. and N.B.).

REFERENCES

Peterson, A.W. and L.M. Kieffaber, Nature, 242, 321 (1973).

Peterson, A.W. and L.M. Kieffaber, Nature, 257, 649 (1975).

Krassovsky, V.I., K.I. Kuzmin, N.A. Piterskaya, A.I. Semenov, M.V. Shagaev,
N.N. Shefov, T.I. Toroshelidze, Plan. and Spa. Sci., 23, 896 (1975).

Hughes, D.W., Space Research XI, Akademie Verlag, Berlin, p.319 (1971).

Rantanen, R.O. and E.B. Rees, Martin Marietta Technical Report MCR75-13
(1975).

DISCUSSION

J. Burger: You mentioned the problem of the water ejected by the flash evaporators of the Orbiter. Have you made any calculations that show where this water is going to move?

P. Rothwell: I understand that water is ejected through a flash evaporator in order to prevent it going into orbit close to Spacelab as lumps of ice. We have not carried out any calculations ourselves, but I suppose the hope is that the water vapour will be left behind the Shuttle as a wake. We would like to use the TV camera to determine experimentally what happens to this wake.

S.A. Bowhill: What advantages exist for making visual meteor observations from Spacelab rather than from the ground?

P. Rothwell: The main advantages are (a) Clear skies;

(b) Earth provides a darker background for the TV camera to look down on (especially when Spacelab is over oceans) than the star field which a ground-based camera must look up at, and so it should be possible to detect more of the faint meteors.

P.M. Banks: (a) What physical processes lead to the patchiness you observe?

(b) Can you describe the intensity variation with regard to its dependence upon magnetospheric activity?

(c) Was the 85 km region sunlit at any time? If so, what effects were seen?

P. Rothwell: (a) According to Krassovsky, passage of gravity waves through an emission layer affects both the intensity and the rotational temperature of the atmospheric emissions. I suppose that successive compressions and rarefactions of the emitting layer as the gravity waves pass through it could produce the variations in intensity which we observe.

(b) We do not have enough information from this one Assess mission to describe dependence of intensity variation upon magnetospheric activity, or to determine whether it is more dependent on atmospheric or magnetospheric activity.

(c) The sky background was much too bright for us to make airglow observations when the 85 km region was sunlit.

L.H. Meredith: What do your measurements of OH patchiness imply in terms of the spatial resolution required for sounding measurements of atmospheric composition?

P. Rothwell: The dimensions of the airglow patches that we observed were about 20 km across.

J.E. Harries: (Comment): As a comment on the Chairman's question about spatial resolution required to detect the variations in OH (or other) constituents in the mesosphere and thermosphere, limb-sounding is used in order to increase vertical resolution, and Dr. Rothwell's observations do not give any information on vertical structure. In order to study horizontal structure, nadir sounding is more appropriate.

EUV – VISUAL MEASUREMENTS FROM SPACELAB

Gerhard Schmidtke

Institut für physikalische Weltraumforschung,
Fraunhofer-Gesellschaft,
Heidenhofstr. 8,
D-78 Freiburg/W.-Germany.

Among others, two areas of atmospheric physics
are not yet studied satisfactorily: the unattenuated
solar photon flux at the top of the Earth's atmosphere and atmospheric emissions.

The determination of the solar spectral irradiance ("solar constance", which is not constant) is
still controversial. Though the accuracy for radiometers of modern design in space is quoted by $\pm$ 0.3 %,
long-term variations of the irradiance with solar
cycle are not yet established (current estimate: 1 –
0.1 %). In the UV spectral region through the Schumann-Runge range relative variations, e. g. with solar
rotation, are determined with high accuracy (better
than 1 %). However, long-term variations, e. g. with
solar cycle, are expected to be below the level of
experimental accuracy. Typically, down to 125 nm the
intercomparison of relevant solar flux data, which is
underway on behalf of COSPAR being coordinated by the
author, does not yet take into account the time scale.
This parameter will be used for the measurements of
HI, 121.6 nm, and below.

Since Hinteregger has revised his fluxes (Hinteregger, 1970), in the EUV spectral region a good agreement seems to be achieved now for the period from end
of 1972 through 1976 by Hinteregger (1976), Schmidtke
(1976) and Timothy (1976). The higher fluxes given by
Heath (1976) will be intercompared soon. Outside this
period the temporal changes are still speculative.

J. J. Burger et al. (eds.), Atmospheric Physics from Spacelab, 351–359. All Rights Reserved.
Copyright © 1976 by D. Reidel Publishing Company, Dordrecht-Holland.

Though strong indications are given for an enhanced solar emission with higher solar activity
(Schmidtke, 1976; Timothy, 1976), the self consist-
ency of the data from the AFGL (former AFCRL) group
neither supports nor excludes a stronger variation.

Summarizing this brief review: Considering the
error sources in the measurements, the determination
of the solar spectral irradiance and the temporal
variations could be achieved from Spacelab only by
absolute measurements from Spacelab and additional
(non-broadband!) monitoring from satellites. This is
urgently needed especially for aeronomy bearing in
mind, that almost no measurements of the EUV fluxes
are planned for the period beyond 1977.

Because solar radiation is controlling most of
the atmospheric emissions, the solar irradiance must
be known for the interpretation of the airglow measurements. More important, the temporal variations of
the irradiance have to be measured e. g. to decide
whether enhanced airglow emissions are caused by
higher densities of the emitting species or by an increase of the solar radiation.

Since Blamont et al. (1974) observed large changes
in the thermospheric temperatures especially at high
latitudes, instrumentation with higher spatial resolution and a wide field of view is requested. Studying
the red arcs, the precipitation of Helium ions in
equatorial regions, the ozone etc. a broad coverage
in wavelengths with higher spectral and temporal resolutions is required. It would also be appropriate to
incorporate extinction measurements, since its analysis still yields the most accurate density profiles
for many atmospheric constituents and allows cross
checks for the interpretation of airglow measurements.
A set of two instruments is proposed for Spacelab to
meet these requirements: a combined telescope spectrometer and a wide angle photometer, both of new design.
The instruments are being presented and discussed at
the 11th ESLAB-Symposium.

References:

H. E. Hinteregger, Ann. Geophys., 26, 547, 1970.
H. E. Hinteregger, J. Atm. Terr. Phys. (in press), 1976.
G. Schmidtke, Geophys. Res. Let. (accepted), 1976.
G. J. Timothy, The Physical Output of the Sun, 1975 (in press), 1976.
D. F. Heath, private communication, 1976.
J. E. Blamont, J. M. Luton and J. S. Nisbet, Radio Sci., 9, 247, 1974.

DISCUSSION

A. Durney: What fraction of the total radiated solar energy is contained in the EUV?

G. Schmidtke: Of the order of 10^{-5}.

A. Durney: So any climatic change caused by variation in the region below 1000 Å would necessarily be catalytic.

AN INVESTIGATION OF MINOR CONSTITUENTS FROM SPACELAB

G. Witt, J. Stegman
Institute of Meterology
University of Stockholm
Arrhenius Laboratory
Fack
S-104 05 STOCKHOLM, Sweden

ABSTRACT

From a simple passive measurement of O_2 emission in the infrared and red atmospheric systems together with the OH Meinel bands it is possible to derive the hydrogen and odd oxygen concentration profiles. The simultaneous measurements of NO delta and gamma bands and the NO_2 continuum afterglow the daytime and nighttime concentration profiles of odd nitrogen may be obtained. Such a measurement made from Spacelab by means of a cluster of five photometers and a concave-grating polychromator, operating in a limb scanning mode will yield valuable information pertaining to the transport of minor constituents between the mesosphere, the upper stratosphere and the lower thermosphere.

1   INTRODUCTION

It is now well recognized that transport processes play a dominant role in the determining the balance between different groups of minor constituents in the upper atmosphere. One such group involves atomic hydrogen and the odd oxygen components, O and O_3. Another such group is that of odd nitrogen (nitrogen atoms and nitric oxide), which is particularly interesting since its chemistry involves auroral processes. An enhanced production occurs in geographically well defined regions and therefore the worldwide distribution of these constituents is intimately related not only to vertical diffusion but also meridional transport. From simple passive measurements of selective airglow emissions observed at one and the

J. J. Burger et al. (eds.), Atmospheric Physics from Spacelab, 361–365. All Rights Reserved.
Copyright © 1976 by D. Reidel Publishing Company, Dordrecht-Holland.

same time it is possible to derive and monitor the concentration profiles of those minor constituents that are parts of
the above mentioned "photochemical groups".

2   DISCUSSION

The first of the objectives of this investigation is to measure
ozone and atomic hydrogen concentrations in the mesosphere
and upper stratosphere. Existing methods of studying H atom
concentrations in this altitude regime are restricted to a
solution of transport equations in the thermosphere with upper
boundary conditions being derived from the measured Lyman-
alpha flux at 400 km from satellites. The significant role of
H_2 in these models is only now recognized (Hunten and Strobel
1974) although it was proposed many years ago (Tinsley 1969),
thus the determination of the H atom concentration is of great
significance. It has also been shown that the total H atom
concentration above the stratosphere is related to the total
mixing ratio of hydrogen above the tropopause (Bowman and
Thomas 1973) so that the measurement of concentrations in
the mesosphere provides data at many other altitudes. The
second objective of the same investigation is to determine
the concentration of odd nitrogen in the lower thermosphere.
Extensive model calculations have been published (Strobel
et al. 1975) which fairly well explain existing rocket and satellite observations of NO. As to atomic nitrogen, only one
single indirect airglow observation has been reported hitherto
(Feldman and Takacs 1974). Simultaneous and continuous
measurements of both N and NO are altogether lacking.

Recent satellite (Rusch 1973, Rusch and Barth 1975, Barth
et al. 1975) and rocket-borne (Zipf et al. 1970, Witt and
Stegman 1976) observations have indicated that auroral processes can lead to an enhancement in the abundance of nitric
oxide in the lower thermosphere, not only locally but even
at great distances from the auroral zone, although the exact
transport mechanisms are still far from being understood.
The continuous observation of both N and NO provides significant information concerning both vertical and lateral transport processes at the upper boundary of the mesosphere and
the objective of this investigation will therefore complement
the ozone-hydrogen study.

The measurement can be derived from passive observations
of airglow emissions, made with a cluster of five photometers
and one concave-grating polychromator operating in a limb
scanning mode. The relevant emission features are the following:

0-0 band of O_2 ($a^1 \Delta_g - X^3 \Sigma_g^-$) at 1.27 μ (ozone photolysis)

0-0 band of O_2 ($b^1 \Sigma_g^+ - X^3 \Sigma_g^-$) at 762 nm (O-atom recombination, O^1D quenching and resonance fluorescence)

3-1 ($v' - v''$) band of OH Meinel System

$(X^2 \Pi_i)$ at 1.505 μ (H + O_3 reaction)

8-3 ($v' - v''$) band of OH Meinel System

$(X^2 \Pi_i)$ at 727.5 nm (H + O_3 reaction)

NO_2 continuum at 540 nm (NO + O reaction)

NO delta bands at 180-190 nm (N + O reaction)

NO gamma bands at 205-237 nm (resonance fluorescence)

Studies of the infrared atmospheric system (Evans and Llewellyn 1972) have shown that the emission in the daytime and twilight may be directly inverted to give the ozone concentrations height profiles. If these derived concentrations are combined with measurements of the OH Meinel System then H atom concentrations profiles may also be obtained. (Evans and Llewellyn 1973). The inclusion of the oxygen atmospheric system ($b^1 \Sigma_g^+ - X^3 \Sigma_g^-$) is to facilitate the evaluation of the O atom concentrations in the mesosphere and taken in conjunction with the other derived concentrations may be used to estimate the temperature in the mesosphere. Alternately these observations may, taken in conjunction with Nimbus Satellite data to confirm reaction rate coefficients. The purpose of the double OH system is to compare H atom concentrations derived from vibrational levels which are only populated by radiative cascade (Charters et al. 1971). The comparison of many profiles derived in this way will permit an extensive investigation of the effect of vibrational excitation on reaction rates (Potter et al. 1971, Fiocco and Visconti 1974); this may be extremely important for stratospheric modelling of the terrestrial atmosphere (Crutzen 1974). The latitudinal variation of the derived H atom concentrations may be used to estimate water vapor mixing ratios at the tropopause and thus develop improved transport parameters for atmospheric models (Hunten and Strobel 1974, Donahue and Liu 1974, Thomas and Bowman 1974). The purpose of the NO_2 continuum photometer is to provide further information of the transport parameters.

Nitric oxide reveals its presence in daytime through the resonant scattering of sunlight in the UV spectrum. As to the
nighttime distribution of NO, only indirect methods are available; primarily the observation of the airglow continuum
produced by the reaction of NO with atomic oxygen. However,
during nighttime another indirect method can be employed to
determine the N atom concentration, namely the chemilumi-
niscent emission in the delta bands of NO arising from the
recombination of N and O atoms (Feldman and Takacs 1974).
Thus, the ultraviolet spectrum of nitric oxide in conjunction
with the visible "air afterglow" continuum measurement can
provide significant information about odd nitrogen both in the
sunlit and dark period of the day.

3   CONCLUSIONS

It must be emphasized that the essential feature of this investigation is the simultaneity in the measurement of the different
constituents. This is necessary for the proper understanding
of the transport processes which carry these constituents
from region to region in the upper atmosphere. The technique
described here is well established and the measurements can
be carried out by relatively simple means but might yield a
wealth desirable pieces of information about the atmosphere.

4   ACKNOWLEDGEMENTS

The authors would like to thank Dr E.J. Llewellyn, University of Saskatchewan, for valuable ideas and discussions and
for having provided the inspiration for this contribution.

5   REFERENCES

Barth, C.A., Gerard, J.C., Kley, D., 1975: J. Geophys.
      Res. 80, 25, 3419.

Bowman, M.R., Thomas, L., 1973: J. Atm. Terr. Thys.
      35, 347.

Charters, P.E., Macdonald, R.G., Polanyi, J.C., 1971:
      App. Optics 10, 8, 1747.

Crutzen, P., 1974: Canad. J. Chem. 52, 1569.

Evans, W.F.J., Llewellyn, E.J., 1972: Planet. Space Sci.
      20, 624.

Evans, W.F.J., Llewellyn, E.J., 1973: J. Geophys. Res. 78, 1, 323.

Feldman, P.D., Takacs, P.Z,: 1974: Geophys. Res. Lett. 1, 4, 169.

Fiocco, G., Visconti, G., 1974: J. Atm. Terr. Phys. 36, 583.

Hunten, D.M., Strobel, D.F., 1974: J. Atm. Sci. 31, 2, 305.

Liu, S.C., Donahue, T.M., 1974: J. Atm. Sci. 31, 1118.

Potter, A.E. et al., 1971: J. Chem. Phys. 54, 992.

Rusch, D.W., 1973: J. Geophys. Res. 78, 5676.

Rusch, D.W. and Barth, C.A., 1975: J. Geophys. Res. 80, 25, 3719.

Strobel, D.F. et al., 1975: J. Geophys. Res. 80, 22, 3068.

Thomas, L., Bowman, M.R., 1974: J. Atm. Terr. Phys. 36, 1421.

Tinsley, B.A., Pl. Space Sci. 17, 769. 1969.

Witt, G., Stegman, J., 1976: to be published.

Zipf, E.C., Borst, W.L., Donahue, T.M., 1970: J. Geophys. Res. 75, 6371.

DISCUSSION

C. Muller: Do you think that simple photometers could be used in order to observe all the emissions described?

J. Stegman: The photometer array will employ a single common telescope with a field of view (half angle) of about $0.07°$ and aperture 10 cm. The signal is modulated and phase-sensitive detection for the photoconductive detectors will be used. The visible and UV images will be recorded by standard photon counting techniques. According to our estimates, the instruments described would be sensitive enough to observe the emissions.

ULTRAVIOLET OBSERVATIONS OF THE UPPER ATMOSPHERE FROM SPACELAB

Francesco Paresce

Space Sciences Laboratory, University of Calif.
Berkeley, California 94720

Introduction:

Ultraviolet spectrophotometry in the 100-3000 Å region is an old, reliable, proven technique used repeatedly and systematically up to now to study the upper atmosphere of the earth and of other planets. Therein, ironically, lies one of the main problems in convincing the scientific community of the usefulness of performing uv spectrophotometry from Spacelab. Have we not done all this before? Should we not think of something really new and exciting to do for such a futuristic experiment as Spacelab? These are but a few of the understandable first reactions of many scientists to the proposal. A second drawback at first glance stems from the common belief that uv spectrophotometry is useful only for the study of the atmosphere above 120 km. altitude while emphasis in Spacelab planning has been on the region below that level. Consequently, it would seem a difficult task indeed to assign a role, if any, to the ultraviolet in the 1980's.

Under serious examination, however, the situation takes on a much more positive outlook. This optimistic reassessment of the role of uv spectrophotometry for the next decade in atmospheric research follows from a consideration of the two basic characteristics of a typical Spacelab mission: the manned and the multi-disciplinary aspect of the shuttle flights.

J. J. Burger et al. (eds.), Atmospheric Physics from Spacelab, 367–388. All Rights Reserved.
Copyright © 1976 by D. Reidel Publishing Company, Dordrecht-Holland.

Although there is some controversy as to man's role in Spacelab his full utilization for scientific purposes is, at least in principle, foreseen after the first few proving flights. Thus we should expect to have on board an experimenter well versed in the operation of the uv and complementary instruments who will be able to dedicate a substantial part of his time to the kind of operations any experimenter on the ground is able to execute in the laboratory environment. Anybody that has had experience in an earth based laboratory knows how crucial it is to be able to react quickly to unexpected phenomena or to test directly various conflicting hypotheses about either the functioning or the results of his instruments. In this spirit the payload specialist could react to rapidly changing situations in the aurora or the tropical airglow or look for the occurrence of sporadically occurring atmospheric phenomena. This facet of his work will be particularly suited to the typical real time data output formats that the new generation of uv instruments will certainly embody. These tasks could be performed by monitoring a few simple CRT displays of the data stream from a console in the Spacelab environment and reacting to them with a limited number of operations on the IPS or on the instrument itself.

Man's role is enhanced even further if we consider the second fundamental characteristic of multidisciplinarity of the future shuttle missions. The accent now and in the future is on the study of the atmosphere and magnetosphere as a whole not on the single isolated elements that have been relatively well studied in the past. In simpler terms, it is more than a hope that Spacelab will permit the observation of the complex phenomena that link together such diverse regions as the mesosphere and the thermosphere, the ionosphere and the magnetosphere and permutations thereof. For this task it is essential that on board any truly atmospheric flight there be an integrated set of instruments ranging from radiometers and lidars to mass and uv spectrometers working nearly simultaneously to study the earth's atmosphere as a whole not as a disjointed set of regions. A striking example of this is the still open question of the production and escape of atomic hydrogen in the atmosphere. Its elucidation is of vital concern to our understanding of the origin and evolution of our atmosphere. The abundance of atomic hydrogen above 120 km. depends crucially on the chemistry involving OH and H_2 in the mesosphere and H_2

and H_2O_2 in the stratosphere and the transport mechanisms of this element through the lower atmosphere. It is rather obvious that we shall make headway in understanding these processes only when simultaneous measurements of the hydrogen bearing constituent is made at each altitude region from the stratosphere all the way into the exosphere. Up to now we have only scattered measurements at various altitudes of some of these parameters that are in no way associated with each other either in space or in time.

Only a manned mission will allow such a complete study of the atmosphere. Man's role is crucial here as he selects instruments, conditions and spacecraft attitudes to carry out the global study of the earth's atmosphere that is called for. Uv spectophotometry plays a fundamental role because it permits remote sounding of the upper atmosphere in the whole wavelength range from 100 to 3000 Å and of the lower atmosphere down to the ozonosphere principally by means of occultation spectrometry in the 1000 to 3000 Å range. It is in the feature of complementarity to other techniques used simultaneously from Spacelab that uv spectrophotometry will realize its full potential. We may venture to predict that uv spectrophotometers will form an important part of every shuttle mission dedicated to atmospheric science.

That this is not an overly optimistic assessment of the possibilities open to ultraviolet observations from Spacelab can be judged from even a quick glance at the experiment-instrument matrix set up by the atmospheric sciences section of the AMPS science working group /1/. In this matrix uv spectrophotometry in one form or another (emission and occultation spectrometers and photometers) are deemed necessary for the fulfillment of the objectives of 12 out of the 17 atmospheric science experiments proposed. This is a lower limit to the actual experiments in which uv spectroscopy would be involved since it is certain also to play a vital role as a diagnostic tool in a number of the gas release and electron accelerator experiments now envisaged by other sections of the AMPS working group.

Theory:

As we have anticipated in the foregoing, uv spectroscopy should play a crucial role in the Space-

lab era. The reasons for this are best considered
by briefly summarizing in the following some of the
problems in atmospheric physics that, in our estimates, will still be outstanding in the 1980's and
that can be studied by Spacelab missions carrying uv
spectrophotometers and complementary equipment. A
number of simultaneous rocket and balloon flights to
obtain cross calibrations of instruments and vertical
profiles of some airglow features will also, as in the
past, probably prove invaluable to these ends. Tethered satellites might also in the future be considered
in this connection.

     Production, loss and transport processes of the
earth's minor constituents throughout the stratosphere,
mesosphere and thermosphere are perhaps the most vital
to understand fully and experiments aimed at elucidat-
ing their behavior will certainly have the highest
priority. As we mentioned earlier H and H bearing
products form an important link in this chain. The
intense emission lines of atomic hydrogen at 1216,
1025 and 972 Å due to resonance scattering of solar
line radiation by atmospheric H can be used to determine the abundance of this constituent from about 80
km. altitude to the outer reaches of the geocorona at
many tens of earth radii. For this, both the spectral
and vertical profile of the line intensities will be
useful with a Spacelab instrument assuming the burden
of providing the photometric accuracy (optimally by
inflight calibrations and correlations with a mass
spectrometer) and the global and temporal coverage
required. Since the transport and escape of hydrogen
from the upper atmosphere and its possible interaction
with solar radiation is far from clear, asymmetries
and variabilities in its distribution especially near
the poles and in the antisolar regions, will form a
special object of investigation.

     Of even greater significance, however, is the
task of fitting the hydrogen profile above the lower
thermosphere to that of it and the major H bearing
constituents such as H_2, H_2O and OH in the stratosphere and mesosphere. In this endeavor uv occultation spectroscopy using the sun or bright stars should
prove invaluable especially when used in conjunction
with cryogenically cooled infrared limb scanning radiometers, near and far infrared spectrometers and lidars, in short, almost the full complement of presently proposed Spacelab instrumentation. Uv

occultation and emission spectroscopy should provide data on H_2, OH, NH_3 and CH_4 abundances and vertical and horizontal profiles that are either unique or that can be profitably compared to those obtained by the other methods just mentioned.

Of course it is only one part of the complex photochemistry of the minor constituents in the atmosphere. Oxygen also plays an important role and what we have just described for the case of hydrogen could very easily be dedicated to this crucial constituents Atomic oxygen can be observed through its emissions at 1304, 1356, 1026, 999, 989, 936 and 879 Å from the lower thermosphere to the exosphere. Its parent species below the thermosphere can be simultaneously monitored by limb viewing or occultation techniques (OH, NO, NO_2, O_2, or O_3) or emission spectroscopy (NO gamma bands at 2150 Å, CO fourth positive and Cameron bands, O_3 by observing absorption of Rayleigh backscatter light). As in the case of H, these techniques greatly benefit from simultaneous observations of these and other constituents by tunable lidars, mass spectrometers and the normal complement of infrared spectrometers. Similar arguments can be made for nitrogen and carbon whose presence is detectable by emissions at 1200 Å and 1135 Å for atomic nitrogen, a variety of bands such as the Birge-Hopfield and Lyman-Birge-Hopfield in the EUV and far uv for N_2 and the 1561 and 1657 Å lines for C.

Another probably unsolved problem will be that of the terrestrial helium budget. It shows a large discrepancy between the rate of production and classical thermal escape such that the present abundance of He is too low by almost a factor to ten. The solution to this problem is relevant to the question of the overall evolution of a planetary atmosphere. It has been postulated that He, like H and perhaps O and N, may escape after ionization along open geomagnetic field lines. Although this effect seems almost certain to be operating, there are no measurements available to estimate its magnitude. An observation of neutral helium and He^+, the product of its ionization by solar Euv radiation through their resonance emission lines at 584 and 304 Å respectively, especially over the poles, should shed some light on this mechanism. High resolution profiles of the He^+ line in particular should reveal in great detail the dynamical history of the escaping gas. This application is a vivid example

of the power of remote sensing techniques such as uv
spectrophotometry that permit one to probe at great
distances from a rather stationary vehicle such as
the Spacelab. Local measurement techniques would be
invaluable, however, to correlate with the remote
measurements.

It should be clear from the foregoing that uv
spectroscopy in one form or another forms an integral
part of any atmospheric science package intended to
study the normal minor constituent distribution in the
atmosphere from the ozonosphere to the outer limits of
the neutral atmosphere. Of similar interest, however,
are the related questions of variability or abnormali-
ties in the minor constituent distribution such as
those that occur during or after   meteor showers,
geomagnetic storms or particle precipitation. Meteor
showers may produce NO and introduce metal atoms into
the lower atmosphere that may have important effects
on the complex and fragile density of these regions.
The NO abundance can be closely followed by monitoring,
for example, the strong gamma band at 2150 Å. The
elucidation of the effects of geomagnetic storms and
particle precipitation on the neutral atmosphere as
part of the overall question of magnetosphere -
ionosphere - neutral atmosphere coupling is perhaps
the most intriguing outstanding task for a Spacelab
atmospheric mission. Herein may lie, in our opinion,
the key to possible influence of the sun on terrestrial
climate. The solution to this crucial problem is not
to be found in simply more correlations of some poorly
known parameters with some other poorly known parameter
over many solar cycles but in the discovery and understanding of the mechanism or mechanisms through which
such an influence can be propagated. Again this task
is peculiarly adapted to Spacelab with its manned and
multidisciplinary approach coupled to the continuous
and global coverage it can afford.

A major objective of an atmospheric Spacelab
mission would be then to identify the relative importance of neutral atmosphere changes and particle precipitation effects at all relevant latitudes. Perhaps
40% of the precipitating particle energy is released
in the form of electromagnetic radiation in the vacuum
ultraviolet band of the spectrum. Thus accurate accounting of the energy budget during a precipitation
event of any type requires a good knowledge of the vuv
spectrum emitted. Secondary effects of the neutral

atmosphere, primarily due to photoionization of the
ambient gas, depend intimately on the intensity of
specific lines in the EUV. Transport of minor constituents, especially of NO, both in altitude and
laterally can be monitored during and especially after
the events by the relevant airglow emissions of N, N_2,
O^+, N^+, NO, etc. all in the ultraviolet. In situ
measurements of the incoming and secondary particle
populations during the precipitation will, of course,
be of particular significance for the unraveling of
the complex interactions between the particles, the
radiation field and the neutral and ionized constituents.

Ultraviolet sensitive imaging systems equipped
with broad or narrow band filters to observe the whole
region in which precipitation is occurring and to
monitor changes in its appearance would have tremendous advantages over similar systems used widely in
the visible. Reflection from the earth's surface is
absent and scattered sunlight in daytime would be at a
minimum below 1500 Å. Different filters select different spectral regions of the precipitating particle
spectrum as well as different particle populations and
therefore yield useful information on the overall morphology of the event and its evolution in time.

The ionosphere or plasmasphere is no less an
interesting domain for research by uv spectrophotometry
in the 1980's. O^+ is the major constituent of the
upper ionosphere yet it has only been observed very
rarely by remote sounding techniques. Its resonance
line at 834 Å has been seen once at night /2/ and once
in daytime /3/ but we have no information on its expected variation over large ranges of geomagnetic
latitudes and longitudes and in time. Its radiative
recombination with electrons, especially intense in
the tropical airglow belts, yields emissions of neutral oxygen at 1304 and 1356 A and possibly a strong
broad line at 910 Å. The asymmetry of the peak intensities of the tropical airglow belts in these lines
with respect to the geomagnetic equator is a sensitive
indicator of horizontal winds transporting plasma
across the equator at thermospheric altitudes.

The distribution of other heavy ions such as O^{++},
N^+ and N^{++} over the whole ionosphere and plasmasphere
can really be studied profitably only by airglow techniques using the resonance lines at 834, 703, 508, 600
and 304 Å for O^{++}, 776, 746, 635 Å for N^+ and 991, 764,

686, and 772 Å for N^{++}. These constituents are expected [4] to be present even at high altitudes and
together with the lighter ion He^+ act as effective
tracers of the plasmaspheric cold plasma. By measuring the backscattered intensity of the appropriate
resonance line as shown schematically in Fig. 1, one
can obtain instantaneous snapshots of the plasmasphere
in differing constituents.

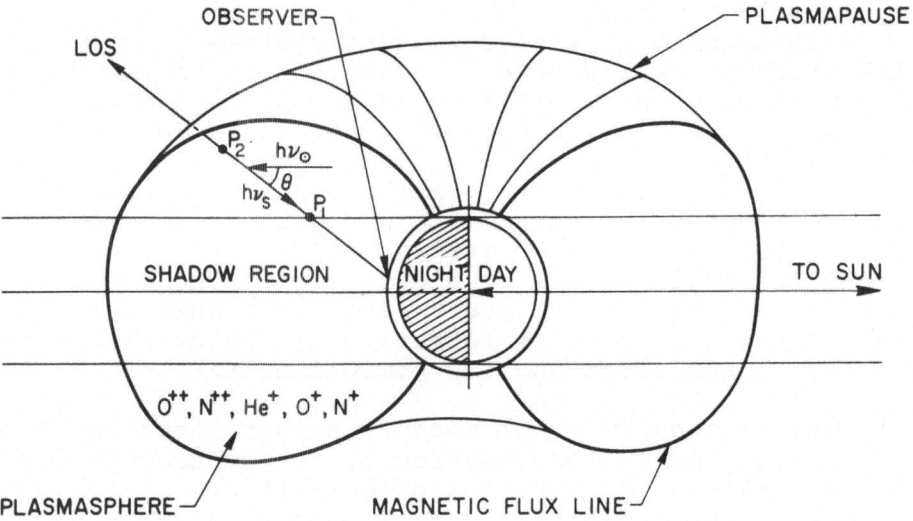

Fig. 1   Observing geometry of the scattered ionic
           emission lines.

These kinds of maps would allow one, for example,
to follow the instantaneous reaction of the plasmasphere to geomagnetic storms and its subsequent evolutions. Localized in situ measurements made at
different times and locations are obviously difficult
to interpret in this connection but remote measurements encompassing large regions of space at one time
should prove useful in understanding this complex
magnetosphere - ionosphere interaction whose main
characteristics are as yet unexplored observationally.
One cannot exclude the possibility of being able to
observe by this technique cold plasma, if it exists,
outside the plasmasphere in the plasma sheet or mantle.
These observations are best made in the directions

where the plasmaspheric component of the backscattered
signal vanishes as in the shadow region or at the
poles (Figure 1).

One final outstanding ionospheric problem that is
easy to predict will still exist in the Spacelab era
is that of the maintenance of the nighttime ionosphere
at all latitudes. According to simple recombination
theory the daytime ionosphere should decay quickly to
very low levels of ionization a few hours after sunset
whereas the level of ionization observed at night is
only one to two orders of magnitude less than in daytime. Where this plasma comes from is a central question at this point. Backscattered EUV radiation from
the plasmasphere, geocorona and interplanetary medium
(this important component due to He I, 584 Å radiation
was discovered in 1970 /57/ might account for part or
all the ionization required but transport processes
from the sunlit ionosphere also play an important role
probably. What is needed to tackle this problem correctly are simultaneous measurements of the EUV radiation field within the night ionosphere as a function
of local time and direction, of the ion and neutral
population and their dynamics. Again only the multi-
disciplinary approach so often lacking in the necessarily restricted present day or past payloads but an
important characteristic of Spacelab will allow any
kind of significant breakthrough in the solution to
this important problem.

In the foregoing we have tried to briefly sum-
marize some of the leading problems that we predict
will be of great interest at the time of the Spacelab
missions and that can be grappled with by using uv
spectrophotometric techniques. But this list does not
by any means exhaust the number of predictable significant problems that will have to be dealt with by
Spacelab atmospheric flights let alone the most probably even larger number of unpredictable ones that
will surely present themselves in the next decade.
Time limitations prevent us from speaking of the exciting possibilities offered by uv observations of
gaseous releases of helium or lithium in the tail or
cleft regions of the magnetosphere to follow interplanetary plasma as it enters the geomagnetic cavity
nor to consider the implications of observations of
artificial aurorae in the uv or the importance of the
measurement of the basic atomic and molecular parameters on board the Spacelab itself. If anything it

is clear that a major task confronting flight planners
will be to carefully select and monitor a viable plan
of operations for uv instruments amongst the almost
infinite number of possible experiments scientists can
easily envisage for this type of instrumentation.

Instrumentation:

At this point we would like to describe some of
the salient characteristics of basic uv instrumentation that we feel will be necessary to carry out the
experiments outlined summarily in the previous section.
None of the instruments we shall describe are the pi-
ous wishes of overly optimistic scientists but are
actually, for the most part, well tested and proven
flight worthy instruments that can be flown today or
new types in advanced stages of construction whose
problems should be well ironed out between now and the
first Spacelab flights.

It very quickly becomes apparent during a careful
study of the basic requirements of uv spectroscopy on
Spacelab that a single instrument cannot begin to cover
effectively the complete wavelength range 100 to 3000
Å and beyond. The range shortward of about 1200 Å can
be covered only by single reflection type instruments
such as the Rowland circle mount /6/ or the Wadsworth
mount spectrometer because of the very low reflection
efficiencies of mirrors at EUV wavelengths. Below
500 Å grazing incidence rather than normal incidence
might be called for for higher sensitivity in this
region. The normal incidence spectrometers have the
advantage, however, of minimizing distortions and
aberrations which become crucial at high spectral or
spatial resolutions. The normal incidence Rowland
circle mount is by far the simplest instrument for the
EUV and has been developed intensively for rocket and
satellite observations from 300 to almost 1500 Å. It
is shown schematically in Figure 2 and it is a simple
and easy to baffle instrument. The Wadsworth type of
mounting permits higher collecting areas but smaller
fields of view and as such is very convenient for limb
viewing of planets to obtain vertical limb profiles.
Because of the large area and bulky collimator, however, it is very difficult to baffle effectively and
this type of instrument is notorious for its stray
light problems in the EUV.

The wavelength region longward of 1200 Å approximately is most efficiently covered by an Ebert-Fastie

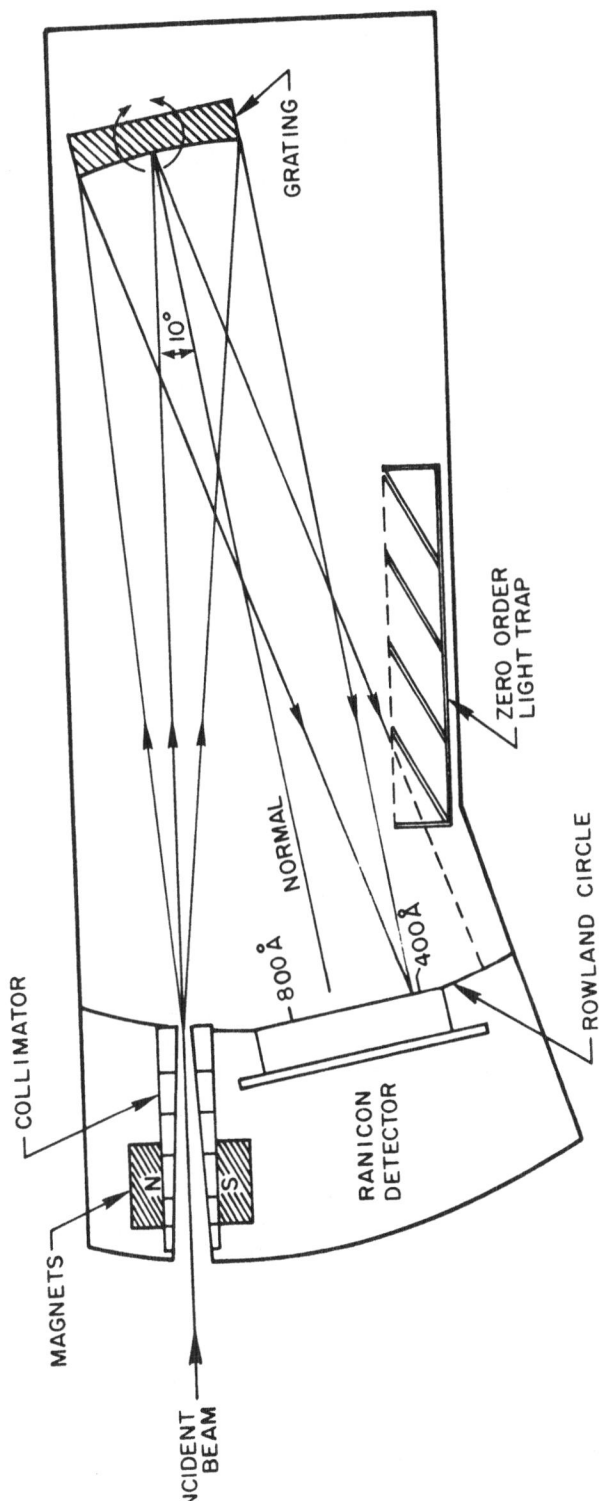

Figure 2.   Schematic diagram of a normal incidence Rowland mount spectrometer.

type spectrophotometer. Even in this range, however,
several spectrometers are necessary, each optimized
by choice of suitable grating, ruling and detector
sensitivity to provide maximum performance and efficiency over roughly an octave of the spectrum. Thus
one can envisage a cluster of small one quarter meter
Ebert spectrometers to cover the ranges 1100-1800,
1800-3400, 3400-6000 and even 6000-10000 Å.

All of the spectrometers just mentioned can be
used either with or without auxiliary optics depending on the application. In the case of no auxiliary
optics high sensitivity to diffuse emissions and appropriate image sizes are achieved with grating focal
ratios of about f/5 (12° x 12°) for the Ebert spectrometer and about f/12 (5° x 5°) for the concave grating
spectometers except the Wadsworth instrument where the
field size is determined by the entrance collimator.
High field of view photon collectors of the type described by Dr. Schmidtke at this symposium might be
particularly appropriate in increasing the collecting
area of the EUV spectrometers for accurate observations of very faint airglow emissions lines or in conjunction with very high resolution spectrometers. For
observations requiring spatial as well as wavelength
resolutions small Cassegrain type feeder telescopes
can be used in front of the Ebert spectrometers in
which case the field of view is determined by the
angular size of the slit at the focus of the telescope.
Typical spatial resolutions in this case could be as
low as 0.5 x 5 arc seconds with spectral resolutions
varying between about 0.5 and 50 A at sensitivities of
typically a few counts per second per Rayleigh. Linear resolutions at the source of about 1 km. in a limb
viewing mode are probably achievable in these configurations.

The recent rapid development of electronic photon-
counting detectors and holographically ruled gratings
has revolutionized the design of modern uv spectrophotometers and has made possible an enourmous improvement in their basic characteristics. It is these two
basic components that the most spectacular changes
have occurred in the past few years. In contrast the
basic mounts in which they are used have remained essentially unchanged since the time they were developed.
Holographic ruling of master diffraction grating blanks
both of the concave and plane variety has liberated
spectrometer designers from the limitations imposed by

the small number of available classically ruled gratings. Where in the past spectrometers were designed primarily around a grating now they can be designed and the grating tailored to fit the scientific requirements. Almost any size, shape, ruling constant and radius within reasonable limits are available. More exotic substrate surfaces such as cylindrical, elliptical, toroidal can now be ruled to minimize aberrations. The possibility of making a stigmatic grating at one or more wavelengths in the desired interval adds even greater significance to this important breakthrough in grating technology.

Up to just a few years ago most of the instruments we described earlier were used with single detectors behind fixed exit slits with the spectrum being sequentially scanned by an appropriate rotation of the diffraction grating. Since with the new detectors with imaging capabilities we shall shortly describe the whole spectrum may be viewed at once, a large gain in sensitivity and time response for any single feature is immediately achieved. Secondly, one can do away with continuous rotation of the grating as it can be left fixed or with at most a few controllable positions for diagnostic purposes. Thirdly, a second dimension that can be used either for imaging of a planetary limb for example or for stacking of different spectra in different orders or for comparison is made available by the new devices. Lastly but not leastly they permit a wide variety of data handling operations that had never been available before. For example, it is a straightforward task to vary the effective picture element dimensions in an imaging spectrometer both vertically (with spatial resolutions along the slit) and horizontally (spectral resolution element) to provide the optimum combination of resolution and sensitivity for each particular application. One might wish in some cases for very high spatial resolutions of the earth's limb while relaxing the spectral resolution for intense emission features or vice versa for the weak features. Part or parts of the spectrum can be selected for study at any time with the rest being ignored in the data stream.

For efficient use in many typical applications the detector at the spectrometer focus, apart from the obvious ones of compactness, simplicity and reliability, must meet the following important requirements: high and stable quantum efficiency over the spectral

band of interest, low internal background, photon
counting capability, large working field sizes, modest
power requirements and 50 x 50 or more pixel formats.
There exist quite a few devices available on the mar-
ket today or under intense laboratory development that
meet with varying degrees of success some or all these
requirements. We shall not try to describe them all
here but rather we prefer to consider in some detail
those devices that make use of the microchannel plate
(MCP) as the primary sensing element. This, because
we feel the MCP is a striking example of the clever-
ness and ingenuity that has been typical of detector
development in the last few years and because it em-
bodies, in our opinion, the most suitable characteristics for detection of uv photons. We stress here that
it should not be considered the only device available
and interesting alternatives (such as the charge
coupled devices for example) could be more useful for
some specific applications.

Although it has no imaging capability, the single
channel electron multiplier (CEM) was one of the first
devices to allow individual uv photons to be counted
with reasonable efficiency and very low background.
The CEM features low power consumption, compact size,
and a large output signal for single event counting
applications. These beneficial features may be kept
and spatial imaging capability obtained, by connecting
a large number of small CEMs in parallel. The sizes
of the individual channels may be reduced to micro-
scopic dimensions without serious consequences because
the electron gain characteristic is dependent only on
the ratio of the channel length to diameter. Microchannel plates made in this way found numerous applications in visible image intensification [7] for which
purposes they are sealed into a vacuum envelope with
a photocathode and a phosphor output screen.

For space applications, an electrical rather than
optical output is required. One way to achieve this
electrical signal is to optically image the phosphor
screen onto an electro-optical sensor. A one dimensional system of this type [8] employed a two-stage
MCP, a phosphor screen, a fiber-optic coupler and a
Reticon MOS photodiode array. Single photon events
were distinguished in this manner at rates up to several hundred counts per second. The linear spacing
of the MOS array elements was 0.025 mm. because the
phosphor light flashes illuminated 5-10 elements in

spite of the fiber optic coupler. Moreover since the phosphor intensity varied from event to event, unambiguous detection of two or more events in the same portion of the MOS array frame was not possible. Hence the count rates had to be maintained below the frame rates to achieve pulse counting performance. Another scheme /9/ combines a MCP, a phosphor screen, a fiber-optic coupler and a SEC Vidicon to achieve 12 lp/mm. resolution. Due to vidicon limitations photon counting sensitivity was not achieved, however.

A far simpler way to achieve an electrical signal output from a MCP detector is to utilize the output electrons directly. Systems incorporating a MCP and proximity coupled multi-element anodes with inter-element spacings of 50 microns or less have been extensively developed /10/, /11/, /12/. An inherent problem with these methods is their complexity: the number of image pixels cannot exceed the number of electronic counting circuits employed. Since these can be several hundred, several hundred circuits are needed. Furthermore, crosstalk between adjacent anodes is always a complication which can cause two or more events to be counted when in fact only one occurred. This leads to severe limitations on the maximum counting rates achievable (less than 25c/s/element in /11/ for example).

Many of the limitations of detectors previously discussed can be overcome by using a readout method which encodes the position of each event. Suitable electronic circuitry external to the detector then combines the functions of decoding and accumulation. The result is an electrical position coordinate being made available in essentially real time. Several methods exist for passively combining the outputs of many anode elements while retaining position information. An especially good one is to couple adjacent anode grid wires with capacitors with signals appearing at the two ends of the resulting capacitor string with a ratio that depends on position. Position information is then recovered with a ratio circuit /13/.

These methods have in common the need for numerous, finely spaced anode wires to supply signals to a complicated passive position-encoding network. The methods can, with some additional difficulty, be made two-dimensional by overlaying a second orthogonal anode grid and network, and arranging the grid bias

potentials so that event pulse currents are shared
between the two systems.

A hybrid readout scheme is to proximity focus
the MCP output onto a pair of orthogonal grids /14/.
Adjacent wires in each grid are resistively coupled,
and every eighth wire connects to an amplifier, dis-
criminator, and an input port of a switchable analog
ratio circuit. A digital computer controls the event
measurement process. In operation, each detected
event is coarsely located by identifying the three X-
axis signals and three Y-axis signals which are largest; then, under computer control, the analog ratio
circuit is switched to the appropriate signal wires
and performs a fine interpolation of the event position.

An enormous simplification is made possible when
a continuous resistive anode is employed for position
encoding: the anode as its own encoding network. The
fact that it is inherently continuous means that the
system resolution need not be limited by the complexity of the anode structure and circuitry.

The two-dimensional resistive anode is an extension of the resistive wire technique into a planar
geometry. The anode is a flat rectangular or circular
resistor equipped with electrical contacts at three or
more points on its periphery. When charge is deposited
at any point on the anode, it is removed via the contacts in relative amounts which depend on event location.

The MCP/resistive anode combination offers the
advantages of compactness, ruggedness, simplicity,
and format versatility in all applications. Resistive
anode image converters or Ranicons for one or two dimensional sensing of ultraviolet images have been only
very recently developed and discussed in the literature /15/, /16/. A Ranicon we have developed is shown
schematically in Figure 3.

For two-dimensional work, four contacts around
the periphery of the anode are connected to four
charge amplifiers, which in turn feed a two-axis pulse
position analyzer (PPA). In operation, the electron
cloud from a detected MCP event ($\simeq 10^7$ electrons) is
collected by the resistive anode. This charge drains
off the anode into the four amplifiers. The relative

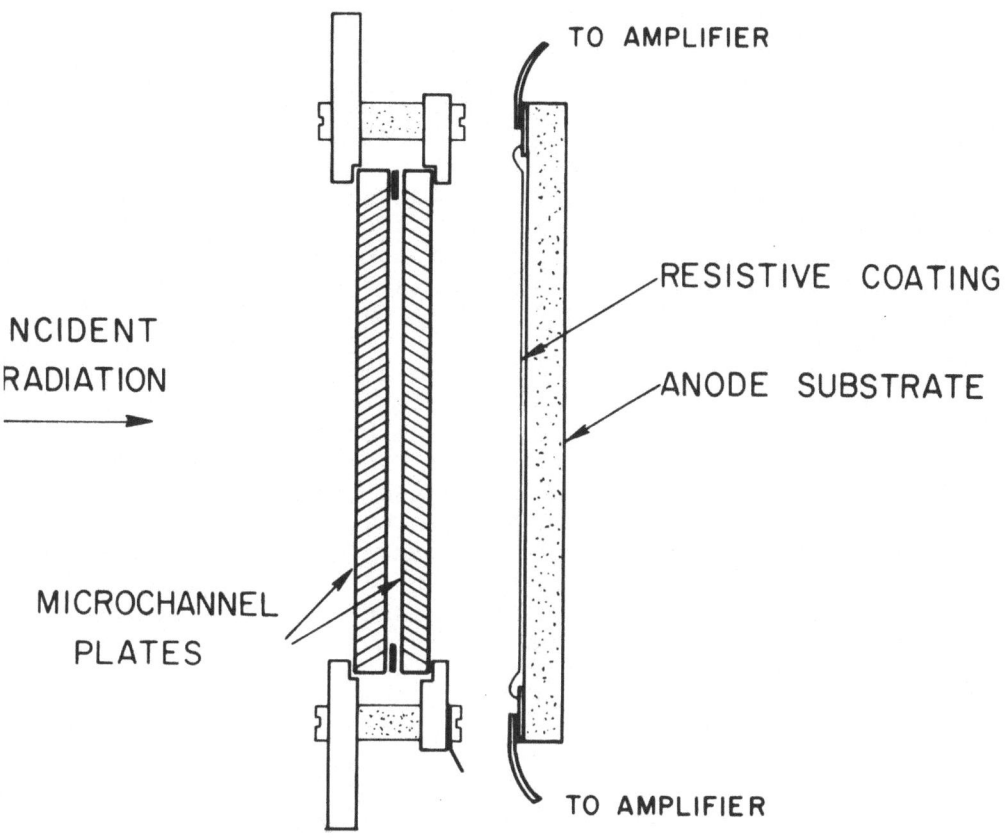

Figure 3.   Schematic diagram of the ranicon.   The
             microchannel plates provide photodetection
             and electron gain.   The resistive anode
             permits the event location to be determined.

proportions of these four signals depend upon the location of the event.   Hence, the x and y coordinates
of each event can be determined from the ratios of
the charges collected at opposing resistor contacts.
The distributed capacitance of the anode resistor
causes a systematic dependence of charge collection
time upon event location.   Hence the event coordinates

can also be obtained from the rise time differences
of the amplified signals. The relative merits of
these PPA techniques are discussed in (16). The readout
technique is continuous in the sense that infinitesi-
mally small motions of the electron image result in
non-zero changes in the charge ratios. This fact dis-
tinguishes the resistive anode method from discrete
location readout devices. The detector's resolution
is fundamentally limited by the channel-to-channel
spacing D of the MCP, and by the thermal noise charge
fluctuations within the resistive anode. Each of these
causes acts to perturb the location of the apparent
centroid of the detected electron cloud. The channel
spacing limitation sets a maximum one dimensional
resolution equal to W/D, where W is the width of the
working field. Typical MCP's have field widths of 40
mm, D between 12 and 40 microns, and permit ≈1000
elements to be resolved along each axis. The thermal
noise resolution limit is a function of the electrostatic capacitance of the anode, which in combination
with the amplifier time constants sets a limit to the
system's charge noise. This precise figure depends upon
certain bandwidth integrals of the processing electronics, but is of the order of 3×10^3 electrons rms.
Individual channel plate signals of 10^7 electrons
suffer random position errors of less than one part
per thousand, corresponding again to a one-dimensional
resolution figure of better than 1000.
The proximity focussed electron cloud arriving at the
resistive anode can be several millimeters in size but
does not limit system resolution because it is the
centroid of each cloud which is sensed by the subsequent
electronic processing. Statistical fluctuations in
the centroid location do arise from the finite number
of electrons in the cloud, but in practice set a resolution limit far finer than the two perturbations
described above. Thus, in addition to being much simpler
than multianode detectors, the ranicon is not subject to
the crosstalk problems characteristic of multianode
systems.

We have used the ranicon for uv spectroscopy by
placing it at the focus of a normal incidence Rowland
circle mount as shown in Figure 2 and have obtained
very satisfactory operating characteristics even with
very weak lines. A typical one-dimensional image of
a laboratory source in the 400-800 Å range is shown
in Fig. 4. The spectrometer permitted 10 Å resolution
in this case (1 mm. slits) and the Ranicon had no
difficulties whatever in resolving lines to this ac-

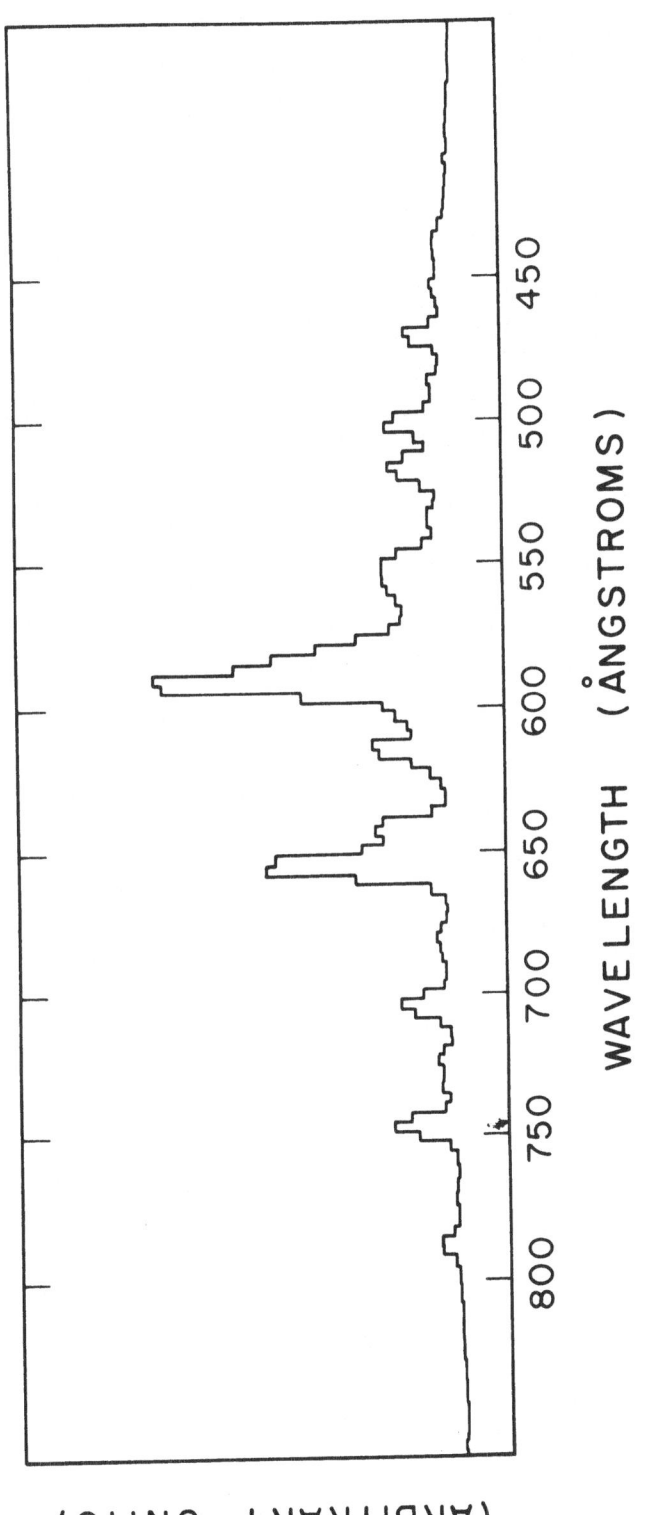

COUNTS (ARBITRARY UNITS)

WAVELENGTH (ÅNGSTROMS)

Figure 4. Spectrum of a hollow cathode discharge source in the 400–800 Å range obtained with a one-dimensional Ranicon.

curacy. Subsequent tests with both one dimensional
and two-dimensional devices using X-ray resolution
test foils indicated that at least in this very simple setup a resolution of at least 5 lp/min was easily achieved up to a total count rate of approximately
20000 counts per second.

The great power of this technique should be quite
apparent. Apart from its inherent advantages for any
application it is particularly suited for Spacelab
since the operator can, for the first time, interact
directly and in real time with the image the instrument or instruments are observing. Thus he can manip-
ulate quickly and smoothly all the envisaged dia-
phragms, filters, slit widths, etc. with immediate
response to their effects on the image a reality. No
fussing with film or guessing correct exposures, he
can actually see his image build up in time and create
the best observing conditions for his instrument. All
these and a few more advantages we have not had time
to delve into ought to make uv spectrophotometry on a
multidisciplinary manned Spacelab mission, as we anticipated in our introduction, a very important and
useful tool for atmospheric research.

There are some applications, of course, in which
devices using electronographic techniques and sensitive film might be more appropriate. This is particularly true when the enormous information storage
capability of the film is necessary and/or when very
intense sources are being viewed as in the case of
solar occultation studies for example. If one ex-
trapolates the incredibly rapid growth of the electro-
optical image converters we have discussed into the
future it is rather tempting to predict, however, that
they might well equal or even surpass some of these
more classical techniques in these areas as well.

Acknowledgements:

Illuminating discussions with Dr. Michael Lampton
on all aspects of detector research and development
have been fundamental for this talk and are gratefully
acknowledged. I also wish to thank all the members of
the Atmospheric science section of the AMPS science
working group and especially Drs. Paul Feldman and
George Carruthers for stimulating conversations on
many facets of uv spectrophotometry.

REFERENCES

/1/. Scientific objectives for atmospheric physics
from AMPS, Atmospheric Science Section of the
AMPS Science Working Group.

/2/. Kumar, S., Bowyer, S., and Paresce, F., Geophys.
Res. Letters, 1, 109, 1974.

/3/. Carruthers, G. and Page, T., Science, 177, 788,
1972.

/4/. Schunk, R. and Walker, J.C.G., Planet. Space
Sci., 17, 853, 1969.

/5/. Paresce, F., Bowyer, S., and Kumar, S., Ap. J.,
183, L87, 1973.

/6/. Kumar, S., Paresce, F., Bowyer, S., and Lampton,
M., Applied Optics, 13, 575, 1974.

/7/. Ruggieri, D., IEEE Trans. Nuc. Sci., NS-19, 3,
75, 1972.

/8/. Riegler, G., and More, K.A., IEEE Trans. Nuc.
Sci., NS 20, 192, 1973.

/9/. Hunter, W.R., and Harlow, F.E., Applied Optics,
12, 968, 1973.

/10/. Smith, D.G., and Pounds, K.A., IEEE Trans. Nuc.
Sci., NS-15, 541, 1968.

/11/. Timothy, J.G., and Bybee, R.L., Applied Optics,
in press, 1976.

/12/. Catchpole, C.E., and Johnson, C.B., PASP, 84,
134, 1972.

/13/. Gott, R., Parkes, W., and Pounds, K.A., IEEE
Trans. Nuc. Sci., NS-17, 367, 1970.

/14/. Kellogg, E., Henry, P., Murray, S., Van Spey-
broeck, L., and Bjorkholm, P., Rev. Sci. Instr.,
in press 1976.

/15/. Lampton, M., and Paresce, F., Rev. Sci. Instr.,
45, 1098, 1974.

/16/. Lawrence, G., and Stone, E., Rev. Sci. Instr.,
        46, 432, 1975.

DISCUSSION

G. Haskell: Would you please elaborate on the functions that you
     foresee for the Spacelab Payload specialists?

F. Paresce: The same as those usually carried out by a scientist
     in his laboratory. I feel these are crucial to the success
     of any good experiment. It is not going to be any different
     on Spacelab.

P. Rothwell: With reference to the importance of manned inter-
     vention with instruments to be flown on Spacelab, our
     experience with the Spacelab ASSESS mission (in which four
     experiment operators had to run a number of experiments on
     the NASA CV990 for a period similar to a Spacelab flight)
     suggests that one cannot count on continuous attention for
     any one experiment. It is safest to design instruments for
     semi-automatic operation, since the experiment operator may
     have to assign first priority to troubleshooting and routine
     operation of all experiments in his charge.

PASSIVE SOUNDING EXPERIMENTS AND EXPERIMENTAL REQUIREMENTS FOR THE DETERMINATION OF TRACE GAS CONCENTRATIONS

Redemann, E. and H.-J. Bolle, Univ. of Munich

D. Offermann, Universität of Wuppertal

ABSTRACT

Three classes of passive optical sounding experiments are necessary to accomplish the scientific goals of the research tasks proposed for the Middle Atmosphere Program (MAP): Sun occultation, limb emission, and airglow emission experiments. Experiences with existing instruments which represent these classes are presented, and necessary future developments for Spacelab missions are discussed.

1. INTRODUCTION

The investigation of the three-dimensional distribution of minor constituents and their concentration variations in time is realized as one of the major problems in atmospheric physics. These quantities regulate the energetics and by this the dynamics of these layers and may even affect climate at the ground. During the last years measurements by instruments mounted on aircrafts, balloons and rockets have been made, and valuable information about concentration and vertical profiles of some constituents has been gained. But these measurements are confined to limited geographical regions and to few constituents only so that we are still far away from a complete understanding of the global fields of atmospheric composition andmotion, of the interaction between different levels of the atmosphere and of diurnal and seasonal

J. J. Burger et al. (eds.), Atmospheric Physics from Spacelab, 389–400. All Rights Reserved.
Copyright © 1976 by D. Reidel Publishing Company, Dordrecht-Holland.

variations in the concentration of some species. It is evident, that remote sounding of the atmosphere from space platforms can substantially contribute to obtain the global coverage needed in order to detect trends in concentration variations and to investigate transport phenomena. In the following we shall discuss a few types of passive experiments which can contribute to this research program.

2. PASSIVE SOUNDING EXPERIMENTS

Three "classes" of passive experiments can be defined

- absorption measurements observing the sun's radiation attenuated in its passage through the atmosphere (sun occultation)

- thermal emission measurements in order to detect either the upelling radiation from the earth's surface and the atmosphere (nadir mode) or the atmospheric emission when the instrument is pointed to the limb (limb mode)

- non-thermal emission measurements of some excited species such as OH with a concentration maximum in the upper atmosphere.

Each of these classes has some advantages but is also restricted in its application. In general, the major advantages of the limb mode are the long optical paths - and therefore a very large amount of absorbing or emitting gas along the line of sight so that even small concentrations can be detected - and the good altitude resolution of about 2 - 4 km. This makes limb measurements more favourable compared to nadir measurements for determining vertical profiles in altitude regions above 10 km height. Limb measurements are less applicable for the detection of concentration profiles below the tropopause where clouds can influence the signal, and possible horizontal concentration fluctuations cannot be detected since the signal depends on the total mass averaged over an atmospheric path of about 600 km.

On the other hand, the sun occultation technique provides relatively simple data reduction which does not need very accurate information about the temperature distribution in the optical path. For constituents with no strong diurnal variations in concentra-

tion mean vertical profiles can be obtained. For
those gases, the concentration of which varies during
the day, occulation measurements can only give addi-
tionall information under the situation near sunrise
and sunset, but which are, however, interesting from
the energetic point of view. Furthermore, if the
orbit is not choosen accordingly, the information
can be limited to a very narrow geographical one.

The constraintsof the occultation method can be
overcome by limb emission measurements which allow
observations continuously in time and also under
different azimuth's of the horizon, so that even information about large scale horizontal concentration
variations is available.

During the last seven years prototypes of instruments measuring either the atmospheric transmission or the thermal or non-thermal emission have
been designed in Germany. A most simple and cheap
instrument is a correlation radiometer consisting of
two small optical systems focussing the solar radiation on one detector after passing two separate
filters or in a different version after reflection
at two diffraction gratings mounted under slightly
different angles. One spectral channel is sorting out
an absorption band, the other being a nearby reference
wavelength outside the absorption band. The comparison
of the two signals allows the determination of atmospheric gas concentrations in the optical path. If
the filters are replaced by gas cells of which one
contains the gas under inspection, the same instruments can be used as a gas correlation radiometer
(Fig. 1). A first version of this type of instrument
has been tested during a balloon flight 1971, and
vertical profiles of CH_4 and H_2O have been obtained
(Burkert et al., 1973). An improved version is proposed to be flown on SPACELAB mounted to a sun pointing platform. At a later stage this platform can be
used for more advanced instruments like a SISAM interferometer (Bolle et al., 1972 a).

To the second class belongs a filter radiometer
(Fig. 2) with a small FOV which consists of a 30 cm
Cassegrain system with a built-in black body for absolute calibration (Bolle et al., 1972 b). This instrument allows continuous measurements of atmospheric
emission by looking downwards to the earth's surface
or scanning the horizon. First experience has been

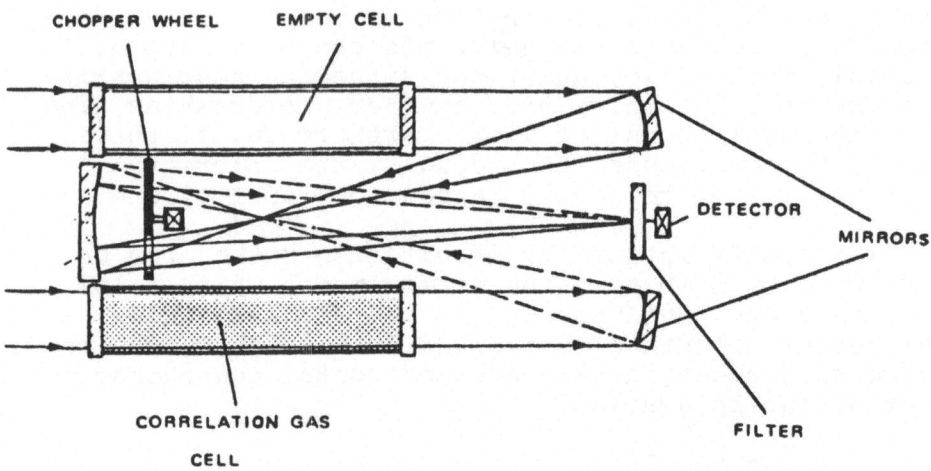

Fig. 1: A simple gas correlation radiometer

gained during a balloon flight 1975 for determining
H_2O and CH_4 vertical concentration profiles. It is
expected that this instrument is also able to determine the stratospheric CO concentration.

In order to detect thermally or non-thermally
excited airglow radiation in the upper atmosphere it
is necessary to measure radiances comparable with the
emission of optical surfaces at room temperature.
Since in atmospheric research we deal with volume
emitters and not with point sources as in astronomy
this emission of optical components cannot be compensated by space chopping. The only way to arrive
at radiances in absolute units is very often to cool
the whole instrument. It has therefore very early
been started to develop cooled radiometers. The first
instruments were cooled by liquid Freon to about
200 K for OH emission measurements in the NIR (Bangert
et al. 1972). Very recently a LHe cooled Fastie-Ebert
spectrometer has been developed (Offermann, unpub-
lished) for far IR spectroscopic investigation on
atomic oxygen emission. This instrument which can be
regarded as a prototype for future more advanced
experiments, for which high detectivity is mandatory,
will be flown on a rocket during a winter campaign
1976/77.

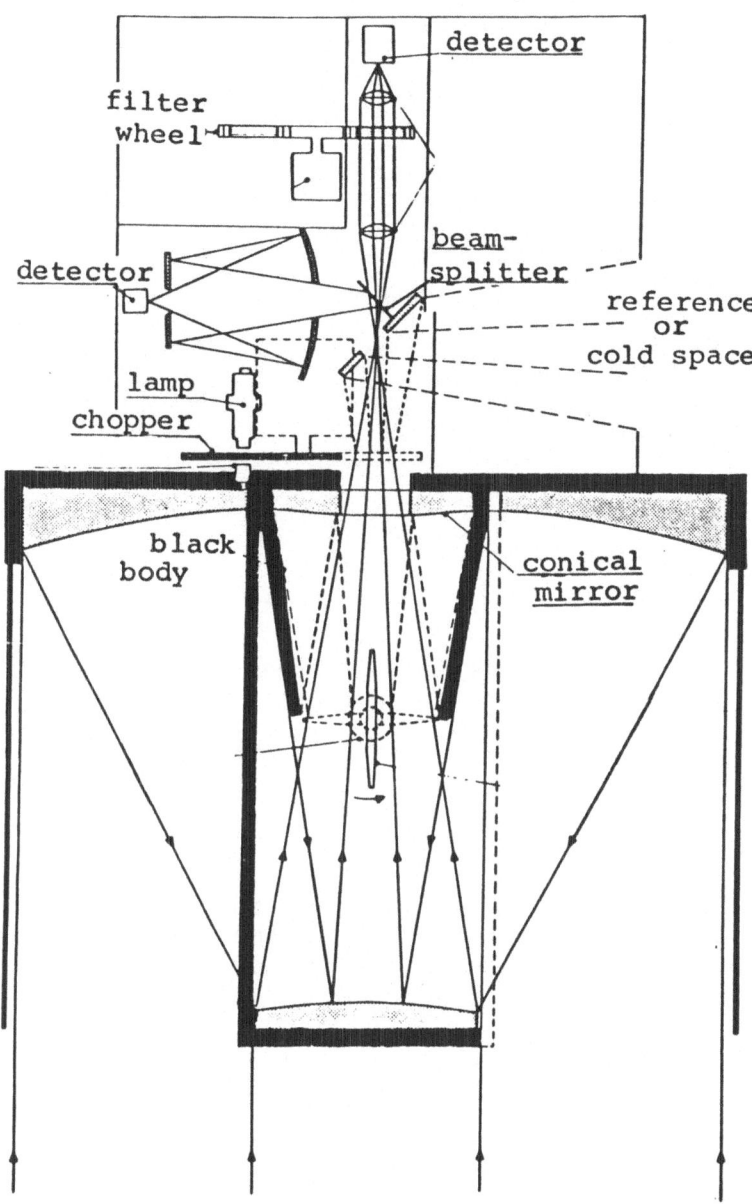

Fig. 2: Uncooled filter radiometer for emission
measurements

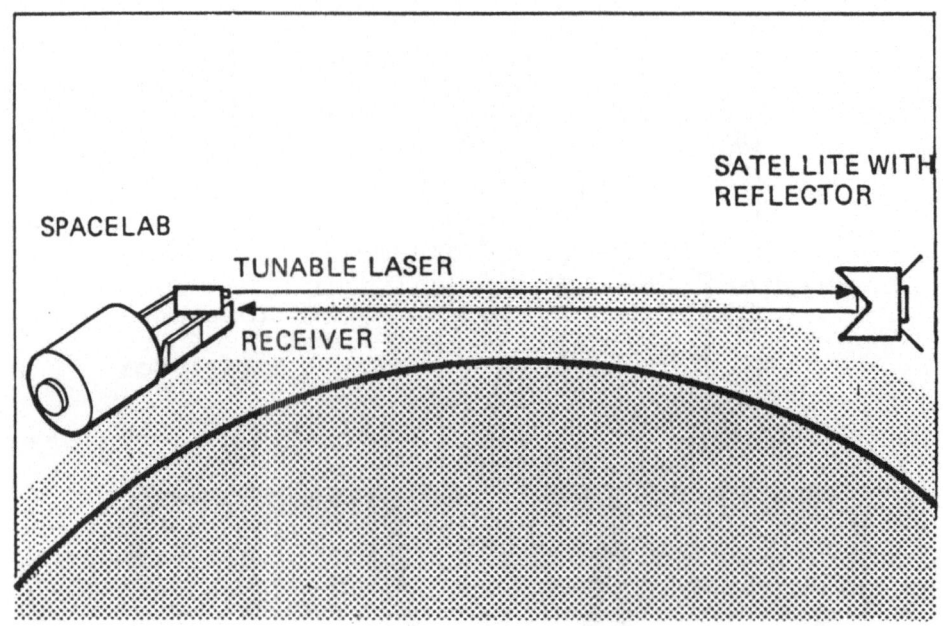

Fig. 3: Active transmission measurements between
        SPACELAB and a slaved satellite with a
        retro-reflector

3. FUTURE EXPERIMENTAL STRATEGY

    In order to accomplish the scientific goals in
atmospheric physics which are outlined in MAP it will
be necessary to make simultaneous measurements of
both photochemically related gases and energetically
important constituents in all layers of the atmosphere.
This means, that measurements will have to be made
with as high accuracy as possible even for gases with
concentration levels in the ppb range. Therefore,
emphasis will have to be laid on the enhancement of
spectral resolution and signal/noise ratio. Two ways
can be followed and are schematically described in
Fig. 4.

    As mentioned above the sun occultation method has
the advantage of high accuracy data reduction as only
little information about the actual temperature profile
is sufficient and as the tangent height of the line
of sight is well known. In addition, this method does
not require any scan mechanism as the line of sight
is moving along with the satellite orbit interval

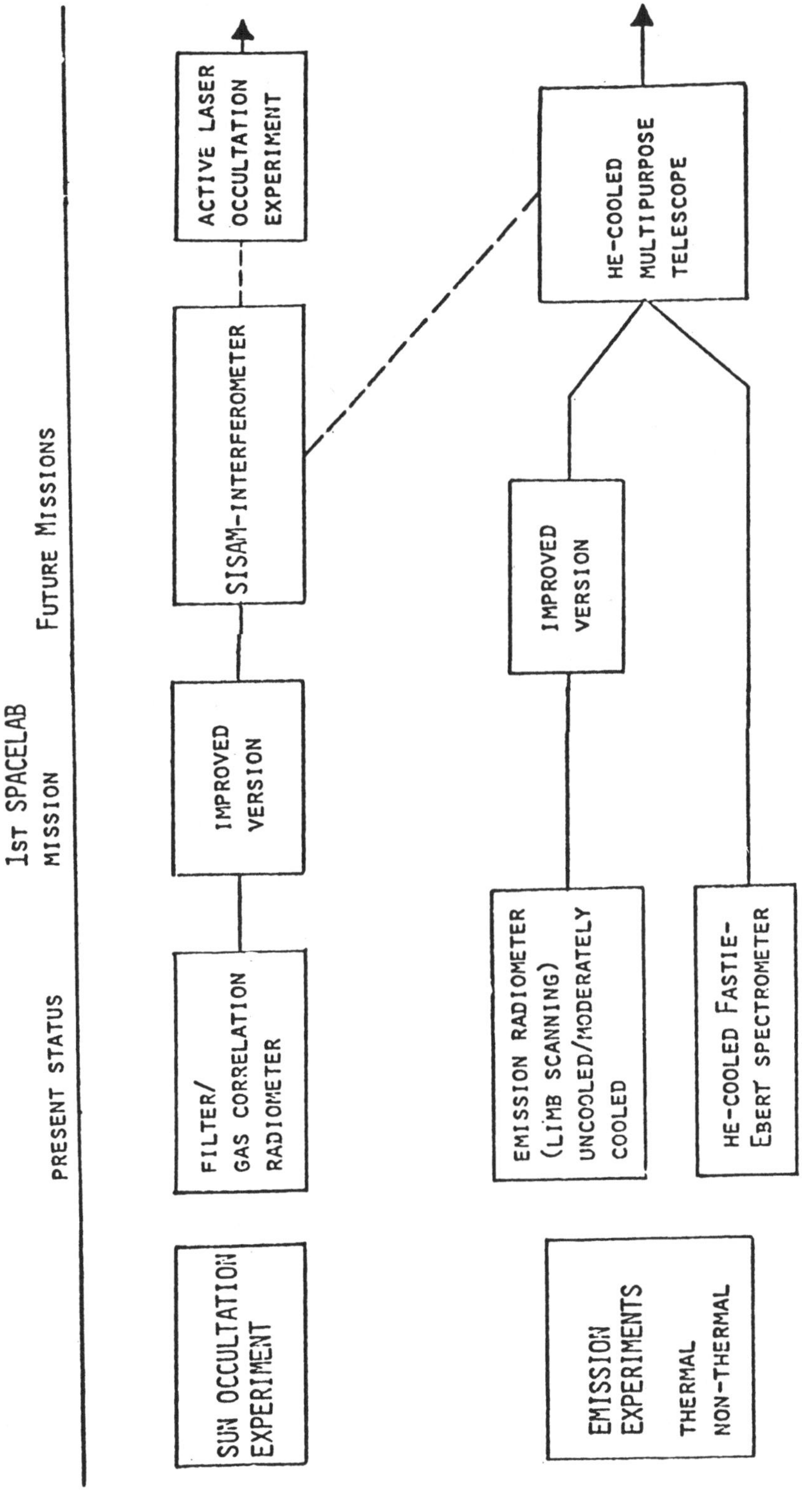

Fig. 4: Schematic description of future experimental strategy

through different atmospheric levels. The high radiant
power allows the use of high dispersive systems. For
the detection of constituent concentrations of about
1 ppb or less spectral resolution better than 0.1 cm⁻¹
will be necessary. This can so far be obtained by gas
correlation spectroscopy.

A very promising improvement - and a very difficult task too - would be to complete the occultation
experiment to a continously measuring experiment if
the sun could be replaced by one or more tunable lasers
mounted on SPACELAB (Fig. 3). The emitted radiance will
be reflected by a subsatellite with retroreflectors
standing near the horizon, and from the signal attenuated after passing the atmosphere twice the amount of
absorbing species along the optical path can be determined.

Compared to the occultation method a disadvantage
of emission measurements is the much lower radiation
power, but in this case the total received signal is
correlated to the emitting air mass instead of a small
signal variation. A rough estimation shows that if we
assume a spectral resolution of 0.2 cm⁻¹ and that 1%
absorption as lower limit can be measured a concentration of about 3 ppb should be detectable if we use the
occultation method. The lower limit for emission measurements depends on the NEP of the detector

$$P = N \cdot A \cdot \Omega$$

with

N = radiation in Watt $cm^{-2} sr^{-1}$
$A = 10^3$ cm^2 (telescope area)
$\Omega = 5 \times 10^{-6}$ sr

Assuming that

$$N = \int N_\nu \, d\nu = \overline{B_\nu(\overline{T})} \cdot S \cdot m$$

with

$B_\nu(T)$ = Planck function
$S \cdot m$ = equivalent width of a spectral line
          (weak line approximation)

we obtain in a first approximation under the same conditions as for the occultation method

$$m \geqslant \frac{NEP}{5 \cdot 10^{-3} \cdot \overline{B}_\nu \cdot S} = 4 \times 10^7 \cdot NEP$$

So if the NEP of the detector can be reduced to 5×10^{-11} Watt or less the limb emission technique should give better results for the detection of very small concentrations than the sun occultation technique. Nevertheless, since the data interpretation of limb emission measurements is morecomplex compared to occultation measurements both experiments have to be combined and the occultation measurements can be used for calibration. Thus, as a first step towards an integrated package for long term observations simultaneous measurements using the occultation and the emission technique should be made. On a second step a large helium-cooled telescope has to be developed which can be used as a general facility with instruments providing high spectral resolution attached to it (Bolle, 1974). The package has to combine different instruments able to measure the emitted radiance be generated in different atmospheric layers. It is expected that such a co-ordinated instrument package can also provide reliable information about the coupling and transport processes between these layers.

Literature

Bangert,W., E. Krieg and R. Scheidle, 1972.
An Interference Filter Radiometer
with Cooled Optics and a Cooled PbS
Detector for Rocket Application.
In: Infrared Detection Techniques
for Space Research. Ed. by V. Manno
D. Reidel Publ.Comp., Dordrecht

Bolle,H.-J., M. Bottema, W. Völker and A. Zickler,1972
A SISAM Interferometer and a Simple
Michelson-Interferometer with Sphe-
rical Mirrors for Space Application.
In: Infrared Detection Techniques
for Space Research.
D. Reidel Publ.Comp., Dordrecht

Bolle, H.-J., P. Müller and E. Rothe, 1972.
Spektralradiometer hoher räumlicher
Auflösung für indirekte Sondierung
der Atmosphäre
Forschungsbericht BMBW - FB W 72-09

Bolle, H.-J., 1974. Payload Considerations for SPACE-
LAB Missions to study Air Quality
and Atmospheric Effects on Earth
Science Data
ESRO Report

Burkert,P., D. Rabus and H.-J. Bolle, 1973.
Profiles of water vapor and methane
mixing ratios from measurements with
balloon-borne instruments
Final Scientific Report
AFOSR-72-2282, AFCRL

DISCUSSION

B. Feuerbacher: When using gas-cell radiometers on Spacelab, don't you run into problems due to the Doppler shift arising from spacecraft motion, unless you have pointing in the degree range?

E. Redemann: It is true that because of the high velocity of Spacelab, the Doppler shift of the absorption lines has to be considered. This is a problem which affects each experiment in the field of gas correlation spectroscopy from Spacelab and must be studied very carefully.

T.M. Donahue: The spacecraft velocity is eight times the average thermal velocity of atmospheric atoms and the Doppler effect will shift the absorption lines of the material in the cells completely out of the emission-line profiles unless observations are made close to the perpendicular to the velocity vector.

F. Paresce: Could you elaborate on the Ebert-Fastie spectrometers you mentioned and what O-line are you looking at?

E. Redemann: This instrument has been developed by Dr. Offermann. I have been told he is using the O-line close to 62µm.

T.M. Donahue (to F. Paresce): In order to obtain the atomic oxygen density from airglow observations of resonance radiation, it is necessary to obtain a complete altitude profile from about 100 to about 500 km since the medium is optically very thick and excitation is primarily by photo-electrons and not resonance scattering of sunlight. Is it necessary to have data at more than one altitude in order to unfold the source function from the intensity - and to know the photoelectron distribution in altitude and in energy - also to know the O_2 density in order to take into account absorption?

F. Paresce: Yes, that is true. We would need a supporting

rocket or tethered subsatellite measurement to get an
altitude profile. The UV spectrometer on Spacelab could be
well calibrated on board by means of calibrated lamps so
that it could obtain a very precise intensity measurement,
while all the rocket would have to do is obtain a relative
altitude variation. O_2 density would also be measured
simultaneously by UV absorption spectroscopy.

L. Thomas (to T.M. Donahue): Would one expect to get information
on the height distribution of atomic oxygen down to turbopause heights from observations of scattered 1304 Å radiations?

T.M. Donahue: Yes it is possible if a vertical profile is
obtained and the effect of O_2 absorption can be taken into
account properly below 120 km.

OPTICAL ABSORPTION TECHNIQUES

T.M. Donahue

Atmospheric & Oceanic Science Department,
University of Michigan, Ann Arbor, USA

A resonance absorption - resonance fluorescence experiment
conducted during the Apollo-Soyuz Test Project was designed to
measure atomic O and atomic N densities at 225 km. Resonance
radiation from Apollo was returned by a collection of retro-
reflections to a spectrophotometer on the Apollo, with total path
lengths of 300, 1000 and 2000 m. The results obtained were
discussed and the applicability of similar experiments from
Spacelab analysed.

J. J. Burger et al. (eds.), Atmospheric Physics from Spacelab, 401–402. All Rights Reserved.
Copyright © 1976 by D. Reidel Publishing Company, Dordrecht-Holland.

DISCUSSION

F. Paresce: How do you know there was no contamination at night
   too?

T.M. Donahue: We don't know for certain.   However, the good
   agreement with the Atmospheric Explorer measurements is
   encouraging.   The agreement between resonance fluorescence
   and absorption-defined densities is also a straw in the wind.

L. Thomas: In the measurements of the line shape of the lamp
   outputs, was there any evidence of Lorentz broadening rather
   than Doppler broadening conditions?

T.M. Donahue: We don't know what the source of the broadening
   was.   It was a function of the frequency of excitation, not
   of pressure - it does not occur if microwave excitation is
   used.

GENERAL DISCUSSION (SUMMARY)

IONOSPHERE/ATMOSPHERE

INTRODUCTION, by S.A. Bowhill

The ionospheric/atmospheric problems and the relevance of Spacelab
will be discussed under the following headings:
1. Leading problems in ionospheric physics.
2. The relation of these problems to and their solution on
   Spacelab.
3. Questions regarding the Spacelab programme.

1. Under the first heading we may list major problems deserving
intense investigation:
(a) Plasma instabilities and their relevance to electrojets,
    artificial heating, parametric instabilities, spread F, etc.
(b) Electrodynamics of the ionosphere i.e. thermospheric motions,
    conductivity along field lines, parallel and perpendicular
    field-line currents, etc.
(c) Thermospheric circulation under quiet and disturbed conditions
(d) Magnetosphere-ionosphere coupling, particularly the transfer
    of electric fields from the magnetosphere and ionosphere, and
    the transport of energetic particles to low latitudes.
(e) Chemistry of the lower ionosphere.

Other problems also deserving attention are the generation of
small-scale irregularities in the E-region and the variation of
winter ionospheric absorption and the correlation with stratospheric warming. Spacelab should be capable of studying the
dynamics and chemistry of the middle atmosphere.

2. The relation of these problems to Spacelab prompts the question
'Is there sufficient emphasis on *dynamics* of the atmosphere in the
Spacelab programme?'

J. J. Burger et al. (eds.), Atmospheric Physics from Spacelab, 403–408. All Rights Reserved.
Copyright © 1976 by D. Reidel Publishing Company, Dordrecht-Holland.

Remote sensing is predominantly a chemical detecting technique and
is mostly applicable to the middle atmosphere. Remote measurements
of temperature profiles may be useful in this context, but do we
have sufficient resolution and density of points to be useful for
dynamical purposes? A number of techniques *hope* to measure winds,
but it appears that the accuracy may not be high enough to yield
useful results. In this context, it is important to identify the
limitations of Spacelab and to ask: 'What is it that Spacelab
cannot measure?'.
(a) Using limb scanning, we cannot obtain horizontal resolutions
     better than tens of kilometres.
(b) It is difficult to assess the resolution of passive sounding,
     the lidar, etc. and it appears that what is needed is more of
     the P. Rothwell-type of experiment (television).
(c) Multibeam sensing from Spacelab was not mentioned at this
     Symposium.
(d) No experiment with the aim of studying turbulent motions was
     discussed.

3.   Questions regarding Spacelab:
(a) What has happened to the concept of 'suitcase'-type experi-
     ments?
(b) What is to be the role of man? - this was not emphasised
     sufficiently.
(c) Where do the atmospheric experiments (A) fit in with the
     magnetospheric (M) and plasma physics (P) experiments on
     AMPS?
(d) Where do ground-based experiments fit into the system?

DISCUSSION

A. Nagy: There are connections between A on one side and M, P on
the other side on AMPS. The electron gun is an example of an
instrument that can be used for investigation of atmospheric,
magnetospheric and plasma physics.

Concerning the problem of measuring turbulence, it must be kept in
mind that the eddy diffusion coefficient is some sort of 'fudge
factor' and not easy to measure. This problem requires more
attention.

Another point, I do not believe we should worry too much at this
stage about how to link ground-based observations with Spacelab.

G. Haskell: In connection with an ESA study on future sounding-
rocket programmes, we found very few cases where rockets and
Spacelab could complement each other with simultaneous measurements
near the same location. Still, rockets have a value of their own
in providing vertical profiles above one point and such measure-

ments are of course of general interest also for interpreting
Spacelab observations.

S.A. Bowhill: For many problems I do not see that it is necessary
to have rockets and Spacelab close together.

J.J. Burger: Returning to your question on the 'suitcase' concept,
the design of Spacelab is such that this concept can still be
practised. However, we should distinguish between the test-flights
(first and second Spacelab flights) and the operational phase,
where integration of the carry-on experiment will be much
simplified.
D. Shapland: Clearly there is the aspect of integration and even
with a 'suitcase' experiment, the integration time could be of the
order of three months.

L. Thomas: I want to add to Prof. Bowhill's list of problems:
measurements of excited states are important for atmospheric
chemistry, where a lot of ground laboratory work is still also
needed.

J. Gregory: It must be remembered that Spacelab flights can only
make short sample measurements in a variable medium and to obtain
accurate measurements requires many flights and lots of statistics.

TROPOSPHERE/STRATOSPHERE

INTRODUCTION, by G. Hunt

Two basic questions:
1. What is justification for atmospheric research?
2. How does one assess results and progress?

We require some definition of climate, e.g. why does it change in
the long-term, say 50 years? It is clear that Spacelab cannot be
used for such long-time-scale studies, but let us consider ways in
which it can contribute. We need to know the *natural* variations
in the atmosphere and also the effect of *artificial* effects (man-
made). We need to determine the *important* chemical reactions
involved and not waste time on insignificant reactions. More
interaction is required between the *modellers* and the *measurers*.
Most models are one-dimensional and these have to be extended,
which requires better measurements. There is a great need for
simultaneous measurements by different instruments in order that
instrumental problems may be removed. Spacelab can contribute
here by tackling these problems piecemeal. In climatology, the
driving factor is the *solar constant*, which has to be known to
0.1% accuracy and Spacelab can contribute here to intercalibrate

monitoring experiments, and also to determine in more *detail* how
the amount of radiation reaching the Earth's surface varies. Even
if the solar constant does not vary, astronomical parameters like
perihelion occurrence drift on a one yearly scale. The very
important parameter is the radiation balance, of which we know in
general that it is also controlled by clouds in the troposphere,
where small variations in cloud altitude can create large differences, e.g. 0.5 km height change is equivalent to $\sim$2% change in
solar constant. More detailed investigations are thus necessary.

DISCUSSION

T.M. Donahue: Models are complicated, but no numbers game. All
reactions included in the model calculations are important and
cannot be neglected. Due to the complications, new reactions are
put into models reluctantly. We also wish to work out two-
dimensional models but this is even much more complicated and
cannot be done easily. In this situation I emphasise that one-
dimensional models are still quite useful in many respects.

G. Hunt: As an example, it is difficult to include interaction
with oceans in the models.

A. Nagy: If natural variations are compared to induced ones, the
former are still the most significant.

G. Hunt: That depends on the place of application and processes
concerned.

E. Raschke: What do you see as the value of Spacelab?

G. Hunt: To test and improve instruments and not to make 'climate'
measurements.

A. Pedersen: I have the impression from presentations and discussions following these presentations that Spacelab could be
used to study long-term variations by making point measurements
once or twice per year, and for absolute calibration of monitoring
experiments.

S.A. Bowhill: Meteorologists tend to look upon Spacelab as a
super Nimbus that flies for a week. However, one has to think
about other important aspects where Spacelab can contribute, for
example, to
 (i) active probing
(ii) correlation with the M and P of AMPS.

D. Shapland: Spacelab will be important for equipment testing and
in the development of meteorological instrumentation. It is also

a good short-reaction device, e.g. for studies during/after a
volcanic eruption.

Experiments on Spacelab

A. Pedersen: Regarding instruments on Spacelab, one of the
questions is how passive and active experiments do complement each
other.

L. Thomas: It must be stressed that passive systems are very
versatile, the prime limiting factor being the resolution attainable. They also are limited for observations in the troposphere.
Here they can be complemented by lidar techniques, which are
particularly powerful for measuring minor constituents and winds
on a global scale. Especially for wind measurements, good prospects
are foreseen, since one can work with aerosols.

R.J. Murgatroyd: The chemistry and composition of the troposphere
is sufficiently known to assess the associated sinks and sources
in the models, so that direct measurements are very important.

E. Raschke: But how does one perform these measurements from
Spacelab when clouds obscure the view? Spacelab is not the answer.

G. Hunt: Cloud cover is only 50%.

T.M. Donahue: Spacelab can, for example, do a lot for the understanding of NO chemistry, and ground measurements can be used to
track changes in NO sources.

Role of Man on Spacelab

P. Rothwell: Spacelab has only four operators. At any one time
only two of them are available. The load on the individual is
very high.

S.A. Bowhill: How much help can operators obtain? They must not
only ensure that data flows from right instrument to right buffer.
More is required.

P. Rothwell: Operators have tremendous responsibility. But their
capabilities are limited. Equipment must be made semi-automatic.

J.J. Burger: ASSESS showed that automation of experiment operations
is required, in order to free operator time for creativity. People
must be trained so that they can operate the experiments intelli-
gently and interpret data in real time and react to unpredictable
opportunities.

D. Shapland: An equilibrium point between man and equipment must
be found.

J.J. Burger: There are roughly three schools of thought on the
role of man: (i) Everything can be automated and no men are
necessary, (ii) Man might be useful in a standby mode for correcting
technical shortcomings, and (iii) The availability of man is going
to be the most precious resource on Spacelab.   It seems extremely
difficult in this situation to strike the right balance.

F. Paresce: Astronauts involved in the Apollo and Skylab missions
did good experimentation, and thus there is hope for significant
contributions from man in space.

INDEX OF AUTHORS

ASTROPHYSICS AND SPACE SCIENCE LIBRARY

Edited by

J.E. Blamont, R.L.F. Boyd, L. Goldberg, C. de Jager, Z. Kopal, G.H. Ludwig, R. Lüst,
B.M. McCormac, H.E. Newell, L.I. Sedov, Z. Švestka, and W. de Graaff

1. C. de Jager (ed.), *The Solar Spectrum. Proceedings of the Symposium held at the University of Utrecht, 26–31 August, 1963.* 1965, XIV + 417 pp.
2. J. Ortner and H. Maseland (eds.), *Introduction to Solar Terrestrial Relations, Proceedings of the Summer School in Space Physics held in Alpbach, Austria, July 15–August 10, 1963 and Organized by the European Preparatory Commission for Space Research.* 1965, IX + 506 pp.
3. C.C. Chang and S.S. Huang (eds.), *Proceedings of the Plasma Space Science Symposium, held at the Catholic University of America, Washington, D.C., June 11–14, 1963.* 1965, IX + 377 pp.
4. Zdeněk Kopal, *An Introduction to the Study of the Moon.* 1966, XII + 464 pp.
5. B.M. McCormac (ed.), *Radiation Trapped in the Earth's Magnetic Field. Proceedings of the Advanced Study Institute, held at the Chr. Michelsen Institute, Bergen, Norway, August 16–September 3, 1965.* 1966, XII + 901 pp.
6. A.B. Underhill, *The Early Type Stars.* 1966, XII + 282 pp.
7. Jean Kovalevsky, *Introduction to Celestial Mechanics.* 1967, VIII + 427 pp.
8. Zdeněk Kopal and Constantine L. Goudas (eds.), *Measure of the Moon. Proceedings of the 2nd International Conference on Selenodesy and Lunar Topography, held in the University of Manchester, England, May 30–June 4, 1966.* 1967, XVIII + 479 pp.
9. J.G. Emming (ed.), *Electromagnetic Radiation in Space. Proceedings of the 3rd ESRO Summer School in Space Physics, held in Alpbach, Austria, from 19 July to 13 August, 1965.* 1968, VIII + 307 pp.
10. R.L. Carovillano, John, F. McClay, and Henry R. Radoski (eds.), *Physics of the Magnetosphere, Based upon the Proceedings of the Conference held at Boston College, June 19–28, 1967.* 1968, X + 686 pp.
11. Syun-Ichi Akasofu, *Polar and Magnetospheric Substorms.* 1968, XVIII + 280 pp.
12. Peter M. Millman (ed.), *Meteorite Research. Proceedings of a Symposium on Meteorite Research, held in Vienna, Austria, 7–13 August, 1968.* 1969, XV + 941 pp.
13. Margherita Hack (ed.), *Mass Loss from Stars. Proceedings of the 2nd Trieste Colloquium on Astrophysics, 12–17 September, 1968.* 1969, XII + 345 pp.
14. N. D'Angelo (ed.), *Low-Frequency Waves and Irregularities in the Ionosphere. Proceedings of the 2nd ESRIN-ESLAB Symposium, held in Frascati, Italy, 23–27 September, 1968.* 1969, VII + 218 pp.
15. G.A. Partel (ed.), *Space Engineering. Proceedings of the 2nd International Conference on Space Engineering, held at the Fondazione Giorgio Cini, Isola di San Giorgio, Venice, Italy, May 7–10, 1969.* 1970, XI + 728 pp.
16. S. Fred Singer (ed.), *Manned Laboratories in Space. Second International Orbital Laboratory Symposium.* 1969, XIII + 133 pp.
17. B.M. McCormac (ed.), *Particles and Fields in the Magnetosphere. Symposium Organized by the Summer Advanced Study Institute, held at the University of California, Santa Barbara, Calif., August 4–15, 1969.* 1970, XI + 450 pp.
18. Jean-Claude Pecker, *Experimental Astronomy.* 1970, X + 105 pp.
19. V. Manno and D.E. Page (eds.), *Intercorrelated Satellite Observations related to Solar Events. Proceedings of the 3rd ESLAB/ESRIN Symposium held in Noordwijk, The Netherlands, September 16–19, 1969.* 1970, XVI + 627 pp.
20. L. Mansinha, D.E. Smylie, and A.E. Beck, *Earthquake Displacement Fields and the Rotation of the Earth. A NATO Advanced Study Institute Conference Organized by the Department of Geophysics, University of Western Ontario, London, Canada, June 22–28, 1969.* 1970, XI + 308 pp.
21. Jean-Claude Pecker, *Space Observatories.* 1970, XI + 120 pp.
22. L.N. Mavridis (ed.), *Structure and Evolution of the Galaxy. Proceedings of the NATO Advanced Study Institute, held in Athens, September 8–19, 1969.* 1971, VII + 312 pp.
23. A. Muller (ed.), *The Magellanic Clouds. A European Southern Observatory Presentation: Principal Prospects, Current Observational and Theoretical Approaches, and Prospects for Future Research, Based on the Symposium on the Magellanic Clouds, held in Santiago de Chile, March 1969, on the Occasion of the Dedication of the European Southern Observatory.* 1971, XII + 189 pp.

24. B. M. McCormac (ed.), *The Radiating Atmosphere. Proceedings of a Symposium Organized by the Summer Advanced Study Institute, held at Queen's University, Kingston, Ontario, August 3–14, 1970.* 1971, XI + 455 pp.
25. G. Fiocco (ed.), *Mesospheric Models and Related Experiments. Proceedings of the 4th ESRIN-ESLAB Symposium, held at Frascati, Italy, July 6–10, 1970.* 1971, VIII + 298 pp.
26. I. Atanasijević, *Selected Exercises in Galactic Astronomy.* 1971, XII + 144 pp.
27. C. J. Macris (ed.), *Physics of the Solar Corona. Proceedings of the NATO Advanced Study Institute on Physics of the Solar Corona, held at Cavouri-Vouliagmeni, Athens, Greece, 6–17 September 1970.* 1971, XII + 345 pp.
28. F. Delobeau, *The Environment of the Earth.* 1971, IX + 113 pp.
29. E. R. Dyer (general ed.), *Solar-Terrestrial Physics/1970. Proceedings of the International Symposium on Solar-Terrestrial Physics, held in Leningrad, U.S.S.R., 12–19 May 1970.* 1972, VIII + 938 pp.
30. V. Manno and J. Ring (eds.), *Infrared Detection Techniques for Space Research. Proceedings of the 5th ESLAB-ESRIN Symposium, held in Noordwijk, The Netherlands, June 8–11, 1971.* 1972, XII + 344 pp.
31. M. Lecar (ed.), *Gravitational N-Body Problem. Proceedings of IAU Colloquium No. 10, held in Cambridge, England, August 12–15, 1970.* 1972, XI + 441 pp.
32. B. M. McCormac (ed.), *Earth's Magnetospheric Processes. Proceedings of a Symposium Organized by the Summer Advanced Study Institue and Ninth ESRO Summer School, held in Cortina, Italy, August 30–September 10, 1971.* 1972, VIII + 417 pp.
33. Antonin Rükl, *Maps of Lunar Hemispheres.* 1972, V + 24 pp.
34. V. Kourganoff, *Introduction to the Physics of Stellar Interiors.* 1973, XI + 115 pp.
35. B. M. McCormac (ed.), *Physics and Chemistry of Upper Atmospheres. Proceedings of a Symposium Organized by the Summer Advanced Study Institute, held at the University of Orléans, France, July 31–August 11, 1972.* 1973, VIII + 389 pp.
36. J. D. Fernie (ed.), *Variable Stars in Globular Clusters and in Related Systems. Proceedings of the IAU Colloquium No. 21, held at the University of Toronto, Toronto, Canada, August 29–31, 1972.* 1973, IX + 234 pp.
37. R. J. L. Grard (ed.), *Photon and Particle Interaction with Surfaces in Space. Proceedings of the 6th ESLAB Symposium, held at Noordwijk, The Netherlands, 26–29 September, 1972.* 1973, XV + 577 pp.
38. Werner Israel (ed.), *Relativity, Astrophysics and Cosmology. Proceedings of the Summer School, held 14–26 August, 1972, at the BANFF Centre, BANFF, Alberta, Canada.* 1973, IX + 323 pp.
39. B. D. Tapley and V. Szebehely (eds.), *Recent Advances in Dynamical Astronomy. Proceedings of the NATO Advanced Study Institute in Dynamical Astronomy, held in Cortina d'Ampezzo, Italy, August 9–12, 1972.* 1973, XIII + 468 pp.
40. A. G. W. Cameron (ed.), *Cosmochemistry. Proceedings of the Symposium on Cosmochemistry, held at the Smithsonian Astrophysical Observatory, Cambridge, Mass., August 14–16, 1972.* 1973, X + 173 pp.
41. M. Golay, *Introduction to Astronomical Photometry.* 1974, IX + 364 pp.
42. D. E. Page (ed.), *Correlated Interplanetary and Magnetospheric Observations. Proceedings of the 7th ESLAB Symposium, held at Saulgau, W. Germany, 22–25 May, 1973.* 1974, XIV + 662 pp.
43. Riccardo Giacconi and Herbert Gursky (eds.), *X-Ray Astronomy.* 1974, X + 450 pp.
44. B. M. McCormac (ed.), *Magnetospheric Physics. Proceedings of the Advanced Summer Institute, held in Sheffield, U.K., August 1973.* 1974, VII + 399 pp.
45. C. B. Cosmovici (ed.), *Supernovae and Supernova Remnants. Proceedings of the International Conference on Supernovae, held in Lecce, Italy, May 7–11, 1973.* 1974, XVII + 387 pp.
46. A. P. Mitra, *Ionospheric Effects of Solar Flares.* 1974, XI + 294 pp.
48. H. Gursky and R. Ruffini (eds.), *Neutron Stars, Black Holes and Binary X-Ray Sources.* 1975, XII + 441 pp.
49. Z. Švestka and P. Simon (eds.), *Catalog of Solar Particle Events 1955–1969. Prepared under the Auspices of Working Group 2 of the Inter-Union Commission on Solar-Terrestrial Physics.* 1975, IX + 428 pp.
50. Zdeněk Kopal and Robert W. Carder, *Mapping of the Moon.* 1974, VIII + 237 pp.
51. B. M. McCormac (ed.), *Atmospheres of Earth and the Planets. Proceedings of the Summer Advanced Study Institute, held at the University of Liège, Belgium, July 29–August 8, 1974.* 1975, VII + 454 pp.
52. V. Formisano (ed.), *The Magnetospheres of the Earth and Jupiter. Proceedings of the Neil Brice Memorial Symposium, held in Frascati, May 28–June 1, 1974.* 1975, XI + 485 pp.
53. R. Grant Athay, *The Solar Chromosphere and Corona: Quiet Sun.* 1976, XI + 504 pp.

54. C. de Jager and H. Nieuwenhuijzen (eds.), *Image Processing Techniques in Astronomy. Proceedings of a Conference, held in Utrecht on March 25–27, 1975.* 1975, XI + 418 pp.
55. N.C. Wickramasinghe and D.J. Morgan (eds.), *Solid State Astrophysics. Proceedings of a Symposium, held at the University College, Cardiff, Wales, 9–12 July 1974.* 1976, XII + 314 pp.
57. K. Knott and B. Battrick (eds.), *The Scientific Satellite Programme during the International Magnetospheric Study. Proceedings of the 10th ESLAB Symposium, held at Vienna, Austria, 10–13 June 1975.* 1976, XV + 464 pp.